BATTERY OPERATED DEVICES AND SYSTEMS

BATTERY OPERATED DEVICES AND SYSTEMS

From Portable Electronics to Industrial Products

G. Pistoia

ELSEVIER

Amsterdam • Boston • Heidelberg • London • New York • Oxford
Paris • San Diego • San Francisco • Singapore • Sydney • Tokyo

Elsevier
Radarweg 29, PO Box 211, 1000 AE Amsterdam, The Netherlands
Linacre House, Jordan Hill, Oxford OX2 8DP, UK

First edition 2009

British Library Cataloguing in Publication Data
A catalogue record for this book is available from the British Library

Library of Congress Cataloging-in-Publication Data
A catalog record for this book is available from the Library of Congress

ISBN: 978-0-444-53214-5

Printed and bound in Hungary
09 10   10 9 8 7 6 5 4 3 2 1

Contents

Preface

A number of handbooks are available to people working in the battery field, where batteries are the main subject and their applications are treated in much less detail. Conversely, there are no books dealing with the large spectrum of applications powered by batteries. In other words, although some books cover specific topics, for example portable devices, electric vehicles, energy storage, no books that aim to summarize all battery applications have thus far been published.

This book aims at bridging this gap, as many applications are reported in detail and others are mentioned, whereas less emphasis is put on batteries. However, basic characteristics of batteries and information on the latest developments are enclosed in a dedicated chapter. As is obvious, a 400-page single-author book cannot be as exhaustive as a multi-author large handbook. Nevertheless, the reader may find here, in addition to data on many applications, links to further literature through the many references that have been included. For researchers, teachers and graduate students interested in devices and systems drawing power from batteries, this book will be a useful information source.

In Chapter 1, all applications in the portable and industrial areas are introduced. Some market considerations follow, with details on the most important sectors, and a forecast to 2016 for portable devices is enclosed.

In Chapter 2, basic characteristics of all primary and secondary batteries used in the applications described are reviewed. The most recent trends, especially for the ubiquitous lithium ion batteries, are mentioned.

In Chapter 3, portable applications, for example mobile phones, notebooks, cameras, camcorders, several medical instruments, power tools, GPS receivers, are described with details on their electronic aspects. Particular emphasis is put on the devices' power consumption and management for their implications on battery life and device runtime. The basic features of some electronic components, for example microprocessors, voltage regulators and displays, are presented for a better understanding of their energy requirements. Battery management is also dealt with in detail, particularly in so far as the charging methods are concerned. The criteria of battery choice are stressed.

Chapter 4, on industrial applications, is the largest one, as it includes aerospace, telecommunications, emergency systems, load levelling, energy storage, different meters, data loggers, oil drilling, oceanography, meteorology, robotics, etc. The final part of this section is devoted to wireless connectivity, that is Wi-Fi, Bluetooth and Zigbee, exploited in many portable and industrial applications.

Chapter 5 deals with battery usage in vehicular applications. For their specific interest, these industrial applications are described in a section of their own. Full electric and hybrid vehicles are presented, and the role that the battery plays in the vehicle control systems is outlined.

Rome, March 2008
Gianfranco Pistoia

Chapter 1

AREAS OF BATTERY APPLICATIONS

1.1. Introduction

This chapter aims at providing an overview of products and systems using batteries. Here, the term product indicates any device – small or large, portable or not – powered by a battery. The term system indicates a large installation, such as an energy storage plant to back up an electricity grid, or an extended sensor network.

Several criteria may be used to classify the countless applications of batteries reported in Table 1.1. In this book, three major categories have been considered: portable, industrial and traction/automotive. The first category is mainly represented by consumer applications but has to be extended to any application whose weight and volume allows portability. Therefore, even applications that a consumer rarely comes to know about, for example in the medical field, are enclosed in this category. Industrial applications encompass a wide spectrum, from robots to weather satellites, from oil drilling to telecommunications. Finally, traction and automotive applications include electric and hybrid electric cars, as well as their control systems. Strictly speaking, car-related applications should also be enclosed among the industrial ones. However, they are treated in a separate chapter because of their special interest: many people are willing to know more about these cars and their batteries in terms of performance, cost, reliability and development perspectives.

On the basis of these categories, Chapters 3, 4 and 5 will deal with applications typical of portable, industrial and traction/automotive batteries, respectively. However, in this chapter, some tables are anticipated: in Table 1.2, batteries are listed according to homogeneous groups of applications; in Table 1.3, applications or requirements in terms of current/power, duty cycle, dimensions, durability, etc., are reported together with the battery type/characteristic; in Table 1.4, the energy ranges of various battery-powered applications are indicated.

General characteristics of the main battery types are reported in Chapter 2. However, this book is more oriented to device (or system) description; more details on batteries can be found in the references listed at the end of that chapter.

Table 1.1. Applications using batteries (listed in alphabetical order).

Aerospace
Access control devices
Airborne control devices
Aircraft
Alarms – burglar
Alarm – fire
Alarm monitoring
Alarm panels
Alarm – pollution
Alarm refrigerator
Alarm water level
Alarm – seismic
Alert devices
Animal ID readers
Animal tracking
Appliances – portable
Audio video equipment
Automobile electronic systems
Automotive accessories
Automotive electronic memory
Automotive fuel systems
Automotive
  locator for theft
Automotive
  security systems
Avalanche rescue
  transmitters

Backup power
Ball pitching equipment
Bar code scanners – portable
Bone healing aids
Buoy – oceanographic

Cable TV
Calculators
Calorimeters
Camcorders
Cameras
Cargo tracking
Chemical sensors
Cellular telephones
Clocks
Clocks – scientific
Clockwise operated devices
Communications
  diagnostic equipment
Communication – radio
Communication
  telephone systems
Computer – portable
Computer – home
Computer laptop
Computer mainframe
Computer peripherals
Construction lasers
Control equipment
Converters/programmers
Cordless telephones
Cordless toothbrushes
Counting
  Industrial
  Thermostatic
  Timing

Data logging
  Inventory
Dental equipment – portable
Digital cameras
Diving equipment

EKG equipment
Electric cash register
Electric door openers
Electric fans
Electric fences
Electric gates
Electric locks
Electric meter transponders
Electric trolling motors (fishing)
Electric/electronic distributors
Electric/electronic scales
Electric vehicles
Electronic counting systems
Electronic games
Electronic nerve stimulation units
Elevator – escalators
Emergency call boxes

Table 1.1. (*Continued*)

Emergency devices
Emergency lighting
Emergency notification
Entertainment
  Musical instruments
  Public address amps
  Stereo tuners
  Tape recorders
  TV recorders
  VCRs
  Video cameras
Environmental
  test equipment
Exercise bikes and equipment
Exit lights

Facsimile machine
Fiber-optic test equipment
Fire alarm panels
Fire suppression systems
Fish finders
Flashlights
Flow meters (heat, gas and water)
Fragrance dispensers
Freeway call boxes

Game feeders and callers
Garden equipment
Garage door openers
Gas emergency cutoff systems
Gas meter transponders
Gas motor starting
Gas station elec. pump
Geometrics
Geophysical
  Seismic instruments
  Surveying equipment
Golf carts
GPS equipment

Hand-held computers
Hand-held test equipment
Hand-held devices
Hearing aids
Hybrid electric vehicles

Identification
  Finger
  Face
  Hand
Implantable medical devices
Industrial control equipment
Industrial tools
Infrared equipment – portable
Intelligent telephones

Laboratory analytical instruments
LAN power backup
Lanterns
Lasers
Lifts
Lights
  Camera, video, etc.
  Highway safety
  Maintenance
  Photographic
  Railroad
  Underwater
Load levelling

Marine communications
Marine instrumentation
Marine depth finders
Marine
  underwater propulsion
Measuring and controlling devices
Measuring and dispensing pumps
Medical alert equipment
Medical beds
Medical CPR equipment
Medical crash carts
Medical
  Bio-sensors
  Blood oximeters
  Cardiac monitors
  Defibrillators

(*Continued*)

Table 1.1. (*Continued*)

Diagnostic equipment
Dialysis machine
Drug dispensers
Ear thermometers
Glucose meters
Incubators
Infusion pumps
Inhalators
Intravenous pumps
Life support equipment
Sleep apnoea monitor
Telemetry equipment
Therapy equipment
Wheelchairs
Memory backup devices
Metal detectors
Meteorological instruments
Meters
Electricity, gas, water
Consumption
Microwave communications
Missile launch/tracking
Military electronics
Military fire control systems
Military target range equipment
Mini-UPS
Modems
Monitors – portable
Motherboards
Motor starters
Muscle stimulator
Musical instruments – electrical

Ocean current monitors
Oceanographic equipment
Office equipment portable/programm.
Oil refinery backup
Ophthalmic instruments
Optical instruments
Oxygen analysers
Oxygen monitors

Pagers
Parking lot tags
Parking meters – digital
PBX (private branch exchange) backup
PDAs
Personal organizers
Photovoltaic
Portable data entry terminals
Portable lights
Portable power line monitors
Portable measuring instruments
Portable monitoring equipment
Portable public address systems
Portable transceivers
Portable VoIP
Portable welding equipment
Portable X-ray equipment
Power supplies
Power tools
Printers – portable
Probes
Pulse power devices

Radar guns
Radio-controlled devices
Radio frequency ID tags
Railroad signalling
Real-time clocks
Refrigeration units
Rehabilitation devices
Remote level control
Remote site equipment
Rescue transmitters
Respirators
Robots

Satellites
Search and detection equipment
Scales and balance devices
Security gates
Security scanners
Security systems
Seismic measurements
Sequence control equipment

Table 1.1. (*Continued*)

Shopping cart displays	Transponders
Smart cards	Transportation
Smoke alarms and detectors	Turner memories for VCRs
Solar energy storage	Two-way radios
Solar walklights	
Spectrometers	Ultrasound equipment
Speed measurement	Unmanned air systems
Laser and radar	Underwater gliders
Scoring systems	Uninterruptible power supplies (UPS)
Skydiving instruments	Utilities
timing systems	
Stenography machines	Vending machines
Surgeon suits	Vehicle recovery systems
Surveying instruments	Video cameras
Switching systems	VSAT backup power
backup power	
	Watches
Taximeters	Water treatment controls
Telecommunications	Weather instrumentation
Timing devices	Well logging instrumentation
Toll road transceivers	Wheelchair and scooters
Toys	Wind energy storage
Electromechanical	Wireless products
Programmable	Turnstiles
Radio controlled	Headsets
Riding	Test equipment
Traffic delineators	Wi-Fi and bluetooth
Trailer tracking devices	Word processing systems
Transmitters	
	Zigbee

1.2. Application Sectors and Market Considerations

The numerous applications listed in Tables 1.1 and 1.2 can be further grouped into the following sectors from a market standpoint [3].

1.2.1. Computing

This large and well-established sector includes portable computers, personal digital assistants (PDAs) and calculators. Portable computer batteries are typically lithium ion (Li-ion) and, less frequently, nickel metal hydride (Ni-MH). PDAs

Table 1.2. Applications using batteries (listed by homogeneous groups).

Agricultural
- Livestock/game feeders
- Livestock reproduction

Automotive
- Electronic memory
- Accessories
- Fuel systems
- Braking systems
- Automatic crash notification
- Tire pressure monitoring system
- Electric bicycles/scooters
- EV & HEV
- SLI (Starting, Lighting, Ignition)
- Toll collection

Back-up
- LAN
- Memory
- Uninterruptible power supplies (UPS)
- PBX (Private Branch Exchange)
- Mini-UPS
- VSAT (Very Small Aperture [Satellite] Terminal)

Communications
- Radio
- Railroad signalling
- Telephone systems
- Global positioning equipment
- Marine communications
- Microwave
- Portable transceivers
- Two-way radios
- Cordless & cellular phones
- Portable PA (Public Address) systems
- Freeway call boxes
- Automatic assistance system

Computing and Data Acquisition
- Computers & peripheral equipment
- Hand-held data gathering devices
- Data loggers

Control Equipment
- Thermostatic
- Timing
- Electro-mechanical systems

Energy Generation, Transmission and Storage
- Solar generators
- Wind generators
- Load levelling
- Electricity substations
- Gas turbine control

Lighting
- Emergency lighting
- Exit lights
- Hand-held lights
- Highway safety
- Photographic
- Underwater
- Lanterns
- Solar walk lights
- Traffic
- Airport runway lighting

Medical Applications
- Electronic nerve stimulation units
- Emergency devices
- Heart defibrillators
- Breathing-assistance equipment
- Laboratory analytical instruments
- Medical alert equipment
- Medical beds
- Medical CPR (Cardio-pulmonary Resuscitation) equipment
- Medical crash carts

Table 1.2. (*Continued*)

- Diagnostic equipment
- Dialysis machine
- Incubators
- Life support equipment
- Therapy equipment
- Wheelchairs
- Patient moving
- Telemetry equipment
- Infusion pumps
- Optic instruments
- Portable X-ray machines
- Cardiac monitors
- Dental equipment

Military
- Aerospace
- Aircraft instruments
- Missile launching/tracking
- Fire control systems
- Target range equipment
- Gunnery control

Miscellaneous
- Freon leak detectors
- End of train signalling
- Railroad track hot boxes
- Invisible fences
- Bowling alley lane cleaner
- DC power lifts
- Floor scrubbers
- Portable welders
- Industrial torque wrenches
- Traffic counters
- Portable heaters
- Laser products
- Robotics
- Lawn & garden equipment
- Point of sale terminals
- Switching systems
- Elevators
- Power tools
- Vacuum cleaners

Monitoring Equipment
- Airborne instruments
- Seismic instrumentation & alarms
- Surveying equipment
- Pollution alarms
- Transmitters
- Tracking systems
- Meteorological instruments
- Fiber-optic test equipment
- Portable monitors
- Bar code portable readers
- Ocean current monitors
- Portable power line monitors
- Search & detection equipment
- Scales & balance devices
- Scientific instruments
- Oil drilling
- Speed measurement
- Water consumption meters
- Heat consumption meters
- Electricity consumption meters
- AMR (Automatic Meter Readers)
- Gas consumption meters
- Gas flow meters

Recreation
- Sporting goods
- Trolling motors
- Fish finders
- Electronic deep sea fishing reel
- Tennis ball thrower
- Hobby craft
- Toys

Security Systems
- Burglar alarms
- Fire alarms
- Alarm panels
- Monitoring alarms
- Electric fences & gates
- Metal detectors
- Access control devices
- Ride-on

(*Continued*)

Table 1.2. (*Continued*)

Video Equipment

- Televisions
- Camcorders
- Audio-visual devices
- Cameras and video lighting
- Cable television

Table 1.3. Applications, or requirements, and related battery types.

Application/Requirement	Battery Types or Characteristics
Low-power, low-cost consumer applications	Low-power primary and secondary cells. Leclanché, alkaline, Ni-Cd, Ni-MH, primary lithium
Power tools, cordless equipment	Ni-Cd, Ni-MH, Li-ion
Small devices, hearing aids, watches, calculators, memory back up, wireless peripherals	Primary button and coin cells, zinc-air, silver oxide, primary lithium
Medical implants, long life, low self discharge, high reliability	Primary lithium, button and special cells
Automotive (starting, lighting and ignition (SLI))	Lead-acid
Automotive traction batteries	Lead-acid, Ni-MH, Li-ion, Na/$NiCl_2$
Industrial traction batteries	Lead-acid, Ni-MH
Other traction batteries: robots, bicycles, scooters, wheelchairs, lawnmowers	Lead-acid, Nickel-Zinc, Li-ion, Ni-MH
Deep discharge, boats, caravans	Nickel-zinc, lead-acid, special construction
Standby power, UPS (trickle charged)	Lead-acid, Ni-Cd
Emergency power, long shelf life	Lithium, water-activated reserve batteries
Emergency power, stored electrolyte	Reserve batteries
Very high power, load levelling	Vanadium-redox flow batteries, Na/S, lead-acid, Ni-MH, Li-ion
Marine use, emergency power	Water-activated reserve batteries
High-voltage batteries	Multiple cells
High-capacity batteries, long discharge times	Multiple cells, special constructions, special chemistries
Low power, maximum energy density	Li-ion
Remote instrumentation	
Maximum power density	Primary lithium, Li-ion
Booster batteries, HEV applications	Ni-MH, Li-ion, Na/$NiCl_2$

Table 1.3. (*Continued*)

Application/Requirement	Battery Types or Characteristics
Long shelf life, low self discharge	Primary lithium, special chemical additives
Long cycle life	Temperature controls, built-in battery management systems (BMS), recombinant systems, chemical additives
Satellites, aerospace applications	Ni-Cd, Nickel-H_2, Li-ion, primary Li, Silver-zinc
High-energy density, lightweight	Zinc-air, primary lithium, Li-ion
Special shapes	Solid state, Li-ion polymer
Wide temperature range	Chemical additives, built in heaters, liquid cooling
Low maintenance	Sealed cells, recombinant chemistries
Inherently safe	Sealed cells, stored electrolyte, solid electrolyte, special chemistries
Robust	Special constructions
Missiles and munitions, safe storage, single use, robust, short one off discharge	High-temperature batteries
Torpedoes, short one off discharge	Water-activated batteries
Intelligent battery (communications between charger and battery)	Built in electronics to control charging and discharging
AC-powered devices	Built-in electronics (inverter) to provide AC power
Remote charging	Solar cells with deep discharge batteries
Short period power boost	Lithium, Ni-MH

Source: Adapted from Ref. [1]

typically use Li-ion batteries, and to a lesser extent Ni-MH or primary alkaline. Calculators may use alkaline, lithium or silver-zinc primary systems.

As with portable communications (see the next section), trends include an increasing convergence between cell phones and other portable products such as PDAs and cameras.

Driving forces and market developments include the following:

- Explosive growth has ended. Slow, but steady, sales until the next technology turning point.
- Tablet computers are becoming more important (mainly for commercial users). They are a viable alternative for many applications, and this could eventually grow from a niche market to a significant market sector.

Table 1.4. Energy ranges of different battery groups and related applications.

Battery Type	Energy	Applications
Miniature batteries	100 mWh–2 Wh	Electric watches, calculators, implanted medical devices
Batteries for portable equipment	2–100 Wh	Flashlights, toys, power tools, portable radio and TV, mobile phones, camcorders, laptop computers, memory refreshing, instruments, cordless devices, wireless peripherals, emergency beacons
SLI batteries (starting, lighting and ignition)	100–600 Wh	Cars, trucks, buses, lawn mowers, wheel chairs, robots
Vehicle traction batteries	0.5–630 kWh	EV, HEV, forklift trucks, bikes, locomotives, wheel chairs, golf carts
Stationary batteries (except load levelling)	250 Wh–5 MWh	Emergency power, local energy storage, remote relay stations, communication base stations, uninterruptible power supplies (UPS).
Military and aerospace	Wide range	Satellites, munitions, robots, emergency power, communications
Special purpose	3 MWh	Submarines
Load levelling batteries	2–100 MWh	Spinning reserve, peak shaving, load levelling

Source: From Ref. [2].

- Wearable computers are now being commercialized. Most technical issues have been solved, but creative marketing approaches are needed.
- Convergence between cell phones, PDAs, digital cameras, Global Positioning System (GPS), etc., is being realized. These applications need higher performance batteries and chargers.
- High-performance broadband wireless devices for data services, e-mail, e-commerce, etc., are being proposed. In the long run, cell phones may also cut into the laptop market, but the convergence issues need to be considered.
- Lower prices for PDA hardware and services are expected, this bringing about higher unit sales but proportionally lower market value. At a certain point, portable phones will clearly be a valid alternative to residential and business landlines. This could boost unit sales.

Computer memory represents a specific area – see also Section 4.17. Memory chips need to be powered by batteries, so as to protect data during power outages or when the product is deactivated. Small primary button cells predominate; they include a variety of Li, alkaline and other types. Li-based memory preservation solutions should be preferred in the future, but use of other battery systems will decline, mainly due to competition from non-battery systems such as ultracapacitors.

1.2.2. Communications

This sector encompasses the well-established and very large market of cellular phones, now mostly powered by Li-ion batteries, pagers (now a declining technology) and portable transceivers (powered by everything from lead-acid to Li-ion).

Trends include, as mentioned above, an increasing convergence between cell phones, PDAs and cameras. Driving forces and market developments in the portable communications industry include the following:

- The requirement that cell phone numbers be 'portable' makes it easier for consumers to switch service providers; more interest by consumers, possibly more inclination to upgrade hardware when a new service provider is selected; lower price.
- Convergence between cell phones, PDAs, laptops, digital cameras, GPS, etc. – see the previous section.
- High-performance broadband wireless devices with computing capabilities – see the previous section.
- Cordless phones adopt cell phone look and features. Relatively lower prices and higher performance.

1.2.3. Portable Tools

This is a niche market for portable personal grooming, power tools, lawn tools and kitchen tools. This is one of the largest remaining nickel-cadmium (Ni-Cd) battery markets, although the share of Li-ion is rapidly increasing in hobby and professional tools. Lead-acid, primary lithium and alkaline batteries are also used.

Ni-Cd batteries will continue to power low-end tools but will lose ground to Ni-MH for medium-performance systems and to Li-ion for high-end systems. New tools powered by Li-ion feature high power and reduced dimensions.

1.2.4. Medical Applications

Portable medical devices include hearing aids, heart pacemakers, defibrillators, and various diagnostic and therapeutic devices (see Section 3.3). A number of different battery types are used, including Zn-air (mainly for hearing aids), lead-acid, alkaline, nickel-based, primary Li and Li-ion.

Driving forces and market developments include the following:

- Population aging: increasing number of disabled elderly people; continued sales growth for a variety of medical products.
- Possibility for increased subsidies for portable medical products.
- Lower prices for established medical product lines, such as hearing aids; increased unit sales for medium- and high-end (digital) hearing aids.
- Steadily improving heart disease treatment products and new guidelines that increase the number of potential implantable defibrillator patients; continued sales growth for cardiac rhythm management devices.

1.2.5. Other Portable Products

This sector includes lighting, toys, radios, scientific instruments, photographic devices, smart cards, watches and clocks, etc. A wide variety of primary and secondary systems are used, with aqueous or non-aqueous electrolytes.

Driving forces and market developments include the following:

- Increased demand for portable video games; growing unit and market value from an already large base.
- Increased demand for wireless game products; growing unit and market value from a relatively small base.
- Increased interest in all kinds of toy robots; a better defined market niche may begin with increased sales in low-, medium- and high-end products.
- Increased use of high-performance Original Equipment Manufacturer (OEM) Li-ion and Ni-MH batteries.
- Continued demand for GPS systems, including units incorporated in cell phones; steadily growing GPS sales, with some decrease in unit price.
- Slow and steady growth in consumer weather instruments.

In Table 1.5, an evaluation of the world battery market for the portable device sectors mentioned thus far is reported (decade: 2006–2016). All sectors manifest a growth, although with a different pace. Changes in the growth rate may result from significant technology developments.

Table 1.5. World portable device battery market for the decade 2006–2016.

Sector	Year		
	2006	2011	2016
Computing[a]	3500	3600	3750
Communications	2450	2900	3100
Tools	280	340	380
Medical	650	770	880
Other portable	13 400	14 650	15 300

Note: The values represent manufacturer's wholesale and are in 2006 million dollars (no correction for inflation).
Source: Courtesy of BCC Research.
[a] Includes computer memory.

The high value of the market for 'other' portable devices corresponds to a very large number of applications in this area (see Table 1.1). Many of these applications are powered by primary batteries, especially Zn-carbon and alkaline, that represent ~70% of the total batteries sold.

1.2.6. UPS and Backup Batteries

Uninterruptible and emergency power supplies are activated when utility power is interdicted. Large units are used to provide standby power to telecommunications arrays.

Lead-acid and Ni-Cd batteries predominate, but higher performance systems, including sodium/sulphur, vanadium-redox and Li-ion batteries are emerging.

1.2.7. Aerospace and Military Applications

In this area, there is a wide variety of portable and stationary high-profile applications, for example civilian and military robots, manned and unmanned aircraft, satellites, wireless transmission systems, beacons, etc.

Virtually all battery types are used, including nickel-based, primary Li and Li-ion, alkaline and lead-acid. Many types of specialty batteries are used to meet unique performance requirements, but there is a continuing trend towards Li-based systems.

Driving forces include the following:

- Increased number of conflicts in some areas of the world.
- Improved advanced battery-powered devices, for example those of the soldier equipment.

- Development of EV fleets for non-combat missions.
- Adoption of battery-powered fighting or exploration vehicles.
- Robots, including those of much reduced dimensions (microrobots).

1.2.8. Electric Vehicles and Hybrid Electric Vehicles

This is the still relatively small, but potentially attractive market for cars and trucks with an electric engine. This includes some ‘plug-in’ electric vehicles (EVs), where the battery stacks are recharged from the utility power grid and hybrid electric vehicles (HEVs), where an internal combustion engine charges the battery through the generator. ‘Regenerative braking’ uses kinetic energy to recharge the battery when the vehicle slows down.

Lead-acid and, especially, Ni-MH systems are used in most vehicles. Li-ion is another promising option.

The current trend is towards HEV systems, whereas pure battery-powered vehicles are trying to regain momentum (see Chapter 5). There is a potential for competition from fuel cells; vehicles powered by hydrogen fuel cells are especially investigated in Europe.

Industrial vehicles, for example forklifts and burden carriers, use lead-acid batteries, and despite promising non-lead alternatives, there is a little motivation to change.

For EV and HEV commercialization to continue, the main problems to be addressed are:

- Cost-effectiveness, especially compared to gasoline or diesel fuel (but also alternative fuels such as natural gas or ethanol).
- Technical problems (optimization of performance, comfort, cycle life, etc.).
- Safety issues.

1.2.9. Internal Combustion Engine (ICE) Vehicles

These vehicles use lead-acid starting, lighting and ignition (SLI) batteries in all areas of the globe. However, developing countries tend to use less expensive units. Japanese, American and Western European consumers tend to be the early innovators who employ new technology as it is introduced.

Examples of innovation are dual batteries, which are essentially two separate batteries fabricated into a single package: if one battery in the set is inadvertently discharged, the other auxiliary battery can provide cranking power (see Chapter 5). Trends include use of 36/42 V systems in substitution of conventional 12/14 V systems, although cost issues have delayed their acceptance. There is some consideration for portable jump-start batteries, including Li-ion products.

1.3. Application's and Battery's Life

Let us consider an electronic device, for example a notebook or a medical instrument. Given the electronic characteristics, the size and the operating conditions of the device, the battery requirements become obvious and the choice is oriented. While this allows discarding a number of batteries, those chosen can be optimized in their functioning, so that they can reach the performance observed in laboratory tests. High-end batteries are now endowed with a battery management system (BMS), which manages critical parameters such as charge/discharge voltage, temperature, and maximum current, so as to prolong battery life, while ensuring at the same time a high safety level.

However, as is obvious, the device's runtime also depends on its own power characteristics, and care must be exerted to reduce power consumption as much as possible. This can be obtained by a proper component selection and by a judicious management of the device especially in standby mode, when unduly high currents must be minimized.

At the same time, any other feature of the device that may reduce the battery life must be considered. For instance, its thermal behaviour is of paramount importance, as any heat transferred to the battery would shorten the battery life; therefore, proper heat shielding and/or cooling means, when possible, must be put in operation.

On the basis of the above considerations, in Chapter 3 (Portable Applications) particular emphasis will be put on the dual management action for the device and its battery.

Obviously, non-portable high-end applications too are endowed with management features. Therefore, mentions of management actions in industrial and vehicular applications will also be given in Chapters 4 and 5.

An overview of the characteristics of battery management is reported in Ref. [4], with examples of management for batteries used in non-portable applications.

References

1. MPower, "Batteries and Other Energy Storage Devices", 2005.
2. MPower, "Battery Applications", 2005.
3. D. Saxman, in *Industrial Applications of Batteries. From Cars to Aerospace, Energy Storage*, M. Broussely and G. Pistoia, Eds., Elsevier, Amsterdam, 2007.
4. MPower, "Battery Management Systems", 2005.

Chapter 2

BATTERY CATEGORIES AND TYPES

2.1. Introduction

The type(s) of batteries used in specific applications will be mentioned in Chapters 3–5, but a concise review of battery chemistries and their main features will be given in this chapter.

Common classifications of batteries are (1) primary/secondary; (2) aqueous/non-aqueous; (3) low/high power; and (4) according to the size, for example button, prismatic and cylindrical. In this chapter, the division will be made according to the main application categories specified in Chapter 1. Therefore, the following three groups may be identified. (In this book, for a battery designated by the chemical formula of the negative and positive electrode, for example Zn and MnO_2, the notation with a slash will be used: Zn/MnO_2. For a battery designated by a conventional definition, for example zinc–carbon, the notation with a dash will be used: Zn-C.)

1. *Batteries mainly used in portable applications*
   Zinc-carbon
   Alkaline
   Zinc-air (small size)
   Primary zinc/silver oxide
2. *Batteries used in both portable and industrial/vehicular applications*
   Primary lithium
   Lithium ion
   Nickel–cadmium
   Nickel-metal hydride
   Lead-acid (in a few portable applications only)
   Secondary zinc/silver oxide
3. *Batteries mainly used in industrial/vehicular applications*
   Nickel–hydrogen
   Nickel–zinc
   Nickel–iron
   Large zinc-air
   Flow batteries
   Thermal batteries (include Li-metal-polymer)

In Tables 2.1–2.13, the characteristics of several systems, aqueous and non-aqueous, primary and secondary, are listed. It is necessary to treat this kind of data with some care when comparisons are made, as the batteries (cells) may differ in size, construction, technology maturity, etc.

2.2. Batteries for Portable Applications

Up to the 1940s, Zn-C was the only system used for primary batteries. Since then, several other systems have been commercialized: alkaline batteries, in particular, have gained wide acceptance thanks to their improved performance vs the Zn-C ones, as shown in Table 2.1, where the most important aqueous primary batteries are compared.

2.2.1. Zinc-Carbon Batteries

The first Zn/MnO_2 battery was introduced in the middle of the nineteenth century; its electrolyte is immobilized in an inert support, which justifies the name of "dry battery". This cheap battery is still largely used in moderate and light drain applications. However, it cannot compete with alkaline Zn/MnO_2 in terms of performance, and its use is declining except for emerging countries [2].

Dry batteries can use either the Leclanché or the $ZnCl_2$ system (Table 2.1). The former uses an aqueous electrolyte containing NH_4Cl (26%) and $ZnCl_2$ (8.8%), while the latter contains $ZnCl_2$ (15–40%). In both electrolytes, inhibitors of Zn corrosion are added.

The electrodes are basically the same in both systems. The Zn can of the cell is also the anode, while the cathode is a mix of electrochemically active MnO_2 and carbon. In principle, the electrochemistry of the Zn-C cell is quite simple with Zn oxidation to Zn^{2+} and Mn^{4+} reduction to Mn^{3+} (MnOOH or Mn_2O_3). In practice, the reactions are rather complicated and depend on several factors, such as electrolyte concentration, temperature, rate and depth of discharge.

These batteries can have a cylindrical or a flat configuration. In the former, a bobbin containing a mixture of MnO_2, carbon black and electrolyte surrounds the carbon rod, serving as a current collector for the cathode (hence the name Zn-C). The separator between the Zn can and the bobbin is usually paper thinly coated with a paste of gelled flour and starch absorbing the electrolyte. To prevent electrolyte leakage due to perforation of the Zn can, the latter is jacketed with a polymer film or polymer-coated steel.

In the flat configuration, rectangular cells are stacked to give prismatic batteries, for example the popular 9-V size. The construction in this case is quite

Table 2.1. Comparison of the main characteristics of aqueous primary batteries.

	Leclanché (Zn/MnO_2)	Zinc Chloride (Zn/MnO_2)	Alkaline/Manganese Dioxide (Zn/MnO_2)	Silver Oxide (Zn/Ag_2O)	Zinc-Air (Zn/O_2)
System	Zinc/manganese dioxide	Zinc/manganese dioxide	Zinc/alkaline manganese dioxide	Zinc/silver oxide	Zinc/oxygen
Voltage per cell	1.5	1.5	1.5	1.5	1.4
Positive electrode	Manganese dioxide	Manganese dioxide	Manganese dioxide	Monovalent silver oxide	Oxygen
Electrolyte	Aqueous solution of NH_4Cl and $ZnCl_2$	Aqueous solution of $ZnCl_2$ (may contain some NH_4Cl)	Aqueous solution of KOH	Aqueous solution of KOH or NaOH	Aqueous solution of KOH
Overall reaction equations	$2MnO_2 + 2NH_4Cl + Zn \rightarrow ZnCl_2 \cdot 2NH_3 + Mn_2O_3 \cdot H_2O$	$8MnO_2 + 4Zn + ZnCl_2 \cdot 9H_2O \rightarrow 8MnOOH + ZnCl_2 \cdot 4ZnO \cdot 5H_2O$	$Zn + 2MnO_2 + 2H_2O \rightarrow Zn(OH)_2 + 2MnOOH$	$Zn + Ag_2O \rightarrow ZnO + 2Ag$	$2Zn + O_2 \rightarrow 2ZnO$
Typical commercial service capacities	Several hundred mAh	Several hundred mAh to 38 Ah	30 mAh to 45 Ah	5 to 190 mAh	30 to 1100 mAh
Specific energies (Wh/kg)	65 (cylindrical)	85 (cylindrical)	80 (button); 145 (cylindrical)	135 (button)	370 (button); 300 (prismatic)
Energy densities (Wh/L)	100 (cylindrical)	165 (cylindrical)	360 (button); 400 (cylindrical)	530 (button)	1300 (button); 800 (prismatic)
Discharge curve	Sloping	Sloping	Sloping	Flat	Flat
Temperature range (storage)	−40 to 50°C	−40 to 50°C	−40 to 50°C	−40 to 60°C	−40 to 50°C

(*Continued*)

Table 2.1. (*Continued*)

	Leclanché (Zn/MnO_2)	Zinc Chloride (Zn/MnO_2)	Alkaline/Manganese Dioxide (Zn/MnO_2)	Silver Oxide (Zn/Ag_2O)	Zinc-Air (Zn/O_2)
Temperature range (operating)	−5 to 55°C	−18 to 55°C	−18 to 55°C	−10 to 55°C	−10 to 55°C
Effect of temperature on service capacity	Poor low temperature	Good low temperature relative to Leclanché	Good low temperature	Low temperature depends upon construction	Good low temperature
Internal resistance	Moderate	Low	Very low	Low	Low
Gassing	Medium	Higher than Leclanché	Low	Very low	Very low
Cost (initial)	Low	Low to medium	Medium plus	High	High
Cost (operating)	Low	Low to medium	Low to high	High	High
Capacity loss per year @ 0°C	3%	2%	1%	1%	NA
Capacity loss per year @ 20°C	6%	5%	3%	3%	5% (sealed)
Capacity loss per year @ 40°C	20%	16%	8%	7%	NA

Source: Adapted from Ref. [1].

different from that of cylindrical cells. The Zn anode is coated with a carbon layer, to act as an electron conductor for the cathode side of the adjacent cell. In the 9 V battery, six flat single cells are stacked in series. In these cells, the flat design provides more space to the cathode mixture, so the energy density is higher. The prismatic battery shape, in turn, is more favourable to space saving: the volumetric energy density is twice that of a cylindrical battery. The flat configuration is available in multicell batteries only (from four to several hundred cells in a stack or set of stacks).

Modern cells mostly use either chemical MnO_2 or electrolytic MnO_2, whose percentage of active material is 90–95% (the remainder is mostly H_2O plus several impurities). As a carbon, the highly porous acetylene black is preferred as it remarkably increases the poor conductivity of MnO_2. The Zn anode is ultrapure and is used to alloy with Cd (~0.3%) and Pb (~0.6%) to improve its metallurgical properties and reduce corrosion. The legislation of several countries now prohibits the use of these toxic metals beyond a given (very low) limit, so that their content in modern cells is practically zero. Similarly, the use of Hg as the main corrosion inhibitor has been eliminated. Other materials now considered as inhibitors include Ga, Sn, Bi, glycols or silicates. Zn corrosion is primarily due to the acidic character of both the NH_4Cl and the $ZnCl_2$ solution (the latter is more acidic).

The solution containing $ZnCl_2$ is preferred. Indeed, formation of sparingly soluble Zn salts, which tend to accumulate near the electrode, greatly limits ion diffusion in the Leclanché cell. With the $ZnCl_2$ solution, this phenomenon is reduced, so that faster diffusion and enhanced rates of discharge are allowed. The better performance of the $ZnCl_2$ cell, especially at high currents and moderately low temperature (down to −10°C), is counterbalanced by a higher cost. In terms of performance and cost, this cell lies between the common Leclanché and the alkaline cell [3]. Another advantage of the $ZnCl_2$ cell is given by its lower self-discharge rate (Table 2.1).

2.2.2. Alkaline Batteries

The alkaline Zn/MnO_2 battery was introduced in the early 1960s. Its advantages over the Zn–C system can be summarized as [4]:

- Up to 10 times the service life
- Long runtime at continuous, high drain discharge
- No need for "rest periods"
- Rugged, shock-resistant construction
- Cost-effective on a cost-per-hour-of-service basis
- Good low-temperature performance (down to −20°C vs −5°C for Zn-C)

- Excellent leakage resistance
- Low self-discharge (3% per year at 20°C).

If the cost-per hour of service is considered, especially at high drains and continuous discharge, the alkaline battery becomes cheaper than Zn-C. Its higher capacity and energy vs the standard Zn-C battery (see Table 2.1) is due to the use of high-grade anode and cathode materials, and to the more conductive alkaline electrolyte. A comparison of the performance of the two batteries is shown in Figure 2.1 [4]. The difference is particularly evident in low-resistance devices: the runtime through a 3-Ω resistance is 3 h for a D-size Zn-C cell and 45 h for a D-size alkaline cell.

The anode is essentially high-purity Zn powder. Its higher surface area vs that of a Zn can afford higher discharge rates, while the electrolyte is more uniformly distributed. Furthermore, the combination of a porous anode and a conductive electrolyte reduces the extent of accumulation of reaction products near the electrode, resulting in a higher rate capability. The low impurity level of the zinc powder (especially Fe) has facilitated elimination of Hg, Pb or other heavy metals as gassing suppressors. A gelling agent is instead necessary for

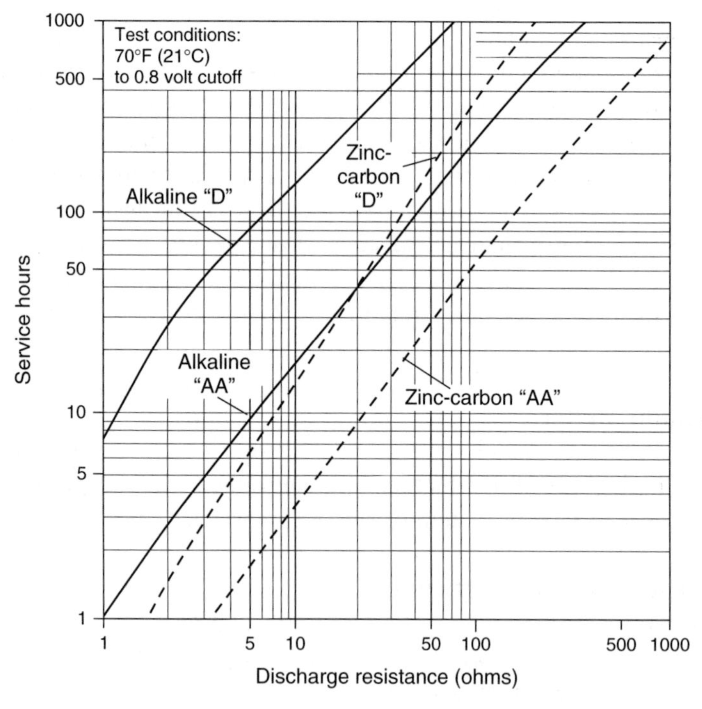

Figure 2.1. Comparison of AA- and D-size Zn-C and alkaline cells.
Source: From Ref. [4].

immobilizing the electrolyte and improving electrode processibility. To this end, starch, cellulose derivatives or polyacrylates are often used. The anode also contains the electrolyte, that is an aqueous KOH solution (35–52%).

The cathode is based on electrolytic MnO_2, as only this form can grant high power and long shelf life. The electronic conductor is carbon in the form of graphite, although some acetylene black may also be added to enhance the surface area.

The separator, which has to be chemically stable in the concentrated alkaline solution, is normally a non-woven fabric, such as cellulose, vinyl polymer, polyolefin or a combination thereof.

The high porosity of cathode, anode and separator allows their saturation with the electrolyte. The homogeneous electrolyte distribution and its high conductivity afford high discharge rates also on continuous drains and at low temperatures.

Zn powder is obviously quite reactive and can decompose H_2O with the production of hydrogen, which can cause MnO_2 self-discharge and generate an overpressure. As mentioned above, reducing the impurity level in the Zn powder greatly limits gassing. Otherwise, additives for the anode are necessary, such as ZnO (or other oxides) or organic inhibitors (polyethylene oxide compounds).

Alkaline batteries can be built with cylindrical, button or prismatic configurations.

In a cylindrical alkaline cell, the can is not an active material, as it is made of steel or nickel-plated steel and acts as the cathode current collector. The cathode is pressed against the steel can either applying a high pressure to the powder when in contact with the can or forming annular pellets, which are then inserted into the can. The Zn powder is allocated in the central cavity, around a brass current collector welded to the cell bottom (negative cap).

A plastic grommet, sealed to the cell can, ensures that the cell is leak-proof. The grommet incorporates a membrane vent for relieving overpressure in case of short circuits or cell abuse.

In a button cell, the Zn powder is in the upper part and contacts the negative cell top, a steel foil usually having an external layer of nickel and an internal layer of Cu or Sn. The can, acting as a container and cathode collector, is made of Ni-plated steel. It is insulated from the cell top by a plastic grommet over which is crimped to seal the cell. The MnO_2 pellet, at the bottom of the cell, is covered by a separator and by an absorber for the electrolyte.

A standard prismatic battery is multicell and constructed as described for the Zn-C battery.

In 1999, premium alkaline cells were commercialized. They have a better performance at high discharge rates than the standard models. This was made possible by a further reduction of the cell resistance through (1) coating both the negative and positive current collector, (2) using a finer graphite grade and (3) packing more MnO_2 into the space available for the cathode. Coating reduces the build up of corrosion products on the current collectors, while a finer graphite powder improves the electronic conductivity.

Button cells have capacities of 25–60 mAh, cylindrical cells of 0.5–22 Ah and prismatic batteries of 0.16–44 Ah. This wide capacity range makes alkaline batteries suitable for several applications, from consumer to industrial/military devices. The former are more numerous and include remote controls, photographic equipment, flashlights, radios, watches, calculators, home healthcare devices, etc., while the latter include portable medical and industrial instrumentation, emergency lighting, communication equipment, electrical measurement devices, etc.

2.2.3. Primary Zinc/Silver Oxide Batteries

The Zn/silver oxide system has a high energy and a flat potential. Furthermore, it performs well at low temperatures and has a good shelf life. These characteristics make this system ideal for electronic devices requiring a small, high-capacity, long-lasting and constant-voltage cell. As a primary battery, it is mainly produced in button sizes, while its use in larger batteries is limited by the high cost of silver [5].

Zn/Ag_2O cells were introduced in the early 1960s as power sources for electronic watches, with currents ranging from a few microamperes (LCD displays) to hundreds of microamperes (LED displays). These cells are also used in pocket calculators, hearing aids, cameras, glucometers, etc.

The anode is zinc powder, the cathode is monovalent silver oxide, Ag_2O, and the electrolyte is a KOH or NaOH aqueous solution (20–45%).

The Zn powder has to be highly pure, as already pointed out for alkaline cells. Indeed, impurities (such as Cu, Fe and Sn) favour Zn corrosion and formation of H_2, which results in an overpressure. In commercial cells, the Zn powder is amalgamated with Hg to keep corrosion under control. A low percentage of Hg is permitted in these button cells, in view of the small amount of Zn: indeed, the maximum capacity is 165 mAh. Gelling agents, such as polyacrylic acid and the like, are added to the anode to facilitate electrolyte accessibility.

Ag_2O is now preferred as a cathode in commercial cells. Unlike AgO, used until the early 1990s, it has a stable potential and does not need to be stabilized by heavy metals, as its reactivity with alkalis is low. As Ag_2O is a poor semiconductor, some graphite ($<5\%$) is added. Furthermore, the reduction of Ag_2O produces Ag, which helps to decrease the cathode resistance. Indeed, the total cell reaction is

$$Zn + Ag_2O \rightarrow ZnO + 2Ag$$

The alkaline electrolyte contains some zincate to control gassing. KOH is preferred over NaOH in button cells submitted to high drains, as its conductivity is higher. Instead, KOH is preferred if the cells are to be used at low temperature: with this electrolyte, a working temperature as low as −28°C can be attained.

Button cells have the cathode contained in the can and the anode attached to the cap. A barrier of cellophane or grafted plastic membrane is set on top of the cathode pellet to prevent Ag^+ migration to the anode (Ag_2O is slightly soluble in alkalis). A separator, usually fibrous polyvinyl alcohol, is added on top of the barrier to act as a further protection. Zn/Ag_2O button cells are available in a variety of sizes with capacities from 8 to 165 mAh.

Large batteries (up to hundreds of Ah) can be used in space applications (especially for launch vehicles).

2.2.4. Primary Zinc-Air Batteries

Primary Zn-air batteries, in their button form factor, are especially designed to provide power to hearing aids. In most hearing aid applications, these batteries can be directly substituted for Zn/Ag_2O and give the longest service of any common battery system.

Some of their characteristics, also useful for other electronic devices, are [6]:

- Highest energy density for miniature batteries
- Relatively flat discharge curve (at 1.2–1.3 V)
- Activated by removing covering (adhesive backed tab) from air access hole
- Constant capacity vs load and temperature at standard drains
- Excellent shelf life (prior to activation)
- Low cost.

Zn-air batteries use O_2 from the air as the active cathode material. O_2 diffuses through the cathode and, after reaching the cathode interface with the alkaline electrolyte, is catalytically reduced. because one active electrode material (O_2) is outside the cell, the other (Zn) occupy most of the cell volume, as shown in Figure 2.2. The capacity of the cell only depends on the anode, and, since the amount of Zn that can be stored in a given volume is about twice that of Zn/MnO_2 or Zn/Ag_2O cells, the energy of the Zn-air cell is very high both on a gravimetric and a volumetric basis (see Table 2.1).

The anode is formed by high surface area Zn powder mixed with the electrolyte and, in some cases, a gelling agent. The cathode region is quite

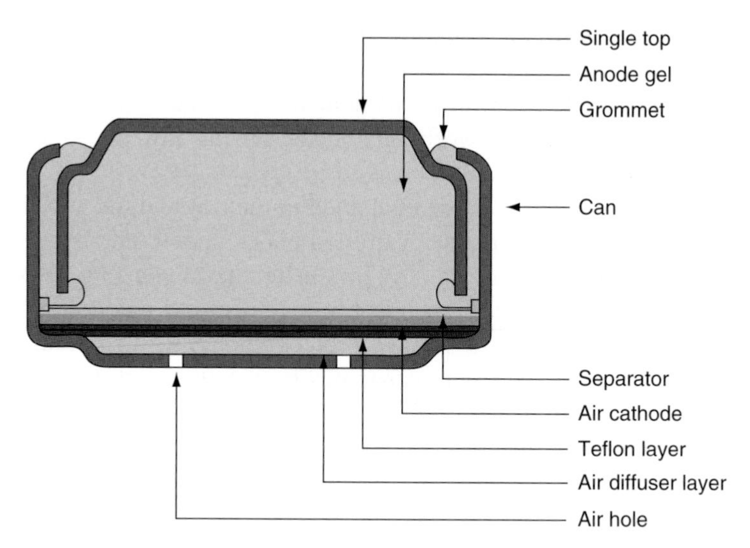

Figure 2.2. Cross-section of a button Zn-air cell.
Source: From Ref. [7].

complex: the holes allow air access; the air diffuser layer distributes O_2 uniformly over the cathode; the hydrophobic Teflon layer is O_2-permeable but limit water vapour access; and the air cathode is formed by a metallic mesh supporting the catalyst layer (carbon blended with Mn oxides and Teflon powder).

The maximum current the cell can support depends on the O_2 availability. If air access were not regulated, excess O_2 could reach the cathode and high discharge rates would be possible. In fact, strict regulation is necessary to limit the inlet of H_2O vapour and other gases that would degrade the cell. The electrolyte is typically a 30% KOH solution in water. In dry days, it tends to lose water, while the reverse occurs in wet days. In both cases, the service life of the cell is affected; this explains why holes and diffusion membrane are necessary.

The overall reaction may be simply written as

$$Zn + ½O_2 \rightarrow ZnO$$

Zn-air button cells have capacities in the range 40–600 mAh, which are delivered at low rates (0.4–2 mA). At these rates, flat discharge curves are obtained, with excellent capacity retention even at 0°C. Operating lives of 1–3 months are typical: this is not surprising if one considers that the cells are in contact with the atmosphere, this favouring direct Zn oxidation, carbonation of the electrolyte and gas transfer. So, they are better used in continuous-drain applications.

Apart from hearing aids, primary Zn-air cell can be used, for instance, in:

- Cardiac telemetry monitors (8.4 V battery)
- Bone growth stimulation (also with a multicell battery)
- Telecommunication receivers (pagers, e-mail devices, etc.).

2.2.5. Strong vs Weak Points and Main Applications of Aqueous Primary Batteries

Table 2.2 lists advantages/disadvantages and main applications of the aqueous primary batteries described in the previous sections. These batteries have a remarkable market share, thanks especially to the Zn/MnO_2 systems. Indeed, the batteries belonging to the group of primary aqueous represent about 35% in terms of value of all batteries sold.

2.3. Batteries Used in Both Portable and Industrial/ Vehicular Applications

2.3.1. Primary Lithium Batteries

This section includes batteries with Li metal as an anode. They feature the highest energies (both on a weight and volume basis) among all primary batteries. Actually, the Zn-air system also features very high energies (see Table 2.1) but this is a peculiar system with very short operating lives, as already outlined in Section 2.2.4.

The first commercial Li batteries were marketed in the 1970s. For cost reasons, their market share lags behind that of aqueous primary batteries. Yet, in some applications, that is those requiring long operation/storage times, extreme temperatures or high power, these are the batteries of choice.

The basic characteristics making primary Li batteries suitable for many applications are [8]:

- High energy (over 200 Wh/kg and 400 Wh/L)
- High and flat potential
- High power (in spirally wound configurations)
- Long shelf life, with capacity losses of $\sim$1% per year at room temperature
- Wide temperature range
- Construction in several form factors (coin, cylindrical, prismatic and very thin batteries) and with capacities ranging from a few mAh to several Ah.

Table 2.2. Pros & cons and main applications of aqueous primary batteries.

Battery Type	Dry Manganese	Alkaline Manganese	Silver Oxide	Zinc Air
Designation	R	LR	SR	PR
Nominal voltage	1.5 V	1.5 V	1.55 V	1.4 V
Pros	• Wide applications • Excellent anti-leakage • No Hg, no Cd • Lowest cost	• For heavy and continuous use • High reliability • No Hg, no Cd • Higher capacity than dry manganese	• Flat discharge curve • Superior long-term reliability • High-energy density • High power	• For heavy and continuous use • Stable discharge curve • Excellent anti-leakage • High-energy density
Cons	• Low-energy density • Poor performance at low temperature and high rate	• More expensive than dry manganese but better cost/performance ratio at high rates	• Expensive but cost-effective in button cells • Good low-temperature performance	• Limited power • Performance affected by environment: flooding, drying out • Short activated life
Applications	• Radios • Headphone stereos • Remote controllers • Transceivers • Flashlights • Clocks • Calculators	• Cordless phones • Headphone stereos • CD players • LCD TVs • Lanterns • Electric shavers • Remote controllers	• Wrist watches • Cameras • Calculators • Hearing aids • Glucometeres	• Hearing aids • Telecommunications • Medical applications

In this section, the batteries listed in Table 2.3 are described. The first two are examples of soluble-cathode batteries: carbon electrodes support their reduction reactions.

2.3.1.1. Lithium/Sulphur Dioxide Batteries

Li/SO_2 cells were the first lithium cells to be commercialized. Their main features are [9]:

- High specific energies and energy densities (~275 Wh/kg and ~430 Wh/L)
- High pulse power density (~650 W/kg)
- Wide temperature range (−55/−60°C to +70°C)
- Long shelf life (more than 10 years at room temperature and 1 year at 70°C).

These cells are produced in spirally wound configurations, to exploit their power capabilities.

The anode is Li foil, while the cathode is Teflon-bonded acetylene black. The solvent is CH_3CN in which LiBr and SO_2 are dissolved. The overall reaction is:

$$2Li + 2SO_2 \rightarrow Li_2S_2O_4$$

$Li_2S_2O_4$ (Li dithionite) precipitates into the pores of the carbon cathode. The stability of this cell (and of other cells with a soluble cathode) is connected with the formation of a passivating film on the Li surface as soon as Li is exposed to SO_2. The film growth rate increases during storage of partially discharged cells.

The presence of SO_2 requires a special construction: hermetic seals are used to prevent SO_2 loss. The cell is pressurized (2 atm) to keep the electrolyte in the liquid state and a safety vent is incorporated in the cell to cope with pressure values exceeding certain limits (e.g. 24 atm). The acetylene black–Teflon mix, supported on Al screen, is characterized by high values of conductivity, surface area and porosity. The last feature ensures that $Li_2S_2O_4$ precipitating on the cathode does not cause its clogging in early reaction stages.

The electrolyte contains 70% SO_2 and has a high conductivity even at −50°C ($2.2 \times 10^{-2}\,\Omega^{-1}\,cm^{-1}$). This affords the use of Li/SO_2 batteries in applications prohibited to other chemistries. They can maintain a high proportion of their capacity even at the 1-h discharge rate, whereas the capacity of aqueous batteries with a Zn anode starts declining at the 20 to 50 h rates.

Table 2.3. Characteristics of primary Li batteries.

Attribute/Cell	Li/SO_2	Li/$SOCl_2$	Li/MnO_2	Li/CF_x
Average voltage (V)	2.7–2.9	3.4–3.6	2.8–3.0	2.6–2.8
Specific energy (Wh/kg)	260–280	450–600 (bobbin) 200–450 (spiral)	250–300 (bobbin) 150–230 (spiral)	200–250 (small) 530–600 (large)
Energy density (Wh/L)	400–450	700–1100 (bobbin) 400–850 (spiral)	580–650 (bobbin) 400–520 (spiral)	580–635 (small) 900–1050 (large)
Power density	High	Low/medium (bobbin) Medium/high (spiral)	Low/medium (bobbin) Medium/high (spiral)	Low
Temperature range (°C)	–55 to –70	–55 to –85 (standard) –50 to –150 (h.t.)	–40 to –85 (laser) –20 to –60 (crimp)	–40 to – 85 –40 to –125 (h.t.)
Shelf-life (years at room temperature)	10	10–15	10	15
Relative market	5	15	100	1
Relative cost (per kWh)	0.9	1.5	1	2

Early design cells had a Li/SO_2 ratio as high as 1.5:1. However, it was ascertained that this ratio greatly impaired the cell safety. Indeed, in deeply discharged cells, when the SO_2 concentration is below 5% and the passivating film removed, the reaction of Li with CH_3CN occurs. Therefore, cells with a Li/SO_2 ratio close to 1 are now preferred: in these balanced cells, Li remains passivated as there is a sufficient amount of SO_2.

Li/SO_2 cell are fabricated in cylindrical sizes with capacities ranging, for standard cells, from 0.86 Ah (1/3C size) to 11.0 Ah (F size). This last cell can stand continuous currents of 8.0 A and pulse currents of 60 A.

A drawback, common to all soluble-cathode cells, is the so-called voltage delay. Extended storage, particularly at high temperatures, favours formation of a thick film on the anode; therefore, discharges at high rates and low temperatures start with a lower voltage. The time needed to resume the standard voltage depends on the length and temperature of storage. This effect is not appreciated in room-temperature operations and can be eliminated by a pulse discharge at high rate to depassivate the anode.

Safety concerns arise on overdischarge and thermal abuse, thus imposing the presence of controlling components in the cells/batteries. If the cell is to deliver high currents, a fuse is necessary; if there is any possibility of cell (or battery) charging, a diode is also needed. The use of a microporous polypropylene separator allows electrolyte flow in its 200-nm channels, while blocking carbon particles that might reach the anode and short the cell. For a more reliable control, an electronic circuit (see Chapter 3) is needed, especially for larger cells on heavy duty.

Li/SO_2 cells have a number of military and civilian applications, including radio communications, lighting/night vision, automotive electronics, professional electronics, meteorology/space, toll-gate systems.

2.3.1.2. Lithium/Thionyl Chloride Batteries

Li/$SOCl_2$ cells have very high energies (see Table 2.3) and their service life can reach 15–20 years. Moreover, these cells can be stored for long (with capacity losses of ~1–2% per year at room temperature) and can be operated in an exceptionally wide temperature range (−80°C (with a special electrolyte) to 150°C) [9].

Low-rate Li/$SOCl_2$ cells are built in the bobbin-type cylindrical configuration, while moderately high-power cells are built in the spirally wound configuration.

The anode is made by a Li metal foil, porous carbon is the cathode support, and $SOCl_2$ is both the active cathode material and the solvent for the electrolyte salt (usually $LiAlCl_4$).

The overall cell reaction can be written as

$$4Li + 2SOCl_2 \rightarrow 4LiCl + S + SO_2$$

LiCl precipitates on the carbon surface and stops cell operation when pore clogging occurs in cathode-limited cells. SO_2 is soluble in the electrolyte, while S is soluble up to ~1 mol/L and can precipitate on the cathode towards the end of discharge. LiCl is also the main component of the passivating film formed on the Li anode.

The cell capacity could be improved by adding excess $AlCl_3$ to the electrolyte. In this case, soluble $LiAlCl_4$ is formed instead of LiCl, so that no pore clogging occurs. However, $AlCl_3$ dissolves the LiCl passivating film on Li, thus favouring its corrosion. For this reason, excess $AlCl_3$ is only used in high-rate reserve cells.

Cells for low-rate applications are essentially constructed with the bobbin-type configuration shown in Figure 2.3. The Li anode contacts a steel can. The porous cathode, Teflon-bonded acetylene black, occupies most of the can volume and includes a metallic cylinder as a current collector for larger cells (see figure) or a pin for smaller cells (i.e. AA size). Bobbin-type cells are cathode-limited, as this is considered safer than the anode-limited type. No hazards have been observed when submitting these cells to short circuits,

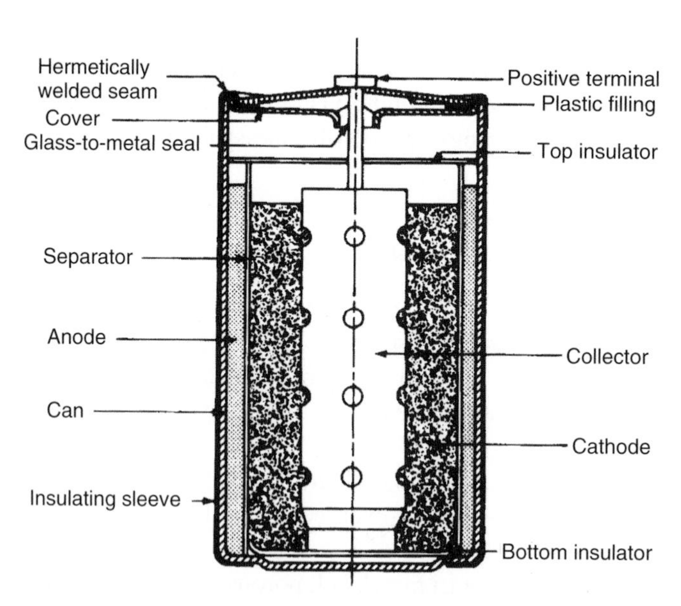

Figure 2.3. Cross-section of a bobbin-type Li/$SOCl_2$ cell.
Source: Courtesy of Tadiran.

overdischarge or overcharge. Their capacities range from 0.36 Ah to an impressive 38 Ah for a DD cell.

The spirally wound configuration allows using the Li/$SOCl_2$ couple in applications requiring medium to moderately high rates. In this case, safety devices, such as a vent and a fuse are incorporated to prevent accidents stemming from overpressures or short circuits. Overpressure can be reached on overdischarge: the temperature may rise to 115°C and the pressure to 140°psig in cathode-limited cells. In the spirally wound configuration, the energy output is reduced (there is more inactive material inside the cell) and the shelf life is also shortened (the reactivity increases with cathode surface area). Furthermore, the upper limit of the operating temperature range cannot overcome ~85°C.

Several additives can be used for a better performance. In particular, the low-temperature performance may be improved by using $LiGaAl_4$ instead of $LiAlCl_4$. With this salt, a working temperature of −80°C is attainable, as demonstrated, for instance, by the cells used in the Mars Microprobe Mission. At the other extreme, Li/$SOCl_2$ cells of the bobbin type can work at temperatures up 150°C, or even 180–200°C in oil exploration (see Section 4.7).

Bobbin Li/$SOCl_2$ cells have a high capacity, but their rate capability is not sufficient for some applications, that is GPS, automatic meter readings (AMRs), etc. Some alternative types of Li/$SOCl_2$ cells can be used if both high energy and power are requested. In an approach proposed by Tadiran (Israel), a bobbin-type cell is coupled to a hybrid layer capacitor. The cell manages ordinary loads, while the capacitor can take over when a high current pulse (up to several amperes) is required.

The voltage delay effect, also present in Li/$SOCl_2$ cells, may be reduced by applying a conductive polymeric film on the Li anode. The double-layer film on uncoated Li has a porous layer whose thickness increases with storage, while the thickness of the conductive polymer film remains constant, thus limiting the ohmic drop at the start of discharge [10].

Ordinary bobbin-type cells (–55 to 85°C) can be used for CMOS memory backup, medical devices, lighting, emergency locators, tracking, automotive electronics, alarm systems, etc.

High-temperature cells (up to 150°C or more) can be used in measurements while drilling (MWD), tyre pressure monitoring systems (TPMS), geothermal applications.

Spirally wound high-power cells can be used in radio communications, space applications, security alarms, GPS and, in general, in military applications.

2.3.1.3. Lithium/Manganese Dioxide Batteries

Commercialized since 1975, this primary Li battery is the most widely used, as it owns several nice features: high voltage, high energy (both on a gravimetric and a volumetric basis), wide operating temperature range, good power (in some designs), long shelf life, safety and low cost. The Li/MnO_2 battery is produced by several manufacturers in coin, cylindrical or prismatic forms (the last one including thin cells) and can be used in a variety of applications.

Li foil is used as the anode, heat-treated MnO_2 (with a residual H_2O content of ~1%) as the cathode, and a preferred solution is $LiClO_4$-PC-DME (PC, propylene carbonate; DME, dimethoxyethane). Heat treating of electrochemical grade MnO_2 at 350–400°C is fundamental, as the non-treated form has a poor performance [11, 12].

The overall reaction may be written as

$$x\text{Li} + \text{MnO}_2 \rightarrow \text{Li}_x\text{MnO}_2$$

During discharge, Li^+ ions are gradually inserted into the channels of the cathode structure, thus giving rise to a so-called solid solution. The formula Li_xMnO_2, $0 < x < 1$, best explains the gradual accommodation of Li^+ ions into the host structure.

Li/MnO_2 cells for portable and industrial applications are mainly manufactured with the following designs: coin, spirally wound cylindrical, bobbin-type cylindrical, 9 V prismatic and thin cells. The first two designs are most widely used. The bobbin cells have a laser seal, while the spirally wound may have a crimp or a laser-welded seal. The laser sealing technology ensures a cell life of 10 years at room temperature and a wider temperature range. Indeed, these cells can be operated in the range –40 to 85°C, while for those with crimp seals the range is –20 to 60°C.

In the bobbin structure, the amount of cathode is maximized to have high capacity and energy. Because of this structure, the cathode surface area is reduced and these cells can only be used at low drains. In the spirally wound structure, a thin layer of the cathode mix is supported on a metal grid; three strips, Li foil, separator and cathode, are tightly wound to form a high surface area structure capable of sustaining high currents.

The cell cans are made of stainless steel and the separator is polypropylene. Bobbin and spirally wound cells are endowed with safety vents, and the latter have a positive temperature coefficient (PTC) device for additional safety. High-power cells can also have relatively high capacities, for example more than 10 Ah in the D size [2].

The Li/MnO_2 system is also available as a prismatic 9 V battery of the same shape as Zn-C or alkaline cells. However, their internal construction is quite different. Each of the three cells in series contains stripes of anode, cathode and separator, just as in spirally wound cylindrical cells, bent to completely fill the space. These consumer-replaceable batteries last up to five times longer than alkaline and 10 times longer than Zn-C batteries [13]. Very thin "paper" cells, with thicknesses of 1–2 mm are also available [13].

Table 2.4. Applications of Li/MnO_2 cells as a function of cell geometry.

Cell Type	Typical Applications
Coin cell	• Memory backup • Watches • Calculators • Medical equipment • Electronic games • Security devices • Small, low power electronic devices • Automatic sensors and transmitters
Bobbin cell	• Utility meters • Memory backup • Real time clock/calendar • Low power electronic devices • Automotive electronics
Spirally wound cell	• Small power tools • Alarms and detection devices • Communications equipment • High-performance flashlights • Medical instruments • Remote sensing devices • Handheld test apparatus • Utility meters • Flash cameras • Computers • Access controls • Laser devices • Bar code readers • Memory backup • Real time clock

Source: From Ref. [2].

As indicated in Table 2.3, the Li/MnO_2 system is the most widely commercialized among Li batteries. It deserves a more detailed list of applications for its most popular types, that is button and cylindrical cells. These are reported in Table 2.4.

2.3.1.4. Lithium/Carbon Monofluoride Batteries

Another interesting primary battery is the one using polycarbon fluoride, $(CF_x)_n$, as a cathode. This compound, synthesized by direct fluorination of carbon with fluorine gas, has an x value in the range 0.9–1.2, and useful cathode materials have $x \geq 1$. In the following, this cathode material will simply be written as CF_x (where $x = 1$ normally) [14].

In large cells, this system has the highest energy, both on a gravimetric and a volumetric basis, of all primary Li batteries, as shown in Table 2.3. This high energy is maintained at high temperatures. The performance is also good at low temperatures, but at −40°C a remarkable voltage drop is observed.

The overall reaction may be written as

$$Li + CF \rightarrow LiF + C$$

The formation of carbon enhances the electronic conductivity of the cathode. Typical electrolytes useful for this system are $LiAsF_6$ in butyrolactone (BL) or $LiBF_4$ in PC-DME.

Apart from the energy, the Li/CF_x system has a number of pleasant features. Its operating voltage is flat and high (~2.8 V), its capacity is quite high at low-moderate drains, its useful temperature range is wide (–40 to +85°C, and up to 125°C for some cells), and its self-discharge rate is the lowest of any primary Li cell. This last feature is particularly notable: a Li/CF_x cell loses <0.5% capacity after 1 year of storage at room temperature, and less than 4% per year at 70°C. In contrast, this system is not characterized by high power and is better used at low rates. Cost is another weak point – twice that of Li/MnO_2 cells, as indicated in Table 2.3.

Pin, coin, cylindrical and prismatic cells are available, with capacities ranging from 25 mAh to 5 Ah. The pin-type cells use the inside-out design with a cylindrical cathode and a central Li anode. Coin-type cells have a Li foil rolled onto a Cu net, while the cathode, also containing Teflon and acetylene black, is supported on a Ni net. The cylindrical cells mainly use the spirally wound configuration [2]. Prismatic cells are built in high-capacity sizes (typically 40 Ah).

The applications of these cells depend on their design:

- Coin type: low-power consuming cordless appliances, memory backup.
- High-temperature coin type: automotive electronic systems, toll-way transponders, radio frequency identification (RFID).
- Cylindrical type: utility meters, emergency signal lights, electric locks, electronic measurement equipment.
- Prismatic type: biomedical and space systems.

2.3.1.5. Comparison of Li Primary Batteries and Market Considerations

In Table 2.5, strong/weak points are presented for the most important Li primary systems, together with their main applications.

Although the market for primary Li is steadily growing, these batteries still represent a minor fraction of all primary batteries. Indeed, if the US market is

Table 2.5. Pros & cons and main applications of lithium primary batteries.

Battery Type	Wound-type Manganese Dioxide	Thionyl Chloride	Sulphur Dioxide	Carbon Monofluoride
Designation	CR	ER	WR	BR
Nominal voltage	3 V	3.6 V	2.9	2.8
Pros	• For heavy duty use • Wide temperature range • Stable discharge curve • Low self-discharge • Excellent anti-leakage	• High 3.6 V voltage • Flat discharge curve • High-energy density • Wide temperature range • Superior long-term reliability	• Low self-discharge • Usable down to −60°C • High rate capability • Storage up to 70°C	• High energy (even at high temperature) • Flat voltage • Wide temperature range • Very low self-discharge
Cons	• Potential safety issues for high-rate cells	• Voltage delay after storage • Potential safety problems	• Voltage delay • Potential safety problems (toxicity) • Lower energy than thionyl chloride	• Low power • More expensive than other Li batteries
Applications	• Cameras • Utility meters • PDAs • Measuring instruments • Sensors	• Memory backup • Security/alarm system • RFID tags • Measuring instruments • Medical devices	• Military • Aerospace • Radio transceivers • Portable surveillance	• Utility meters • Emergency signals lights • Electric locks • Electronic measurement equipment

considered, Freedonia has estimated in consumer applications to 2009 the following shares: 69% for alkaline batteries, 16% for primary lithium and 15% for others (i.e. zinc-air button batteries).

While primary batteries will continue to have the largest market share in portable applications, their growth rate will lag behind secondary batteries. Most of the attention of market share analysts is on the secondary batteries and power packs used in many electronic devices. The Freedonia's study predicts that most of the US demand increase will be associated with secondary batteries, while primary batteries are growing "at a below-average pace since they are generally used in more mature markets and have less potential for technological upgrades"

2.3.2. Rechargeable Lithium Batteries (Lithium Negative Electrode)

A few rechargeable batteries based on Li anode are present in niche markets: Li/V_2O_5, Li/Li_xMnO_2 and Li/Nb_2O_5 [8]. All of them use Li–Al alloys, instead of pure Li, as the negative electrode. Al improves storage characteristics and reduces formation of dendrites. However, a serious drawback with the formation of this alloy is the large volume expansion, which produces electrode pulverization and loss of electrical contact on prolonged cycling. Li–Al alloys cannot be extruded into thin foils and, so, can only be used as disks in coin cells. Indeed, all three batteries listed above are commercialized as coin (button) cells, with liquid electrolytes; memory backup is their ideal application. Memories can be found in a very large number of electronic portable and industrial devices (see Section 4.17). Hence, these batteries are introduced here among those useful for both application types.

2.3.3. Lithium-Ion Batteries

These batteries have represented a turning point in the field of power sources for a variety of applications. This is the result of such strong points as [2]

- High specific energy and energy density
- Low self-discharge
- Long cycle life
- No maintenance
- No memory effect
- Fairly wide operating temperature
- Fairly high rate capability
- Possibility of miniaturization and very thin form factors.

In contrast, some weak points also need to be mentioned:

- Relatively high initial cost
- Need of a protection circuit to avoid overcharge, overdischarge and excessive temperature rise
- Degradation at high temperature
- Lower power than Ni–Cd or Ni–MH, especially at low temperatures.

However, it is to be stressed that some of the above drawbacks are being progressively reduced: the cost is steadily decreasing, some Li-ion batteries (especially the polymeric ones) can work with simplified protection elements, and the power output has been greatly enhanced, thanks to proper battery engineering and new positive electrodes.

A Li-ion cell is based on two electrodes able to insert Li^+ in their structure. The term insertion includes both bi- and tri-dimensional structures. In the case of bi-dimensional (layered) structures, the term intercalation is preferentially used. At present, most commercial Li-ion cells have carbon as a negative, $LiCoO_2$ as a positive, and an organic liquid or polymeric electrolyte. However, after many years of predominance of the C/$LiCoO_2$ couple, new electrode materials have emerged, especially as substitutes for the positive (see later).

Carbons capable of Li^+ intercalation can be roughly classified as graphitic and non-graphitic. Pure graphite is crystalline while non-graphitic carbons contain more or less extended amorphous areas. Both have been utilized as negative electrodes in Li-ion cells.

In pure graphite, up to 1 Li^+ can be intercalated per 6C atoms, that is the limiting composition is LiC_6. The Li^+ intercalation/deintercalation reaction at the negative electrode can be described as

$$Li_xC \rightleftarrows C + xLi^+ + xe$$

During the first charge, electrolyte decomposition and formation of a solid electrolyte interface (SEI) on C occur. Such a process is necessary as the behaviour of the C electrode depends on the characteristics of this layer. Electrolyte decomposition starts at ~0.8 V vs Li/Li^+ and SEI formation continues down to ~0.2 V; at this potential Li^+ intercalation begins. When the voltage approaches 0 V, the charge is stopped to avoid Li plating on C. Because of the SEI formation, the first-charge capacity exceeds the first-discharge capacity. The difference (irreversible capacity) should be minimized to reduce excess of the Li^+ source (the positive electrode) in a real battery [15].

$LiCoO_2$, the most common positive electrode, has a layered (hexagonal) structure from which Li^+ can be deintercalated upon charge and re-intercalated upon discharge, according to the reaction

$$Li_{1-x}CoO_2 + xLi^+ + xe \rightleftarrows LiCoO_2$$

Combining this reaction with that of graphite, one obtains the overall reversible process in a C/$LiCoO_2$ cell:

$$Li_{1-x}CoO_2 + Li_xC \rightleftarrows C + LiCoO_2$$

The maximum practical delithiation for $LiCoO_2$ is ~60% (160 Ah/kg). Beyond this value, structural changes limit reversibility of the above reaction. More Li^+ can be deintercalated from $LiNiO_2$, so that a capacity of about 200 Ah/kg may be obtained. However, this material is difficult to synthesize, has a remarkable capacity loss on cycling, and a limited thermal stability. Much better performance is obtained by partly substituting, in $LiCoO_2$, Co with Ni and another metal (e.g. Al or Mn).

In recent years, new positive electrodes have gained attention, especially in view of building larger and more powerful Li-ion batteries. The properties of several systems are summarized in Table 2.6.

A positive electrode, wherein part of Co is replaced by Ni and Al (NCA), allows extending the specific energy up to 240 Wh/kg and the energy density to 630 Wh/L. If Mg-doped Li–Mn spinel is used, the power capability is increased: these Li-ion cells can be used in power tools and are considered as power sources for HEVs. Another excellent positive is nanosized-doped $LiFePO_4$, capable of very high power outputs. An entirely new system, where both electrodes have been changed with respect to the traditional C/$LiCoO_2$ couple, is Nexelion by Sony, which is based on Sn–Co–C as a negative and $LiCo_xNi_yMn_zO_2 + LiCoO_2$ as a positive [16]. The new negative electrode uses nanoparticles that allow minimizing changes in particle shape during charge/discharge, as was instead the case of other Sn-based anodes. It can accept considerably more lithium in its structure upon charge with respect to a conventional C anode, as depicted in Figure 2.4. The Nexelion cell can be recharged to 90% capacity in 30 min at the 2C rate.

More details on these new systems for Li-ion batteries will be given in Chapter 3.

Li-ion cells can use both liquid and polymeric electrolytes. Among the prerequisites of organic liquid electrolytes to be used in these cells, two are of particular importance due to the specific nature of the electrode materials. First, the electrolyte has to grant a stable and efficient SEI on graphite, thus limiting self-discharge (Li^+ deintercalation) while allowing fast reversible Li^+ transport. Second, the electrochemical window of the electrolyte has to range from 0 V to at least 4.3 V vs Li/Li^+. Electrolytes commonly used in Li-ion batteries are based on $LiPF_6$ and a binary solvent mixture, EC-DMC or EC-DEC

Table 2.6. Basic characteristics of Li-ion batteries with different chemistries.

System	Discharge Voltage (V)	Temperature Range (°C)	Specific Energy (Wh/kg)	Energy Density (Wh/L)	Cycles	Power
$LiCoO_2$	3.6	−20/60	140–190	360–500	800–1200	L-M
NCA[a]	3.5	−20/60	220–240	500–630	800–1200	L-M
NCM[b]	3.7	−20/60	100–150	230–400	500–700	M-H
Mn spinel[c]	3.7	−20/60	130–150	300–320	500–700	H
Fe phosphate[d]	3.3	−30/70	100–140	250–380	>1000	VH
Nexelion[e]	3.5	−20/60	160	480	~1000	M-H

[a] NCA, Ni-Co-Al.
[b] NCM, Ni-Co-Mn.
[c] $LiMn_2O_4$ doped with Mg.
[d] Nano-sized, doped $LiFePO_4$.
[e] Sony's hybrid battery with Sn-Co-C as a negative and $LiCo_xNi_yMn_zO_2 + LiCoO_2$ as a positive (L = low, M = moderate, H = high, MH = moderately high, VH = very high).

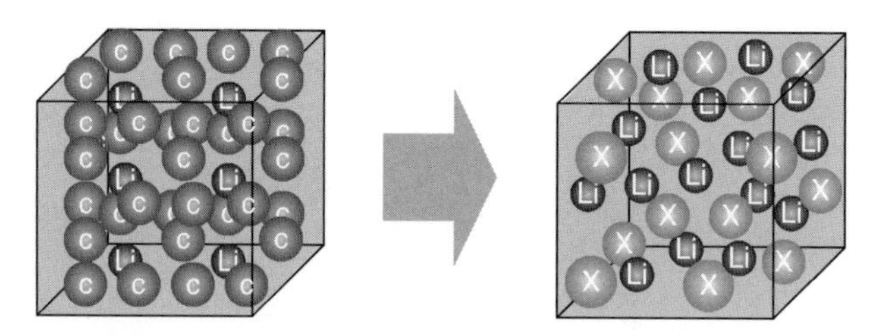

Figure 2.4. Comparison of a C anode with a composite containing Co, Sn and C. *Source*: From Ref. [16].

(EC, ethylene carbonate; DMC, dimethyl carbonate; DEC, diethyl carbonate). These solutions are stable up to 60°C, and can be used down to −20/−30°C (see also Table 2.6). Temperatures above 60°C are problematic for Li-ion cells operation: while at the positive electrode parasitic reactions with the electrolyte are fastened, the SEI on the C surface becomes unstable. It can reform, but irreversible losses have been observed [2]. Additives to improve the SEI are vinylene carbonate, used by SAFT and Sanyo, and methyl cinnamate, used by NEC.

In a few applications, working temperature below −20/−30°C may be reached. In these cases, the conductivity of the binary solutions becomes too low, and ternary or quaternary mixtures of carbonates must be used, for example $LiPF_6$ in EC-DMC-DEC or $LiPF_6$ in EC-DMC-DEC-EMC (EMC, ethyl methyl carbonate) [17].

One of the greatest concerns with liquid electrolytes is their flammability, which becomes a source of risk in case of cell venting. Additives that may lower the flammability (fire retardants) include trimethyl and triethyl phosphate, and other P-containing organic compounds.

The electrolyte is supported on a microporous separator (polyethylene or polypropylene). Commercial cells are mainly available in cylindrical or prismatic form factors. Ribbons of electrodes and separators are wound together, as shown in the cylindrical cell of Figure 2.5. The negative electrode is supported on a thin Cu foil, while the positive electrode is supported on a thin Al foil. The case normally contacts the positive electrode and is made of stainless steel. However, the last generation of prismatic cells uses an Al case to take advantage of its lower weight.

In 1999, batteries with polymeric electrolytes have been commercialized. They can offer some advantages over the conventional ones, for example no leakage, flexibility and very thin form factors. The use of the so-called gel polymer electrolytes has resulted in commercial products for consumer applications with performance characteristics comparable to those of liquid-electrolyte cells. A gel polymer electrolyte (GPE) is formed by immobilizing a liquid electrolyte in a polymeric matrix. An example is provided by the electrolyte

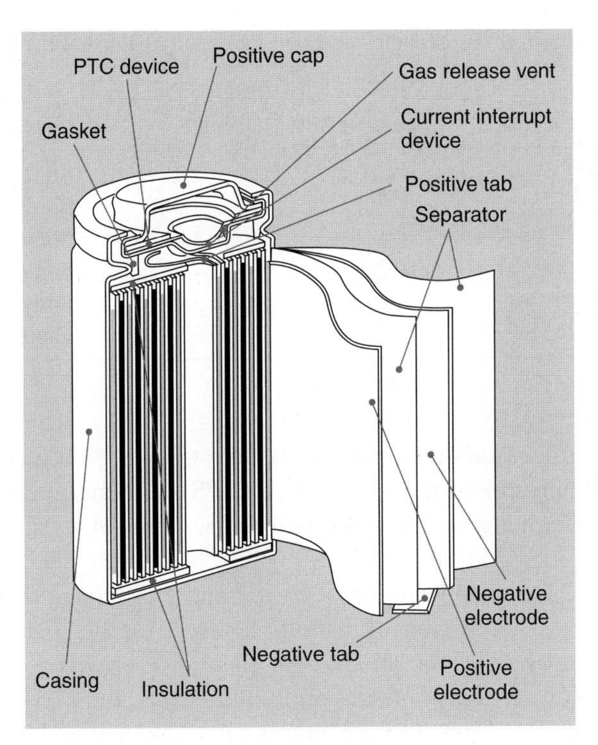

Figure 2.5. Cross-section of a cylindrical Li-ion cell.
Source: Courtesy of Sanyo.

used in Sony's cells: a solution of $LiPF_6$ in PC-EC is added to a copolymer of poly(vinylidene fluoride) (PVDF) and poly(hexafluoropropylene) (PHFP) [18]. The conductivity of this gel electrolyte, $9 \times 10^{-3}\ \Omega^{-1}\,cm^{-1}$ at 25°C, is comparable to those of liquid electrolytes.

The evolution of Li-ion batteries has been remarkable in these years, with increasing capacity, energy, reliability and safety. For the popular 18 650 cell size (diameter: 18 mm, height: 65 mm), the progress in capacity and energy density since 1993 is shown in Table 2.7.

The use of Li-ion batteries in portable (not only consumer) electronics is well known and will be further outlined in Chapter 3. These batteries are also increasingly used in industrial applications: for instance, in Table 2.8, the characteristics of different Li-ion batteries for stationary and traction applications are reported.

Applications only thought possible for aqueous batteries are now at hand for Li-ion systems too. In aerospace applications, where price is not the main issue, these batteries are already used, for example in GEO and LEO satellites [19] and in rovers for planetary missions, such as the Mars Exploration

Table 2.7. Evolution of the capacity and energy density of 18650 Li-ion cells.

Year	Positive Electrode	Capacity (Ah)	Energy Density (Wh/L)	Notes
1993	$LiCoO_2$	1.0	250	Sony
1996	$LiCoO_2$	1.4	280	Sony
1999	$LiCoO_2$	1.7	340	Sony
2002	$LiCoO_2$	2.2	470	Sony
2004	$LiCoO_2$	2.6	545	Sony (G8), 200 Wh/kg
2006	$LiCo_xNi_yAl_zO_2$	2.9	620	Panasonic, 240 Wh/kg

Rover [17]. Forthcoming applications include: traction (advanced vehicles are available), uninterruptible power supplies (UPS) and energy storage systems. Relatively smaller batteries are used in cleaners, motor-assisted bicycles, power tools, etc.

In Figure 2.6, the splitting up of Li-ion batteries among different applications is shown as a function of battery demand for the period 2002–2007. Cellular phones account for the largest share, followed by notebooks. The growing use in power tools is noteworthy: until 2004, only Ni–Cd and Ni–MH were used in this segment, but the new Li-ion chemistries have changed the scenario [20].

Finally, the long-term forecast for Li-ion batteries to be used in portable applications and HEVs is shown in Figure 2.7 [20]. The demand for these batteries features an almost linear growth till the year 2016, with cellular phones and notebook computers representing major shares. The power tool and HEV segments will become increasingly relevant. For the latter, this would be more evident if the battery value, not the number of battery sold, were taken into account.

2.3.4. Rechargeable Aqueous Batteries

Rechargeable batteries with an aqueous electrolyte, which can be used in both portable and industrial applications include Pb-acid, Ni–Cd, Ni–MH and Ag–Zn.

2.3.4.1. Lead-Acid Batteries

Details on electrode construction, separators, electrolyte, discharge characteristics and charge techniques can be found, for example, in Refs. [21, 22]. Here, only basic features will be described.

Table 2.8. Characteristics of Li-ion batteries for stationary and traction applications.

Application and Positive Electrode	Voltage Range (V)	Operating Temperature (°C)	Cycle Life (cycles)	Specific Energy (Wh/kg)	Energy Density (Wh/L)	Specific Power (W/kg)[a]
Stationary ($LiNi_{0.7}Co_{0.3}O_2$)	4.0–2.8	−20 ÷ 60	900[b]	128	197	
Traction ($LiNi_{1-x-y}Co_xMn_yO_2$)	4.0–2.8	−20 ÷ 50	570	150	252	490
Stationary ($Li_{1+x}Mn_2O_4$)	4.0–3.0	−20 ÷ 60	1200[b]	122	255	
Traction ($LiCr_xMn_{2-x}O_4$)	4.0–3.0	−20 ÷ 50	580	155	244	440

[a] Pulse discharge.
[b] Single cells.
Source: From Ref. [2].

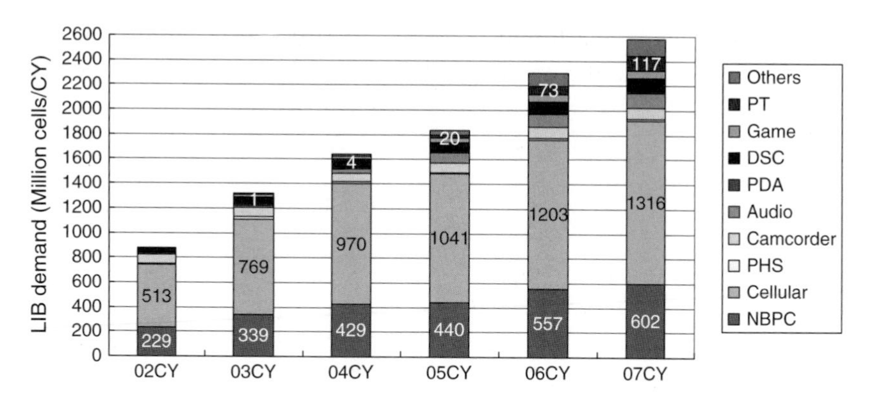

Figure 2.6. Evolution of the market of Li-ion batteries (million cells/year, 2002–2007). (The legend corresponds to the rectangles of each bar, from bottom to top; NBPC, notebooks and portable computers; DSC, digital still cameras; PT, power tools (upper rectangle with a number))
Source: From Ref. [20].

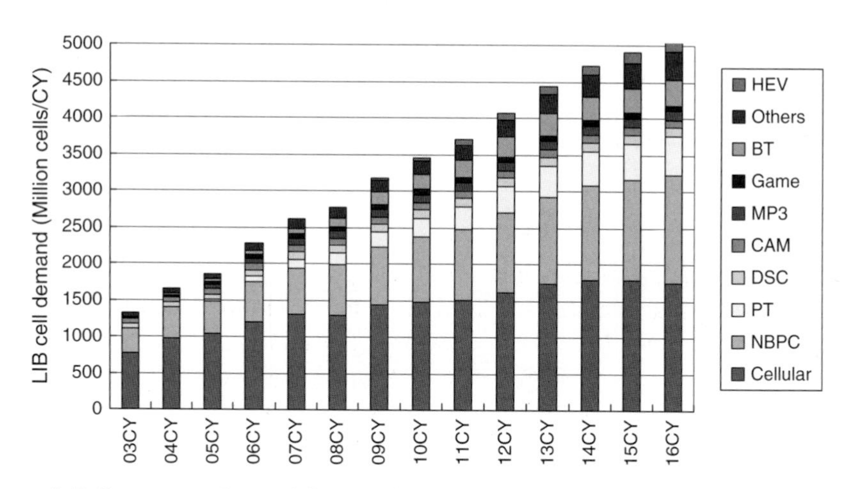

Figure 2.7. Long-term demand forecast for Li-ion batteries in portable applications and HEVs. (see also Figure 2.6; CAM, camcorders; BT, bluetooth devices)
Source: From Ref. [20].

Sealed batteries

Small, sealed lead-acid (SLA) batteries are available to power portable devices. These batteries have capacities in the range 1.2–10 Ah. However, due to their low energy density, they cannot be used in common applications, as computers and cellular phones, but in some applications that will be mentioned later.

These batteries have a limited amount of immobilized electrolyte. Their chemistry is the same as the common Pb-acid battery with excess electrolyte. The negative electrode is Pb, the positive is PbO_2 and the electrolyte is a concentrated H_2SO_4 aqueous solution. The total reaction is

$$Pb + PbO_2 + 2H_2SO_4 \rightleftarrows 2PbSO_4 + 2H_2O$$

The main difference with the common, electrolyte-flooded Pb-acid battery is the electrolyte immobilization and the O_2 recycle permitted by this feature. The aqueous H_2SO_4 solution can either be soaked into an absorptive glass mat (AGM) or gelled by addition of fumed SiO_2. There is enough free space inside the cell to allow for O_2 diffusion in the gaseous phase, which is more than 10^5 times faster than in the dissolved state [23].

The last piece of evidence determines the main characteristics of SLA batteries, that is the internal O_2 cycle. An important secondary reaction of the Pb/PbO_2 system is the one occurring at the positive electrode:

$$2H_2O \rightarrow O_2 + 4H^+ + 4e$$

In a SLA cell, O_2 can diffuse fast in the pores of the AGM not filled with electrolyte or in cracks of the gelled electrolyte. Therefore, it can reach the negative electrode and reform H_2O:

$$O_2 + 4H^+ + 4e \rightarrow 2H_2O$$

This reaction is fast and allows recovering of O_2 generated at the positive, thus maintaining practically unchanged H_2O amount in the electrolyte.

SLA batteries can be made in cylindrical or prismatic configurations. For the latter, a thin version also exists. SLA cylindrical batteries can stand high currents also at low temperatures. The main applications in portable devices can be summarized as:

- Portable TVs
- Measuring instruments
- Lighting equipment
- Various power toys and recreational equipment.

Batteries for Starting, Lighting, Ignition (SLI)

These batteries must possess the requisites of power density, energy density, cycle life, float service life and cost to a satisfactory level [21]. Power is required to start an internal combustion engine (cranking). As the engine runs, the battery is maintained on float, that is at full charge. To maximize the cranking ability, SLI batteries must have electrode grids with very low resistance, thin plates and concentrated acidic solutions.

The introduction of maintenance-free batteries has represented a major advance. They do not require H_2O replenishment, have a better capacity retention and very low corrosion of the electrode terminals. Charge control is needed to avoid H_2O loss. In contrast, as these batteries use low-Sb or Sb-free grids (e.g. Ca–Pb), the rate of H_2O loss on overcharge is reduced.

Traction batteries

Batteries of this type should be able to withstand long cycling, while featuring high energy density and low cost. These requisites overcome the performance of the present Pb-acid batteries, even in their advanced versions. The technical challenges can be summarized as (a) greater specific energy; (b) higher specific power; (c) longer service life (at deep discharge); and (d) faster recharge time [21].

A greater energy can be obtained by minimizing the weight of the inert components (normally ~30%) and by improving the active material utilization. Valve-regulated lead-acid batteries (VRLA, a term interchangeable with SLA) used in EV have specific energies of 30–40 Wh/kg and energy densities of 80–90 Wh/L [24]. It has been calculated that with bipolar or quasi-bipolar designs the specific energy could reach 50 Wh/kg [22]. The last figure is the target set by the Advanced Lead-Acid Battery Consortium (ALABC). In contrast, the United States Advanced Battery Consortium (USABC) has set the values to 80–100 Wh/kg for midterm EV batteries. Present values of specific power for EV Pb-acid batteries are around 190 W/kg. Advanced bipolar designs are claimed to bring power to 800 W/kg at 70% DOD, a value well above the USABC's objective [25].

Stationary batteries

Most of these batteries are of the flooded-electrolyte type and are positive-limited. High energy and power are not requested, but the capability of long living on float or moderate overcharge is necessary. Overcharge causes expansion of the positive grid, and a 10% growth is allowed before the plates touch the container walls. Traditional batteries for telecommunication networks have lifetimes of 5–20 years. To extend this life, a cylindrical battery has been built with circular, saucer-shaped pure Pb grids and horizontally stacked plates. Life was initially predicted to be 30 years or more. Data based on long-term corrosion investigations have allowed extending the life prediction to 68 years [22]. VRLA batteries are now increasingly used, especially in telecom and UPS applications.

A comparison of the basic characteristics of the four types of Pb-acid batteries mentioned above is presented in Table 2.9.

Table 2.9. Characteristics of lead-acid and Ni-Cd batteries.

System	Voltage Range (V)	Operating Temperature (°C)	Cycle Life (cycles)[b]	Specific Energy (Wh/kg)	Energy Density (Wh/L)	Specific Power[c]	Self-Discharge (%/month)
Lead-Acid[a]							
Sealed	2.0–1.8	–40 to 60	250–500	30	90	H	4–6
SLI	2.0–1.8	–40 to 55	200–500	35	70	H	3
Traction	2.0–1.8	–20 to 40	1500	25–30	80	MH	4–6
Stationary	2.0–1.8	–10 to 40	400	20–30	50–70	MH	2
Ni-Cd[a]							
Vented Pocket Plate	1.35–1.1	–20 to 45	500–2000	15–35	35–45	H	5
Vented Sintered Plate	1.35–1.1	–40 to 40	500–2000	30–40	60–100	H	10
FNC	1.35–1.0	–50 to 60	500–3000+	10–40	15–80	L to VH	10–15
Sealed	1.35–1.0	–40 to 45	300–700	15–35	50–120	M to H	15–20

[a] Cell level.
[b] Dependent on DOD.
[c] Pulse discharge (L = low, M = moderate, H = high, MH = moderately high, VH = very high).
Source: From Ref. [26].

2.3.4.2. Nickel–Cadmium Batteries

Ni–Cd batteries belong to a group of five rechargeable batteries, for example Ni–Cd, Ni–MH, Ni–H_2, Ni–Zn, Ni–Fe, having in common a Ni-based positive electrode and an alkaline solution. Even though some known drawbacks can be associated with the Ni–Cd battery, especially low energy density and environmental impact, it can still be used in a number of applications. Ni–Cd batteries can be assigned to two broad categories: vented and sealed. In all types, Cd is used as the negative electrode and β-NiOOH as the positive. The electrolyte is a solution of KOH containing some LiOH (8–20 g/L); the latter improves cycle life and high-temperature performance.

Sealed batteries

Sealed Ni–Cd was the battery used in portable devices up to 1990. Since then, Ni–MH and Li-ion batteries have gained increasing market shares especially in the most developed countries. The overall, reversible reaction is

$$Cd + 2\beta\text{-}NiOOH + 2H_2O \rightleftarrows 2\beta\text{-}Ni(OH)_2 + Cd(OH)_2$$

The reversible reduction of NiOOH to $Ni(OH)_2$ involves the intercalation of H^+ into the layered structure of the oxyhydroxide, thus giving rise to a solid solution.

As for sealed Pb-acid batteries, there is recombination of the oxygen generated at the positive electrode during overcharge. Indeed, the cell is constructed so as to be positive-limited and, on overcharge, O_2 is formed and diffuses through the separator to the negative electrode where it is reduced to OH^-. The $Cd(OH)_2$ formed is converted into metallic Cd during cell charge.

The construction of the positive electrode is of utmost importance in determining its performance. Three technologies that differ for the substrate are possible, which can be based on sintered Ni grid, Ni foam or Ni-plated fibres [8]. The substrates are then impregnated with the active material in its reduced state.

The negative electrode can be, in turn, of the sintered or non-sintered type. In the first case, the micropores of the substrate are filled with $Cd(OH)_2$. The non-sintered type is obtained by coating a Ni-plated steel grid with a CdO-based paste. The first battery charge will form metallic Cd at the negative.

Sealed Ni–Cd batteries are produced with cylindrical, prismatic or button shapes, the first two being more commercialized.

Positive features are high-rate capability, good low- and high-temperature behaviour, good shelf life in a wide temperature range (−40 to +60°C). Some batteries have been stored for up to 10 years and could still deliver, after recharge, almost 100% of their original capacity [27]. On the other hand, these batteries have a high self-discharge (see Table 2.9) and lose rapidly their capacity even at ambient temperature.

In comparison with sealed Pb-acid, these batteries experience minor degradation on storage at high temperatures, have a longer cycle life and perform much better at low temperatures. On the other hand, SLAs are better in operation at high temperatures, on float service and do not manifest the so-called memory effect.

Applications of sealed Ni–Cd batteries in the portable area are now being restricted by Ni–MH and Li-ion cells. Power tools and cordless phones are the major applications, with a minor share being represented by radios, toys and game devices, as substitutes for alkaline cells. Sealed batteries are also used in a number of stationary applications (emergency lighting, communication equipment, energy storage, etc.).

Vented batteries

Vented batteries are for industrial use and can have any of the four electrode constructions listed in the following [26].

Pocket plate. Both electrodes have flat pockets of Ni-plated, perforated steel strips holding in place the active materials. These folded strips are interlocked with each other to form long electrode strips (5–10 m), which are then cut into plates to be used in batteries.

Sintered plate. The sintered electrode (see sealed cells) was developed in order to increase the energy output of the battery. Indeed, with this electrode, the energy density is up to 50% greater than that of the pocket plate. Furthermore, the sintered electrode can be much thinner, thus resulting in a higher rate capability. Other favourable features include flat discharge profile, good capacity retention and low-temperature performance.

FNC (Fibre nickel cadmium). These batteries use fibre-structured electrodes: a mat of synthetic fibres is nickel-plated and subsequently sintered in a H_2 atmosphere at 800°C. The final step is impregnation with the active mass. This method allows the production of electrodes with a thickness ranging from 0.6 to 10 mm. Depending on the electrode thickness, batteries with low to very high power output can be constructed [28].

Plastic-bonded plate. This recent design, mainly used for the negative electrode, has afforded improved performance. The plastic-bonded negative electrode, used in conjunction with a sintered positive electrode, results in reduced battery weight and volume (about 20%), and lower overcharge current that leads to reduced water consumption.

A comparison of the characteristics of different types of NiCd cells is done in Table 2.9. The ability to sustain high-discharge rates is generally good, but with a marked dependence on cell construction. Some fibre electrodes may manifest an outstanding rate capability and cells featuring ultra high rates can be constructed. For example, a 47-Ah, high-rate cell can be discharged for 60 s with currents in the range of 1–2.5 kA.

The rate of capacity loss is related, at a given temperature, to the electrode structure. Pocket-plate electrodes lose 3–5% per month of their capacity at room temperature. Sintered-plate and fibre electrodes lose, at the same temperature, 10–15% per month. The faster self-discharge of these electrodes is connected with their higher surface area: this favours chemical reactions rapidly bringing the battery to a discharged state, especially above ~30°C.

Ni–Cd batteries are capable of long cycling even at high DODs. Cycle numbers in the range 500–5000 are possible. The capability of maintaining an almost stable capacity upon extended cycling can be referred to a very limited variation of the internal resistance. This is normally low (some milliohms for a sintered-plate electrode) and has been found to remain practically constant as a function of state of charge (SOC) and after long cycling [29]. The above-mentioned longer cycle life in comparison to Pb-acid batteries is especially evident when a deep discharge is requested.

Major applications for industrial Ni–Cd batteries include [30]:

- Conventional power stations
- Solar power stations
- UPS plants
- Train lighting and air conditioning
- Marine batteries
- Alarms
- Aircraft
- Traction
- Telecommunications systems
- Signal equipment
- Emergency lighting
- Broadcasting stations
- Switchgear
- Signalling
- Satellites
- Starting (diesel engines, etc.)

2.3.4.3. Nickel–Metal Hydride Batteries

In the Ni–MH system, the negative electrode is a hydrogen-storing alloy (see later for composition), while the positive electrode is β-NiOOH. The solution is the same used in Ni–Cd cells. The overall process consists in the reversible transfer of a proton from an electrode to the other, as indicated by the overall reaction:

$$\beta\text{-NiOOH} + \text{MH} \rightleftarrows \beta\text{-Ni(OH)}_2 + \text{M} \; (E_{\text{cell}} = 1.32\,\text{V})$$

The proton released from the storage alloy during discharge is transferred to NiOOH to form $Ni(OH)_2$. This process is reversed upon charge. Therefore, at both electrodes a homogeneous solid-state process is operating.

At variance with the overall reaction of the Ni–Cd system, where H_2O is consumed on discharge and reformed on charge, here there is no net reaction involving H_2O, so that the concentration and conductivity of the solution does not change.

The cell capacity is limited by the positive electrode, with a negative to positive ratio of 1.5–2 to 1. During overcharge, O_2 is evolved at the positive and diffuses to the negative to form H_2O by reaction with MH. During over-discharge, H_2 is evolved at the positive and again gives rise to H_2O at the negative. Therefore, both H_2 and O_2 recombine to form H_2O, thus assuring true sealed operation of the Ni–MH battery.

The alloys currently used as a negative electrode are of the AB_5 or AB_2 type. An example of the former is $LaNi_5$, whereas ZrV_2 exemplifies the latter. Their main characteristics can be summarized as wide operating temperature range, high capacity and energy, long cycle life, high hydrogen diffusion velocity and environmental friendliness [31]. In the $LaNi_5$ alloy, using a naturally occurring mixture of rare earth elements, La, Nd, Pr and Ce, instead of pure La, enhances the resistance to alkali and reduces costs. Partial Ni substitution with Co prevents alloy pulverization, Mn increases the capacity, and Al improves the resistance to oxidation during the manufacture [32]. AB_5 alloys have capacities of ~300 mAh/g.

AB_2-type alloys (A: Zr, Ti; B: V, Ni plus minor amounts of Cr, Co, Mn, Fe) have higher capacities per unit weight and volume [31]. However, at low and high temperatures and demanding discharge rates, the AB_5 alloy works better. This alloy is also less prone to self-discharge and, as it is cheaper and easier to use, is still preferred.

The positive electrode is basically the same of Ni–Cd batteries. High density, spherical $Ni(OH)_2$ is used (cells are assembled in the discharged state) which also contains Co and Zn. Both Co and Zn greatly limit the formation of γ-NiOOH on overcharge, which is taken as responsible for the memory effect and causes morphological changes in the electrode [33]. Furthermore, these additives improve the charge acceptance and retention at high temperatures (up to 45°C).

Cell construction and performance are indicated in the following.

Small cells (e.g. for electronics) are constructed with the same designs of the Ni–Cd ones. The electrodes have a high surface area so to stand high current rates. In the cylindrical configuration (the most common one), the positive electrode has generally a felt or foam substrate, into which the active material is impregnated or pasted. The negative electrode has also a porous structure (perforated Ni foil or grid) supporting the plastic-bonded hydrogen storage alloy. The electrodes are spirally wound and contained into a Ni-plated steel can. The cell top contains a releasable safety vent, which operates at around 27 atm.

Prismatic batteries are able to deliver up to 20% more capacity. The electrodes are flat and rectangular, other things being the same as in the cylindrical design.

In large cells, the metal alloy is supported by a perforated (or expanded) Ni foil or foam. The positive active material is commonly supported by a Ni foam, but Ni fibres or sintered Ni fibres may also be used, especially for operation at high rates and low temperatures [28, 34].

Both cylindrical and prismatic configurations can be used, the former being preferred for capacities below 10 Ah. Cylindrical cells of more than 20 Ah are difficult to construct, so the prismatic form factor prevails in this case. Prismatic cells use the conventional flat electrode stacks with intermediate separators.

Practically, all Ni–MH batteries are sealed and use a limited amount of electrolyte (6 M KOH plus LiOH as an additive) to allow for fast gas diffusion and recombination.

In cylindrical cells, the metal hydride electrode is connected to the can, which serves as the negative terminal and has to be metallic due the high internal pressures deriving from O_2 evolution on fast charge. Large prismatic batteries may use both metal and plastic cases. The latter ones have an advantage in cost and electrical isolation, this being of prime importance in large packs for EV.

The cells lose little capacity by increasing the current, although at the expenses of the mean voltage. However, repeated discharges at high currents reduce the battery's cycle life. Best results are achieved with rates of 0.2–0.5 C. At the 0.2 C rate, the Ni-MH battery delivers 40% more capacity than Ni–Cd, and a correspondingly higher energy density. In optimized cells, specific energies approaching 100 Wh/kg and energy densities above 300 Wh/L are now available. In comparison to Ni–Cd cells, however, the rate capability is generally lower, self-discharge is faster especially at high temperatures, and tolerance to overcharge is lesser. Furthermore, long-term storage at high temperatures produces permanent damages to seals and separator, and should be avoided. Some tests have shown that continuous exposure to 45°C reduces the cycle life by ~60%.

Ni–MH batteries are commonly charged at constant current, which has to be limited to avoid overheating and incomplete O_2 recombination. The charge temperature has to be controlled and should not overcome 30°C. On the other hand, it has been ascertained that the charge efficiency is very good in the range 10–30°C, so temperatures below 10°C should also be avoided.

The need of a charge control is especially evident in fast charges, where temperature and pressure may reach exceedingly high values with possible cell venting. Methods for charge control are summarized, for example, in Refs. [35, 36]. In addition to conventional charging techniques for Ni–MH cells, considering that large Ni–MH batteries are used in EV and HEV applications, charging by regenerative breaking has also to be mentioned. The energy lost during breaking is used to charge the battery at very high power, 500 W/kg. Not all batteries can accept charge in these conditions, but the Ni–MH one can do so over a relatively wide SOC and temperature range [34].

Under mild conditions, that is charge/discharge at 0.2 C, 20°C, limited overcharge, the Ni–MH battery can deliver in excess of 500 cycles (up to ~1000–1200) before its capacity decreases to 80% of the initial value. In HEV applications, where high-current pulses are applied and the SOC normally varies between 2 and 10%, more than 100 000 cycles are possible. In contrast, if repeated deep cycles are requested at high currents, the performance starts to deteriorate after 200–300 cycles. Shallow rather than deep discharge cycling should be applied at these currents.

The main characteristics of Ni–MH batteries, and the aqueous secondary batteries of next sections, are reported in Table 2.10.

Ni–MH batteries have been introduced in the early 1990s. Since then, they have experienced significant improvements in their characteristics and substituted Ni–Cd batteries in several applications, both portable and industrial.

In portable applications, the challenge by Li-ion batteries is now difficult to tackle, but the lower price of Ni–MH is an advantage. Apart from EV and HEV, examples of industrial applications include standby power systems, aircraft and satellites.

2.3.4.4. Secondary Zinc/Silver Oxide Batteries

In Section 2.2.3, primary Ag-Zn cells have been described. In this section, rechargeable cells based on the same chemistry are mentioned.

These batteries are attractive by virtue of their high specific energy and energy density, proven reliability and safety, and the highest power output per unit weight and volume of all commercially available batteries [37]. They can be even discharged at the 20 C rate. However, notable disadvantages, that is low cycle life (50–100 cycles), poor performance at low temperature and high cost, have limited their use to some specific applications [38].

The characteristics of the negative Zn electrode in an alkaline solution (35–45% KOH) have already been mentioned; it contains small amounts of metal oxides to limit corrosion. Its discharge product is ZnO. The positive electrode is prepared from a sintered Ag powder supported on Ag grid; it contains a mixture of Ag, AgO and Ag_2O. AgO is unstable both in dry charged batteries (reserve batteries) and in activated batteries, as it tends to react with Ag support to give Ag_2O. If chemically prepared AgO is used, this reaction can be minimized [39]. Pure Ag_2O may be used, in spite of its lower capacity, also in view of these advantages: the voltage of its plateau is more stable at high rates and this material does not damage the separator, so that a longer wet life and cycle life is obtained.

Ag–Zn batteries are especially produced with prismatic form factors and with flat electrodes wrapped with multiple layers of separator. As mentioned above, this system is especially notable for its high specific power: conventional high rate cells may give 1.5–1.8 kW/kg, which may become 3.7–4.3 kW/kg, by using thin electrodes and thin separators, or even 5.5 kW/kg in batteries with bipolar electrodes [40]. These unusually high power characteristics are

Table 2.10. Characteristics of aqueous secondary batteries (except Pb-acid and Ni–Cd).

System	V Range (V)	Operating Temperature (°C)	Cycle Life (cycles)[a]	Specific Energy (Wh/kg)	Energy Density (Wh/L)	Specific Power[b]	Self-Discharge (%/month)
Ni-MH[c]	1.4–1.2	−30 to 65	800–1200	60–80	200–270	M to H	15–25
Ni-H_2[c]	1.5–1.2	−10 to 30	>2000	45–55	65–80	M	60
Ni-Fe[c]	1.5–1.2	−10 to 60	2000–3000	30	60	L to M	25–30
Ni-Zn[d]	1.9–1.5	−20 to 50	300–600	60	100–120	H	15–20
Zn-Air[d]	1.2–1.0	0 to 45	20–30	150–200	160–220	L to M	–
Zn/AgO[c]	1.8–1.5	−20 to 60	50–80	90–100	180	VH	5
Zn/Br_2[d]	1.9–1.6	10 to 50	>500	65–70	60–70	L to M	12–15
VRB[d]	1.5–1.1	10 to 50	3000	10–30	10–30	M to H	5–10

[a] Battery.
[b] Pulse discharge (L = low, M = moderate, H = high, MH = moderately high, VH = very high).
[c] Cell.
[d] Dependent on DOD.
Source: From Ref. [26].

especially exploited in space applications. An example is provided by the 150 Ah, 28 V battery for the Atlas V launch vehicle used for the Mars Reconnaissance Orbiter (August 2005). The Ag–Zn battery has also been used by the astronauts of several missions in their extravehicular activities [37, 38].

Secondary Ag–Zn batteries can also be used, by virtue of their high energy and power, in portable applications, such as medical equipment, television cameras, lights, communications equipment, etc.

2.3.4.5. Comparison of the Main Secondary Batteries

In Table 2.11, the best conditions for discharge, charge and storage are summarized for the most important secondary systems.

In Table 2.12, strong/weak points are presented for the same systems, together with their main applications.

In terms of market shares, large Pb-acid batteries have, and will maintain, a dominant position. However, if only portable applications are considered, the Pb-acid battery has a marginal position, and only the other three chemistries have significant shares. As can be appreciated in Figure 2.8, where the trend since 1991 is reported for Ni–Cd, Ni–MH and Li-ion, this last chemistry has been so far the most sold in recent years [20]. Since about 2001, Ni–Cd and Ni–MH have kept an almost constant share. Laminated Li-ion batteries, that is thin batteries with a laminated film bag as a casing to reduce weight, are being increasingly sold.

2.4. Batteries Only Used in Industrial/Vehicular Applications

Table 2.10 lists the characteristics of secondary aqueous batteries, except Pb-acid and Ni–Cd, used in industrial applications (as described above, Ni–MH and Ag–Zn are also used in portable applications). In Table 2.13, secondary non-aqueous batteries other than Li-ion are listed. All of them belong to the class of thermal batteries (see below).

2.4.1. Secondary Aqueous Batteries

2.4.1.1. Nickel-Hydrogen Batteries

The Ni-H_2 cell utilizes H_2 as a negative electrode, while the positive electrode is NiOOH. This system has specifically been developed for aerospace applications and has been in continuous development since the early 1970s.

Table 2.11. Best discharge, charge and storage conditions for the main rechargeable batteries.

	Lead-Acid	Ni–Cd	Ni-MH	Li-ion
Discharge	The limit is ~80% DOD	Can be discharged to 100% DOD	The limit is ~80% DOD; few deeper discharges are allowed	The safety circuit prevents full discharge. ~80% DOD is a safe limit
Typical charge methods	Constant voltage to 2.40 V, followed by float charging at 2.25 V. Float charge can be prolonged. Fast charge is not possible. Slow charge: 14 h. Rapid charge: 10 h	Constant current, followed by trickle charge. Fast-charge preferred to limit self-discharge. Slow charge:16 h. Rapid charge: 3 h. Fast charge: ~1 h	Constant current, followed by trickle charge. Slow charge not recommended. Heating when full charge is approached. Rapid charge: 3 h. Fast charge: ~1 h	Constant current to 4.1–4.2 V, followed by constant voltage. Trickle charge is not necessary. Rapid charge: 3 h. Fast charge recently reported: <1 h
Storage	To be stored at full charge. Storing below 2.10 V produces sulphation	To be stored at ~40% state of charge. Five years of storage (or more) possible at room temperature or below	To be stored at ~40% state of charge. Storage at low temperature is recommended, as this cell easily self-discharges above room temperature	To be stored at an intermediate DOD (3.7–3.8 V). Storing at full charge and above room temperature is to be avoided, as irreversible self-discharge occurs

Table 2.12. Pros & cons and main applications of secondary batteries.

Battery Type	Sealed Lead-Acid	Nickel–Cadmium	Nickel-Metal Hydride	Lithium Ion
Nominal voltage	2.0 V	1.2 V	1.2 V	3.7 V
Pros	• For heavy duty use • Superior long-term reliability • Economical • Easy to recycle	• For heavy duty use • High mechanical strength • High efficiency charge • Charge cycle: 500 times • Easy to recycle	• For heavy duty use • No heavy metals • Relatively high capacity • Charge cycle: 500 times	• For heavy duty use • High 3.7 V voltage • No memory effect • Low self-discharge
Cons	• Relatively low cycle life • Low energy • High self-discharge in flooded batteries	• Low energy • Memory effect • Toxicity • High self-discharge espec. in sealed cells	• More expensive than Ni–Cd • Very high self-discharge	• The most expensive • Potential safety problems • Requires control of charge/disch. limits • Degrades at high temperature
Applications	• Automot. applications • Portable AV equipment • Lighting equipment • Stationary applications	• Portable OA equipment • Portable AV equipment • Power tools • Medical instruments • Stationary applications • Space applications	• Portable OA equipment • Portable AV equipment • Power tools • Medical instruments • Hybrid cars	• Portable OA equipment • Military and space appl. • Many consumer devices • Candidate for next-generation HEV • Power tools

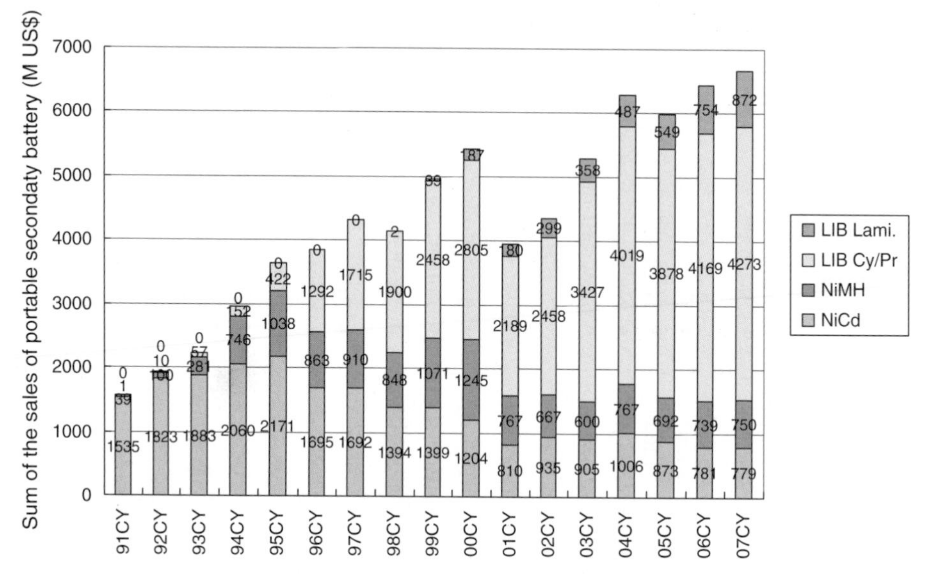

Figure 2.8. Worldwide market (1991 to 2007) of secondary batteries used in portable applications (millions of US Dollars). The legend corresponds to the rectangles of each bar, from bottom to top. LIB Cy/Pr, Li-ion batteries of the cylindrical or prismatic type; LIB Lami., laminate Li-ion batteries (thin batteries with a laminate film as a casing). *Source*: From Ref. [20].

Table 2.13. Characteristics of thermal cells for stationary and vehicular applications.

System	Voltage Range (V)	Operating Temperature (°C)	Cycle Life (cycles)	Specific Energy (Wh/kg)	Energy Density (Wh/L)	Specific Power (W/kg)
Li-Al/FeS[a]	1.7–1.0	375 ÷ 500	1000	130	220	240
Li-Al/FeS_2[a]	2.0–1.5	375 ÷ 450	1000	180	350	400
Na/S[b]	2.0–1.8	300 ÷ 350	6000	155[1]	300[1]	250[2]
				175[2]	350[2]	
Na/$NiCl_2$[c]	2.1–1.7	250 ÷ 300	2500	115	190	260
Li-Metal-Polymer[d]	3.0–2.0	40 ÷ 60[1]	800	140[1]	174[1]	260
		60 ÷ 80[2]		120[2]	160[2]	

[a] From Refs. [41, 42].
[b] From Ref. [43].
[c] From Ref. [42].
[d] From Refs. [44, 45].
[1] Stationary applications.
[2] Traction.
Source: From Ref. [2].

During normal operation the overall reaction is

$$1/2H_2 + NiOOH \rightleftarrows Ni(OH)_2 \quad (E_{cell} = 1.32\,V)$$

One of the major advantages of this system is tolerance to overcharge and reversal. Indeed, on overcharge (and in the final charge stages), O_2 is evolved at the Ni electrode and recombines at the H_2 electrode to form H_2O. During reversal, H_2 is generated at the positive electrode (in a positive-limited cell) and consumed at the negative at the same rate. Therefore there is no pressure build-up or change in electrolyte concentration.

The H_2 electrode consists of a thin-film of Pt black catalyst supported on a Ni foil substrate, backed by a gas diffusion membrane [46]. The preferred Ni electrode consists of a porous sintered Ni powder substrate, supported by a Ni screen, electrochemically impregnated with $Ni(OH)_2$. The separator is a thin and porous ZrO_2 ceramic cloth supporting a concentrated KOH solution.

Ni-H_2 cells are manufactured in various configurations. The individual pressure vessel (IPV) cell contains an electrode stack (single or double) in a cylindrical pressure vessel (voltage: 1.25 V); in the common pressure vessel (CPV) cell, two stacks are in series in a vessel (voltage: 2.5 V); in the single pressure vessel (SPV) battery, a number of cells (typically 22 [46]) are connected in series and placed in a single vessel.

Cell of this type are typically used in geosynchronous earth-orbit (GEO) satellites and in low earth-orbit (LEO) satellites (see Chapter 4). In both cases, the primary power source is represented by solar panels. When the orbit brings the satellite to the earth shadow (eclipse period), the battery starts to deliver energy. The solar panel will subsequently recharge the battery during sunlit periods.

Figure 2.9 shows a 16-Ah battery obtained by connecting in series 11 CPV cells. This battery has been used in LEO satellites, for example the Hubble Space Telescope. Ni-H_2 batteries have also been used in several planetary missions, especially to Mars. Instead, terrestrial applications, for example standby power for emergency or remote sites, are limited due to the high initial cost of these batteries and the drawbacks mentioned below.

To achieve a long life, a Ni-H_2 battery has to work in a limited temperature range, preferably -10 to $+15$°C. Self-discharge for this battery is rather fast even at $+10$°C, for example 10% capacity loss after 3 days. Other weak points are: low volumetric energy, high thermal dissipation at high rates, and safety hazards.

Figure 2.9. 28 V, 16-Ah, 11-cell Ni-H_2 battery for a LEO satellite.
Source: Courtesy of Eagle-Picher Technology.

2.4.1.2. Nickel-Iron Batteries

These batteries, already developed at the beginning of the twentieth century, are based on Fe as a negative electrode and NiOOH as a positive, with the following main total reaction:

$$\text{Fe} + 2\text{NiOOH} + 2\text{H}_2\text{O} \rightleftarrows 2\text{Ni(OH)}_2 + \text{Fe(OH)}_2 \quad (E_{\text{cell}} = 1.37\,\text{V})$$

Even under abuse conditions, such as shocks and vibrations, overcharge/overdischarge, and storage in a fully charged or discharged state, the Ni/Fe battery can undergo ~3000 deep-discharge cycles with a calendar life of about 20 years. A better cycle life is obtained with positive-limited cells. High self-discharge, low coulombic efficiency, poor energy and power density, poor low-temperature performance, higher cost than Pb-acid are the weak points of this battery. On the other hand, the extreme ruggedness, resistance to any abuse and the long cycle life are points in favour of this old system, which has been re-proposed, for example by SAFT, in advanced configurations with better energy and power characteristics [21]. Possible applications are in EVs, material handling and stationary operations.

2.4.1.3. Nickel-Zinc Batteries

A Ni-Zn battery has NiOOH as a positive and Zn as a negative electrode. The overall reaction is

$$Zn + 2NiOOH + H_2O \rightleftarrows ZnO + 2Ni(OH)_2 \quad (E_{cell} = 1.73\,V)$$

Zn has a higher specific energy than the Cd electrode, so that the specific energy of this battery is higher than that of the Ni-Cd system. Other pleasant features are no environmental problems, relatively low cost, good rate capability and good cycle life. However, problems connected with the use of the Zn anode, that is solubility in the concentrated KOH solution, dendrite growth on charge and migration, have so far limited the applications of this system. All these drawbacks could be tackled by reducing the Zn and ZnO solubility. To this end, in addition to the use of a less concentrated KOH solution, approaches based on electrode additives have proven effective. $Ca(OH)_2$ is especially useful when coupled to a ZnO electrode (cells are normally assembled in the discharged state). The use of a microporous polypropylene separator, chemically stable in concentrated alkaline solution, proves an excellent barrier to zinc migration towards the Ni electrode [47].

This battery can be discharged at high currents, for example at the 6C rate, even at low temperatures. However, it manifests a self-discharge of ~20% per month at 25°C

The Ni-Zn battery has been used in motive applications, such as scooters and EVs. EV batteries are claimed to have a performance similar to that of the Ni-MH ones, but at a much lower cost. Large packs are reported to provide an effective range of up to 240 km, and a top speed of 130 km/h. A hybrid all-electric prototype car, equipped with a high-energy Zn-air battery and a high-power Ni-Zn battery (for acceleration), recently proved able to run for 525 km on a single charge, which is the best performance of all times for an EV. However, the limited cyclability of the Ni-Zn system does not allow foreseeing a bright future in this application.

2.4.1.4. Large Zinc-Air Batteries

The chemistry of this system has been mentioned in Section 2.2.4.

As the energy of the cell only depends on that of Zn, which is rather high, the theoretical energy of the Zn-air cell is also high (1350 Wh/kg). Problems connected with the use of a Zn electrode in an alkaline solution have been mentioned. Other problems arise with the use of the O_2 electrode (see Zn-air for portable applications), this justifying the limited commercial success of this system. For industrial applications, both large primary and secondary batteries can be used.

The Zn electrode can be manufactured with different techniques. In primary cells, it is a high-purity foil containing small amounts of metals (e.g. Pb and In) to decrease H_2 evolution. In secondary cells, several types of electrode structure have been proposed (see Refs. [48, 49]); in these cells, Zn can be cycled to nearly 100% of its theoretical capacity.

The air electrode has carbon as the primary structural component. In a typical battery configuration, such as that with two air electrodes of Figure 2.10, the electrode is carbon black mixed with a binder; the current collector is expanded Ni metal; a hydrophobic gas diffusion layer is on the air side and a hydrophilic catalyst layer is on the electrolyte side.

Primary Zn-air batteries are usually very large (more than 1000 Ah) and have been used for many years in such applications as railroad signalling, seismic telemetry, buoys and remote installations [49]. A typical Zn-air primary battery is the so-called Edison Carbonaire, formed by two or three cells in series or parallel, with capacities ranging from 1100 to 3300 Ah.

Secondary Zn-air batteries can be quite different according to the recharge procedure, which can be electrical or mechanical. In turn, the latter may involve replacement of the entire Zn compartment or the Zn powder only.

Electrically rechargeable batteries are based on the bifunctional O_2 electrode, while the negative can be Zn foil, high-porosity Zn or a Zn paste [48, 49]. Large batteries, for example for traction purposes, prefer the last type of negative electrode. Cycle life may largely vary according to the DOD and is generally short at high DOD [48]. During cycling, a steady capacity fade is

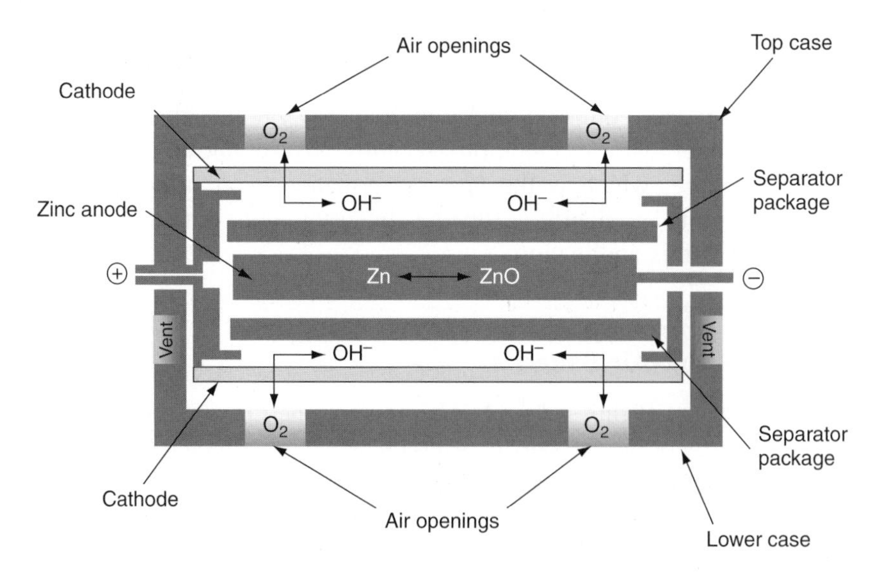

Figure 2.10. Cross-section of a Zn-air cell with a dual air electrode.
Source: From Ref. [48].

observed, which is related to electrolyte carbonation and H_2O loss. Therefore, cells of this type have to provide means for removing CO_2 while allowing a certain humidity level in the incoming air. Other drawbacks are the limited rate capability and the short separator life.

Batteries proposed for electric traction are based upon a mechanical substitution of the exhausted Zn electrodes. Such batteries have been tested on the road, especially in buses and vans. Prototype transit buses have been equipped with 3 trays of 6 modules each, for a total of 312 kWh. Depleted modules are substituted by new ones at servicing stations: this procedure raises obvious concern for logistic and cost reasons.

2.4.1.5. Zinc/Bromine Batteries

This battery and the following one, the vanadium redox battery, belong to the class of flow batteries, that is systems with an aqueous flowing electrolyte.

Two main problems are associated with the Zn/Br_2 system: Zn plating in dendritic form, and solubility of Br_2 in the $ZnBr_2$ solution, this accelerating self-discharge. Only by resorting to a complex system with electrolyte circulation, during battery operation and its confinement in reservoirs, during standby, development could be fulfilled. Details on the battery construction can be found, for instance, in Ref. [50].

The total reaction may be written in the simple form:

$$Zn + Br_2 \rightleftarrows ZnBr_2 \quad (E_{cell} = 1.85\,V)$$

Diffusion of Br_2 towards the Zn deposit is largely impeded by the microporous separator. However, to keep to a low level the concentration of the corrosive Br_2, a complexing agent is used. This is an alkylammonium salt, for example *N*-methyl-*N*-ethylmorpholinium bromide, which associates with the polybromide ions resulting from the reaction of Br_2 and Br^-, thus forming a low-solubility second liquid phase [51]. This complex is stored in a reservoir, where it settles by gravity.

The electrodes, of the bipolar type, are a composite of carbon, high-density polyethylene and glass reinforcement. The separator is made from silica-filled polyethylene [51]. Pumps, reservoirs and tubing are all made from plastic materials to avoid attack from Br_2.

The battery performance (see Table 2.10) can be summarized as: acceptable specific energy; tolerance to deep discharge (the battery can be discharged to 100% DOD without damage); long cycle life (the battery can exceed 2000 full charge/discharge cycles during its operating lifetime [51]).

Two factors may negatively affect the battery performance: self-discharge and energy requested by the auxiliary systems.

The Zn/Br_2 battery is especially intended for stationary applications: utility load management, solar and wind energy storage, emergency backup power and UPS [51]. These systems may have capacities as high as 2 MWh.

Traction batteries, with energies in the range 5–45 kWh, have been tested in small cars and buses.

2.4.1.6. Vanadium Redox-Flow Batteries

These redox-flow batteries are characterized by the fact that all reactants are ionic species. Absence of solid phases and electrolyte circulation confer to these batteries an exceptionally high power capability. Other pleasant features are (1) the capacity can be increased by just increasing the electrolyte volume; (2) full discharge is possible without detrimental effects; (3) the solution can be used for an indefinite time; (4) self-discharge is negligible and (5) the battery response is instantaneous (<1 ms).

The total reaction is (discharge from left to right)

$$V^{2+} + VO^{2+} + 2H^+ \rightleftarrows V^{3+} + VO_2{}^+ + H_2O$$

In practical batteries, the vanadium concentration is around 2 M in 2–4 M H_2SO_4. The actual OCV for a charged cell is ~1.6 V.

The cell scheme is similar to that of the Zn/Br_2 battery and is shown in Figure 2.11. There are stacks of bipolar electrodes, and two electrolyte tanks.

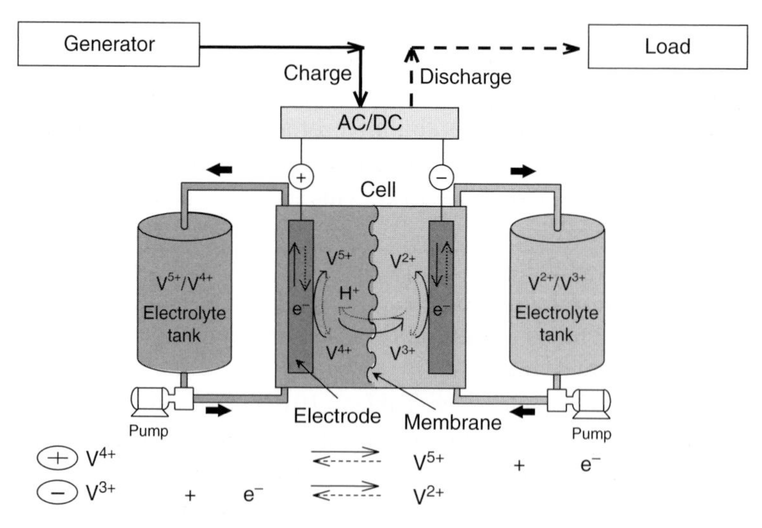

Figure 2.11. Schematic representation of the VRB.
Source: From Ref. [52].

A cation-exchange membrane acts as an efficient separator, as mixing of the electrolytes would correspond to self-discharge.

The battery could give more than 3000 cycles in the operating temperature range of 10–45°C [53]. The positive features of the VRB are counterbalanced by rather low specific energies and energy densities (see Table 2.10).

In general, the power output of a VRB may be raised by increasing the electrolyte flow rate or the number of cells in the stack. At the current stage of development, the stack membrane is the component more prone to deterioration and has to be replaced every 5 years [53].

The VRB can be used in several stationary applications, for example load levelling, momentary voltage sag suppression, emergency power supply, wind/ solar power generation stabilization, power demand and supply control [52, 54].

2.4.2. Thermal Batteries

This group includes batteries with Li- or Na-based negative electrodes. With the exception of the Li-metal-polymer battery, thermal batteries have a molten-salt electrolyte and work well above ambient temperature (see Table 2.13).

2.4.2.1. Li-Al/Iron Sulphide Batteries

These batteries contain lithium alloy as the negative and iron sulphide as the positive. The alloy is multiphase, $(\alpha+\beta)$ Li–Al and $Li_5Al_5Fe_2$, while both FeS and FeS_2 could be used as a positive. The electrolyte is a low-impedance LiCl-LiBr-KBr mixture (LiCl-rich). Using a dense FeS_2 cathode in this electrolyte, more than 1000 cycles have been obtained. Fabrication of electrolyte-starved, bipolar cells or stacks is made possible by the use of chalcogenide ceramic sealants. These materials produce strong bonds between a variety of metals and ceramics. A separator made of pressed MgO powder is normally used.

The electrochemical reactions for FeS and FeS_2 (upper plateau only) can be described as

$$2\text{Li-Al} + \text{FeS} \rightleftarrows \text{Li}_2\text{S} + \text{Fe} + 2\text{Al}$$

$$2\text{Li-Al} + \text{FeS}_2 \rightleftarrows \text{Li}_2\text{FeS}_2 + 2\text{Al}$$

Stopping the discharge at the end of the upper plateau enhances the reversibility of FeS_2 and prolongs its cycle life [21]. FeS_2 has a higher sulphur activity than FeS, hence a higher voltage, but poses corrosion problems. The former material is also a better electronic conductor, thus allowing the use of thicker electrodes.

Positive electrodes are fabricated by compressing a mixture of FeS_x and electrolyte onto a current collector or loading the material into a honeycomb matrix [21]. Graphite may be included and, sometimes, CoS_2 and NiS_2 are also added.

The bipolar cell configuration, where the negative and the positive electrodes contact back to back through a conducting plate, is preferable over the monopolar one. The electrodes are of high surface area and the cell design is compact; moreover, owing to the use of starved electrolyte and MgO separator, it results in a higher power output.

The specific energy and power values of cells (not batteries) are presented in Table 2.13. In terms of energy, both cells meet the midterm USABC's target (80–100 Wh/kg) and the Li/FeS_2 cell approaches the long-term target (200 Wh/kg). However, it has to be noted that the above targets refer to fully engineered batteries. The specific power of the Li/FeS_2 cell is very interesting. Indeed, the mid-term and long-term values set by the USABC are 150–200 and 400 W/kg, respectively. Other positive features of this system are: good cyclability (in excess of 1000 cycles), tolerance to overcharge and overdischarge and to freeze–thaw cycles, and resistance to abuse.

The Li–Al/FeS_x system shows an improved safety over sodium/sulphur (see point 2), while with respect to sodium/nickel chloride (see point 3) it has a higher specific power (but less safety). As any other thermal battery, energy is consumed to keep the cell warm during standby periods; moreover, the bipolar design is likely to have manufacturing problems and high costs.

This system remains a viable candidate for stationary energy storage, while studies for traction applications have been discontinued. Finally, it has to be noted that molten salts electrolytes are being challenged by polymeric electrolytes.

2.4.2.2. Sodium/Sulphur Batteries

High-temperature batteries based on sodium as a negative electrode, sulphur as a positive, and β''-Al_2O_3 as a Na^+-conducting solid electrolyte are now used for energy storage applications. The energy density of the Na/S system is very high (see Table 2.13) and by far exceeds that of aqueous systems (Pb-acid and Ni–Cd) thus far used for energy storage.

Other distinct advantages of the Na/S battery are: good power density, high cycle life, independent of external temperature and moderate cost.

The operating temperature of this system is between 300 and 350°C. In this temperature range, both Na and S are liquid, while the solid electrolyte has a high Na^+ conductivity, thus ensuring good kinetics.

During discharge, Na^+ migrates from Na to S and forms polysulphides; formation of Na_2S_3, at 1.78 V, is taken as the discharge limit. At the C/3 rate, the average voltage is 1.9 V.

During charge, the reactions leading to Na_2S_3 are reversed and, in the final stages, there is a marked resistance increase due to the insulating character of sulphur. Therefore, the charge has to be stopped before complete Na recovery, and subsequent discharges provide 85–90% of the theoretical capacity [43].

As already mentioned, this system has a high cyclability (up to 5000–6000 cycles). This is mainly due to the liquid state of reactants and products: the aging mechanism based on morphological changes of the electrodes does not operate here.

An essential prerequisite for the Na electrode is high purity, that is other metals and Na compounds are not allowed. Contaminants tend to concentrate at the interface with the electrolyte, reducing the electrode-active area or even causing its failure [43].

The sulphur electrode is impregnated into a layer of carbon or graphite felt. The carbon fibres ensure a good electronic conductivity, as sulphur is an insulator for both electrons and ions. Fortunately, Na polysulphides are good ionic conductors.

The electrolyte, β''-Al_2O_3, has a negligible electronic conductivity and is impermeable to molten Na and S. The idealized composition of β''-Al_2O_3 is $Na_2O{\cdot}5.33Al_2O_3$. Pure β''-Al_2O_3 is not easy to prepare, so it has to be stabilized with Mg or Li-ions that substitute for Al ions.

The ionic conductivity of this electrolyte is $\sim$0.5 Ω^{-1} cm^{-1} at 350°C for the polycrystalline form. However, β''-Al_2O_3 is rather sensitive to moisture, this favouring deterioration of its mechanical properties. Therefore, some β-Al_2O_3 (idealized formula, $Na_2O{\cdot}11Al_2O_3$) is enclosed in the mixture, in spite of its lower conductivity, as it is less hygroscopic. A conductivity of $\sim$0.2 Ω^{-1} cm^{-1} is regarded as acceptable for practical electrolytes [21].

Production of Na/S batteries for stationary applications is particularly active in Japan. Very large energy-storage systems can be built with Na/S modules: their storage capacity may be as high as 57 MWh (see Section 4.16.1). They can live in this application for up to 15 years providing thousands of cycles.

2.4.2.3. Sodium/Nickel Chloride (Zebra) Batteries

The negative electrode and the electrolyte of this battery are those of the Na/S battery, but a metal chloride, preferentially $NiCl_2$, is used instead of sulphur as a positive electrode. As the initial research was carried out in South Africa for EV applications, the acronym Zebra (zero emission battery research activity) is often used.

During normal discharge, Ni and NaCl are formed. The voltage, 2.58 V, is higher than that of Na/S, 2.06–1.78 V. Other differences are given below.

Liquid Na is placed in the outer part of the cell, while in practical Na/S cells it is in the central part. Furthermore, Na is not initially included in the cell assembly, but produced *in situ* during charge, as the cell is assembled in the discharged state. The positive electrode is solid and this would pose problems, as the solid-solid interface with the electrolyte would not allow high currents. Therefore, a second electrolyte, $NaAlCl_4$, which is liquid at the battery-operating temperatures, is added.

The tolerance to overcharge and overdischarge is outstanding by virtue of the reversible reactions occurring in these conditions. On overcharge, $NaAlCl_4$ further chlorinates the Ni matrix, while on overdischarge excess Na reacts with the liquid electrolyte. This feature has a practical implication: it is possible to connect in series several cells, without parallel connections, as cell imbalances are levelled out by the above reactions [43].

Another advantage of this battery, with respect to the Na/S system, is represented by its enhanced safety. As a further advantage, the Zebra cell can work in a wider temperature range (i.e. 220–450°C), although the practical range has to be restricted to 270–350°C.

The Zebra system is specifically intended for EV applications, where such factors as safety, easiness in construction, cyclability, and low corrosion, are appreciated.

2.4.2.4. Lithium-Metal-Polymer Batteries

This system is rather different from those enclosed in this section, as it works just above ambient temperature (see Table 2.13) and uses a polymeric electrolyte.

Avestor, a partnership of Hydro-Quebec (Canada) and Anadarko (USA), has produced a battery formed by: Li as a negative, vanadium oxide as a positive electrode, and the polymeric solution LiTFSI-PEO-PPO (LiTFSI, Libis (trifluoromethylsulphonyl) imide; PEO, polyethylene oxide; PPO, polypropylene oxide) [44]. The operating temperature of this battery, often referred to with the acronym LMP, is kept above 40°C, as below this value the electrolyte has a conductivity lower than $1 \times 10^{-4}\ \Omega^{-1}\,cm^{-1}$, the minimum for a satisfactory performance. This battery has been especially developed for standby applications, for example as backup power in telecommunications, and has received attention for a possible use in electric cars. Single cells are connected by a bus bar, so to create voltages of 24 or 48 V, typical of telecom applications. Heating element layers interspersed in the cell stack ensure a temperature range (40–60°C for stationary uses) that is maintained whatever the external temperature, thanks to the insulation surrounding the stack [45].

On storage at room temperature and up to 60°C, the Avestor battery looses only 1% per year of its capacity, so that more than 10 years of shelf life are expected. Tests carried out in operating conditions, that is during float and backup periods, have shown that the battery can be operated for at least 12 years.

Prototypes for traction, working in the 60–80°C temperature range, have demonstrated 800 cycles with energy values of 120 Wh/kg and 160 Wh/L (see Table 2.13).

In spite of the fact that, by August 2006, 20 000 batteries for telecommunications were sold, 2 months later Avestor closed its manufacturing plant in Canada. In March 2007, the French group Bolloré acquired Avestor, and a revival of interest for this battery, especially for traction purposes, seems possible.

References

1. Energizer/Eveready: Typical Characteristics
2. G. Pistoia, in *Industrial Applications of Batteries. From Cars to Aerospace and Energy Storage*, M. Broussely and G. Pistoia, Eds., Chapter 1, Elsevier, Amsterdam, 2007.
3. D.W. McComsey, in *Handbook of Batteries*, D. Linden and T.B. Reddy, Eds., Chapter 8, McGraw-Hill, New York, 2002.
4. Duracell Technical/OEM, Alkaline Manganese Dioxide.
5. S.A. Megahead, J. Passaniti and J.C. Springstead, in *Handbook of Batteries*, D. Linden and T.B. Reddy, Eds., Chapter 12, McGraw-Hill, New York, 2002.
6. Energizer Zinc-Air – Application Manual.
7. Duracell Technical/OEM, Zinc-Air.
8. G. Pistoia, *Batteries for Portable Devices*, Elsevier, Amsterdam, 2005.
9. D. Linden and T.B. Reddy, in *Handbook of Batteries*, D. Linden and T.B. Reddy, Eds., Chapter 14, McGraw-Hill, New York, 2002.
10. K.M. Abraham, *J. Power Sources* 34 (1991) 81.
11. H. Ikeda, T. Saito and H. Tamura, *1st International Symposium on Manganese Dioxide*, Cleveland, OH, 1975, p. 384.
12. H. Ikeda, M. Hara and S. Narukawa, U.S. Patent 4,133,856 (1979).
13. Ultralife Batteries: Products.
14. P.J. Spellman, K.J. Dittberner, J.G. Pilarzyk and M.J. Root, in *Energy Storage Systems for Electronics*, T. Osaka and M. Datta, Eds., Gordon&Breach Science Pub., Amsterdam, 2000.
15. M. Winter, K.-C. Moeller and J.O. Besenhard, in *Lithium Batteries – Science and Technology*, G.A. Nazri and G. Pistoia, Eds., Kluwer Academic Pub., Boston, 2004.
16. Sony Corp., "Sony's New Nexelion Hybrid Lithium Ion Batteries", February 2005.
17. B.V. Ratnakumar and M.C. Smart, in *Industrial Applications of Batteries. From Cars to Aerospace and Energy Storage*, M. Broussely and G. Pistoia, Eds., Elsevier, Amsterdam, 2007.
18. Y. Nishi, in *Advances in Lithium-Ion Batteries*, W. van Schalkwijk and B. Scrosati, Eds., Kluwer Academic Pub., Boston, 2002.
19. M. Broussely, in *Lithium Batteries – Science and Technology*, G.A. Nazri and G. Pistoia, Eds., Kluwer Academic, Boston, 2003.
20. H. Takeshita, "Worldwide Market Update on NiMH, Li Ion and Polymer Batteries for Portable Applications HEVs", *24th International Seminar on Primary/Secondary Batteries*, Fort Lauderdale, FL, March 2007.
21. D.A.J. Rand, R. Woods and R.M. Dell, *Batteries for Electric Vehicles*, Research Study Press, Taunton, U.K., 1998.

22. A.J. Salkind, A.G. Cannone and F.A. Trumbore, in *Handbook of Batteries*, D. Linden and T.B. Reddy, Eds., Chapter 23, McGraw-Hill, New York, 2002.
23. A.J. Salkind, R.O. Hammel, A.G. Cannone and F.A. Trumbore, in *Handbook of Batteries*, D. Linden and T.B. Reddy, Eds., Chapter 24, McGraw-Hill, New York, 2002.
24. D. Linden and T.B. Reddy, in *Handbook of Batteries*, D. Linden and T.B. Reddy, Eds., Chapter 22, McGraw-Hill, New York, 2002.
25. T. Juergens and R.F. Nelson, *J. Power Sources* 53 (1995) 201.
26. G. Pistoia, in *Industrial Applications of Batteries. From Cars to Aerospace and Energy Storage*, M. Broussely and G. Pistoia, Eds., Chapter 2, Elsevier, Amsterdam, 2007.
27. J. Carcone, in *Energy Storage Systems for Electronics*, T. Osaka and M. Datta, Eds., Gordon & Breach Science, Amsterdam, 2000.
28. M. Ohms, G. Kohlhase, G. Benczur-Urmossy and G. Schadlich, *J. Power Sources* 105 (2002) 127.
29. J. Carcone, in *Handbook of Batteries*, D. Linden and T.B. Reddy, Eds., Chapter 28, McGraw-Hill, New York, 2002.
30. H. Franke, in *Battery Technology Handbook*, H.A. Kiehne, Ed., Marcel Dekker, New York, 2003.
31. M. Suzuki and M. Wada, in *Energy Storage Systems for Electronics*, T. Osaka and M. Datta, Eds., Gordon & Breach Science Pub., Amsterdam, 2000.
32. Y. Morioka, S. Narukawa and T. Itou, *J. Power Sources* 100 (2001) 107.
33. Y. Sato, S. Takeuchi and K. Kobayakawa, *J. Power Sources* 93 (2001) 20.
34. M. Fetcenko, in *Handbook of Batteries*, D. Linden and T.B. Reddy, Eds., Chapter 30, McGraw-Hill, New York, 2002.
35. D. Linden and D. Magnusen, in *Handbook of Batteries*, D. Linden and T.B. Reddy, Eds., Chapter 29, McGraw-Hill, New York, 2002.
36. Duracell: Technical/OEM, Ni-MH Rechargeable Batteries.
37. A.P. Karpinski, B. Makotevski, S.J. Russel, J.R. Serenyi and D.C. Williams, *J. Power Sources* 80 (1999) 53.
38. A.P. Karpinski, S.F. Schiffer and P.A. Karpinski, in *Handbook of Batteries*, D. Linden and T.B. Reddy, Eds., Chapter 33, McGraw-Hill, New York, 2002.
39. D.F. Smith and C. Brown, *J. Power Sources* 96 (2001) 121.
40. A.P. Karpinski, S.J. Russell, J.R. Serenyi and J.P. Murphy, *J. Power Sources* 91 (2000) 77.
41. G.L. Henricksen and A.N. Jansen, in *Handbook of Batteries*, D. Linden and T.B. Reddy, Eds., Chapter 41, McGraw-Hill, New York, 2002.
42. P.C. Symons and P.C. Butler, in *Handbook of Batteries*, D. Linden and T.B. Reddy, Eds., Chapter 37, McGraw-Hill, New York, 2002.
43. J.W. Braithwaite and W.L. Auxer, in *Handbook of Batteries*, D. Linden and T.B. Reddy, Eds., Chapter 40, McGraw-Hill, New York, 2002.
44. V. Dorval, C. St-Pierre and A. Vallée, "Lithium-Metal-Polymer Batteries: From the Electrochemical Cell to the Integrated Energy Storage System", Avestor Report, 2004.
45. C. St-Pierre, T. Gauthier, M. Hamel, M. Leclair, M. Parent and M.S. Davis, "Avestor Lithium-Metal-Polymer Batteries Proven Reliability Based on Customer Field Trials", Avestor Report, 2003.
46. D. Coates, D. Caldwell and R. Hudson, *5.5 Inch Diameter Nickel-Hydrogen Cell Development Update*, Eagle-Picher Technologies LLC, 1998 (published by AIAA).
47. D. Coates and A. Charkey, in *Handbook of Batteries*, D. Linden and T.B. Reddy, Eds., Chapter 31, McGraw-Hill, New York, 2002.
48. L.A. Tinker and K.A. Striebel, in *Energy Storage Systems for Electronics*, T. Osaka and M. Datta, Eds., Gordon & Breach Science Pub., Amsterdam, 2000.

49. R.P. Hamlen and T.B. Atwater, in *Handbook of Batteries*, D. Linden and T.B. Reddy, Eds., Chapter 38, McGraw-Hill, New York, 2002.
50. A. Leo, in *Modern Battery Technology*, C.D.S. Tuck, Ed., Ellis Horwood, Chichester, UK, 1991.
51. M.H. Thomas, "Persistence and Progress – The Zinc Bromine Battery in Renewable Energy Systems", *Rega Forum 2003*, Yarra Valley, Victoria, Australia, June 2003.
52. Sumitomo Electric Industries, "Vanadium Redox-Flow Batteries (VRB) for a Variety of Applications".
53. C.J. Rydh, *J. Power Sources* 80 (1999) 21.
54. Redox Flow battery [KEPCO] www.kepco.co.jp.english/rd/solution/solution_2.

Chapter 3

PORTABLE APPLICATIONS

3.1. General Considerations

The numerous portable applications listed in the tables of Chapter 1 can be further divided into consumer and non-consumer applications. Some groups may be identified, as listed below.

Consumer portable applications:

- Radio and other portable audio equipment (e.g. CD and MP3 players)
- Sporting goods (fishing accessories, heart rate monitors, etc.)
- Cordless and cellular phones
- Toys
- Notebook and laptop computers
- Portable TV and DVD players
- Calculators
- Electronic games
- Handheld lights
- Digital still cameras
- Camcorders
- Lawn and garden equipment
- Hobby power tools
- GPS navigators
- Personal digital assistants (PDAs)
- Pagers

Non-consumer portable applications:

- Transceivers
- Professional audio/video equipment
- Medical applications (e.g. glucose meter, pulse oximeter, pacemaker, defibrillator, hearing aid, and telemetry, portable X-ray equipment)
- Portable payment terminals
- Meteorological instruments
- Scientific instruments
- Professional power tools
- Bar code readers
- Mini-UPS

In describing the characteristics of some representative portable devices, particular emphasis has been put into (1) description of their components (also using block diagrams when available) and (2) analysis of the power management needed to maximize the device's runtime. The last point takes into account the designer's solutions to reduce power consumption through a proper component choice, the mode of using the device, and the battery management.

3.2. Video/Audio Applications

3.2.1. Notebooks, Tablet PC and Ultra Mobile PC (UMPC)

In this section, the characteristics of portable PCs of various dimensions and functionalities are presented briefly.

A typical block diagram for the most popular portable PC, the notebook, is shown in Figure 3.1. The central processing unit (CPU) performs all operations according to the instructions contained in the operating system.

The northbridge connects the CPU with the memory chips (SDRAM in the figure), while the southbridge connects peripherals to the northbridge chip. Several interfaces (I/F) allow cable-connectivity of external devices, for example those USB-compatible, while the IrDA allows wireless connectivity, for example with a printer. The low-voltage differential signalling (LVDS)

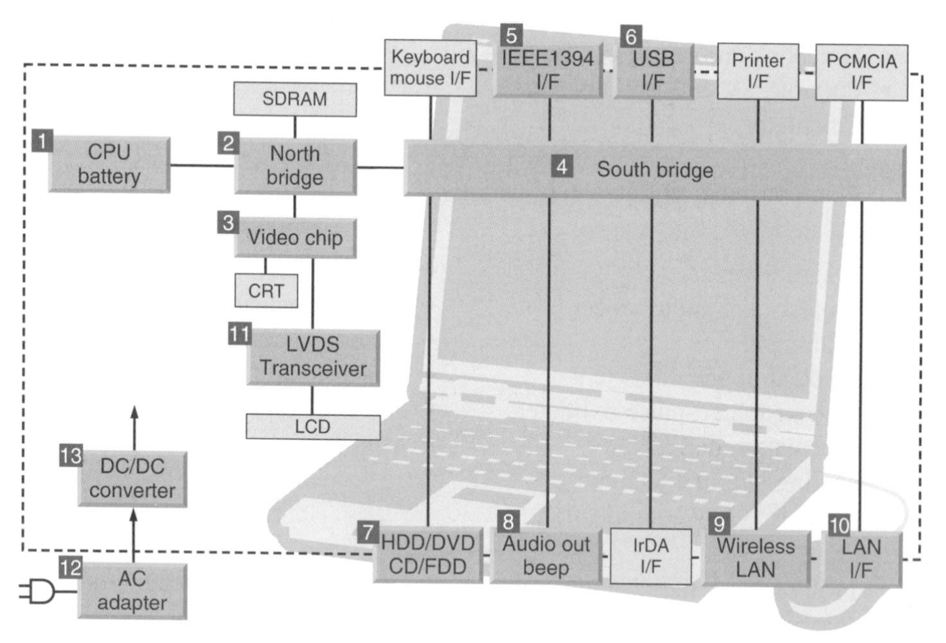

Figure 3.1. Block diagram of a notebook computer.
Source: Courtesy of TDK.

transceiver between video chip and LCD display allows high-speed (Gbits/s) data transmission, much higher than that allowed by a standard I/O signalling. Furthermore, the LVDS allows noise to be filtered easily and effectively. For the significance of other acronyms, see the list.

The AC adapter or the battery may power the notebook: their output voltage is delivered to the various blocks after DC/DC conversion. With reference to the battery, notable elements in the system power are the charger, the battery management/protection and the gas gauge to determine its state of charge.

The power consumption of the major components may be determined by monitoring the power required by the system in various operating modes [1]. Figure 3.2 depicts the total power consumption when switching off the notebook and waiting for its stabilization: a 45-W spike is followed by a levelling off to about 22 W.

In Figure 3.3 (left), the power consumed when viewing a DVD is shown: the power stabilizes just below 40 W, but spikes above 55 W occur when audio/video conditions are particularly demanding. CD's operation is also power consuming, as shown in Figure 3.3 (right): a stable 50–55 W level is observed.

In Figure 3.4 (left), the total system power consumption is shown with a 100% utilization of the CPU: the power minimum at about 25 W corresponds to

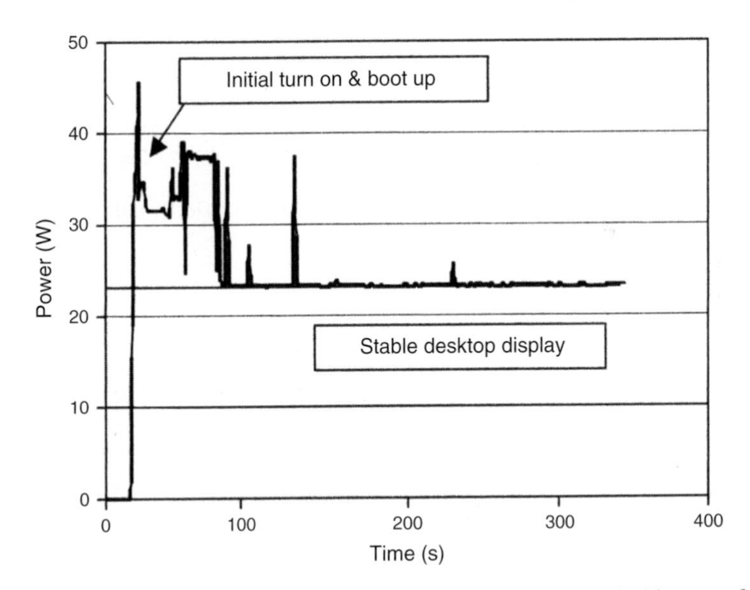

Figure 3.2. Power consumption of a notebook computer upon switching on, followed by stabilization (about 5 min).
Source: From Ref. [1].

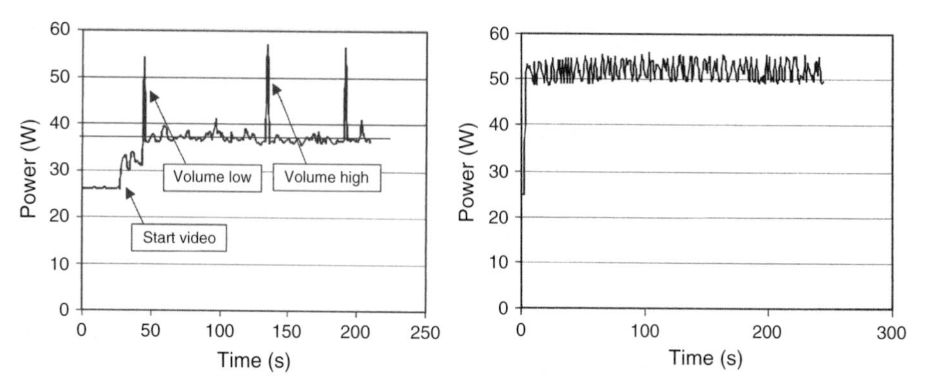

Figure 3.3. Left: DVD power consumption. Right: CD power consumption.
Source: From Ref. [1].

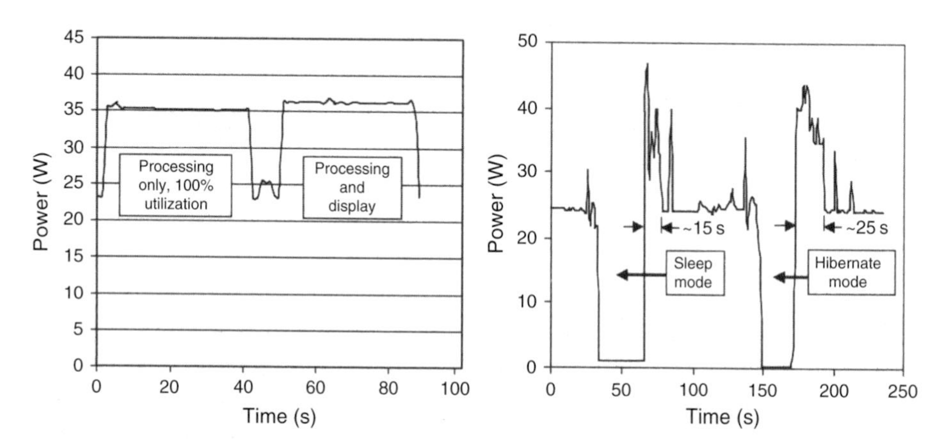

Figure 3.4. Power consumption of processor and display (left) and during sleep and hibernate (right).
Source: From Ref. [1].

a static image, while the 35 W consumption corresponds to a continuously updating image on the display. Figure 3.4 (right) shows how sleep and hibernate modes allow to save power.

High-resolution graphics is also rather demanding (about 50 W). This issue will be later dealt with in more detail. The power requested by the display greatly changes as a function of backlighting. This is depicted by the sequence of Figure 3.5. The total display consumption amounts to 12 W, with 3 W requested by the LCD and its driver, while the backlight and its inverter board consume 9 W.

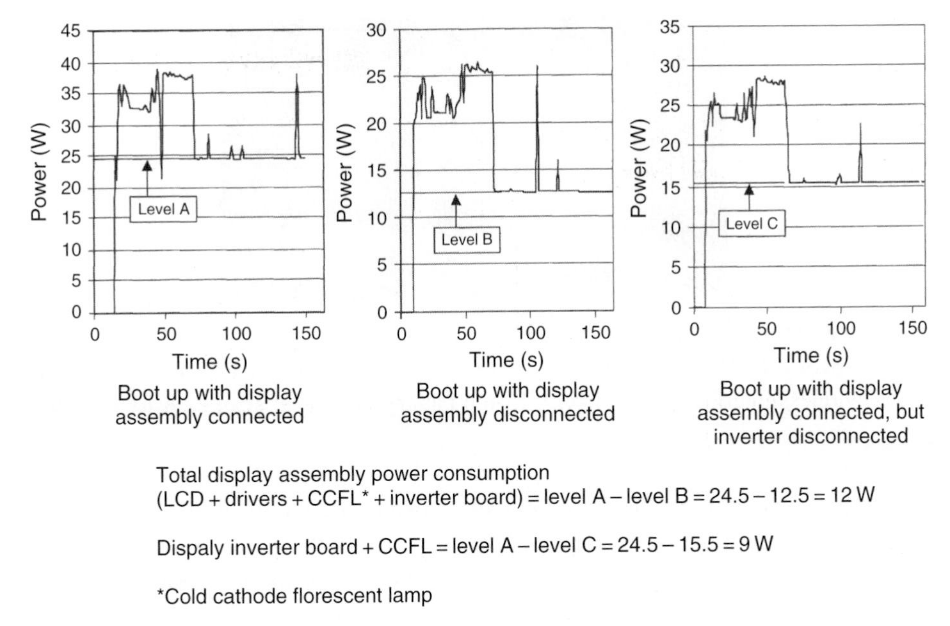

Figure 3.5. Evaluation of the power consumed by a notebook's display.
Source: From Ref. [1].

Other factors affecting power consumption in a portable PC include: the tasks performed, wireless throughput, processor speed, amount of file fragmentation, etc. It can be noted that power consumption due to a wireless connection is not so big. Indeed, with the wireless card *on* and connected to an access point (AP), the battery consumption increases by 2.5%; with the card *on* and continually searching (worst-case scenario) for an AP, the consumption may rise up to 7.5%. These values are acceptable and well below those of other PC's components.

Tablet PCs may be divided into two classes: slate or convertible. The former have no keyboard and the user can operate the touchscreen with a digital pen or a fingertip; the latter have a keyboard, and may be converted into slate mode by rotating the screen so to make it lay on the keyboard. The touchscreen incorporates a digitizer correlating the physical touch on a given point with the information described on that point.

Tablet PCs tend to have lower weight and dimensions, and a smaller screen vs conventional notebooks. Power consumption of a typical tablet PC (Lenovo ThinkPad X61) is between 7.5 and 15 W, lower than that of a notebook (see above). The battery lasts 7 h with a consumption of 10 W/h, while its runtime decreases to 4.6 h for a consumption of 15 W/h.

In the past few years, even smaller types of portable PC have been commercialized. They are known as ultra mobile personal computer (UMPC),

Figure 3.6. Ultra mobile PC (UMPC) by Samsung (Q1 Ultra model). Some characteristics: CPU, Intel A110 (800 MHz); O.S., Windows Vista; display, 7″; RAM, 1 GB; wireless interface (802.11 b/g); Li-ion battery of 30 Wh; size, 22.8/12.4/2.4 cm.

and sometimes the term Origami devices is also used. To officially qualify as an UMPC, a device must have a screen size of 7″ or smaller, a minimum resolution of 800×480 pixels, weigh <900 g, have a landscape display orientation and have a touch-sensitive screen. One such device, produced by Samsung, is shown in Figure 3.6. Other competitors in this niche market include Asus, Sony, OQO, Founder, etc.

Some attention has been devoted to the power consumption of these new devices. Their batteries (Li-ion) have energies around 25 Wh (30 Wh for the product of Figure 3.6), this being about half the energy of a typical notebook. Their CPUs (e.g. Celeron-M or Pentium-M) are based on low-voltage chips and have a maximum power consumption of 5–7 W. Ultra low-voltage CPUs may require lower power: 3.5–5 W.

Even with 100% utilization of the CPU, one-third of total power is consumed by the LCD screen, and with an average CPU use (low load), the screen accounts for more than half the power consumption. This raises the question: is it necessary to use a 7″ screen in such a small device, or a 5–6″ screen would prove sufficient while granting longer runtimes?

An average power consumption for an UMPC is in the range 8–12 W, this corresponding to 2–3 h usage. The graphic chipset has low consumption, but three-dimensional (3D) graphics (e.g. for videogames) needs 2–3 W more or 10–15% of battery runtime. The hard disk's impact on battery life is not very large. However, substitution of hard disks with flash memory devices could result in a power saving of ~5%. This last change has been implemented, for instance, in Sony's VAIO UX1, where a 32-GB flash memory is used for storage.

Intel plans to reduce, by 2010, the power consumed by its UMPC platform to 1/20th of today's value. Furthermore, Intel's new system-on-chip (SoC) for UMPCs is claimed to reduce idle power consumption to 1/100th of the original.

Another niche product in the PC area is the wearable computer, for example the one that is worn on the body. It is particularly useful when the user is engaged in a specific action and, at the same time, needs computer's support. A wearable computer is always *on* and always accessible; it is a fully featured computer that can be reconfigured (programmed).

Applications of wearable computers include the following:

Mediated/augmented reality. It is the ability of the computer to offer enhanced presentations of reality to the user.

Blind vision. It is a personalized radar system that is integrated in a close-fitting vest and is able to process objects in the vicinity of the wearer.

Medical alert. Clothes with embedded wearable computers monitor the wearer's body functions. Should any one of them become critical, a remote medical unit would be notified.

Impression sharing. The wearer can send to another person, or group of persons, his/her impressions, for example something seen in a particular moment or place.

Other application areas include emergency search and rescue, maintenance, law enforcement, logistics, transportation and defence applications.

An example of wearable PC is given in Figure 3.7. It can be worn on the user's wrist, thus allowing hands-free operations. In order to save power, a device like this has the so-called tilt and dead reckoning system, which detects the user's arm position and puts the PC in standby mode when the arm is hanging down beside the user's hip.

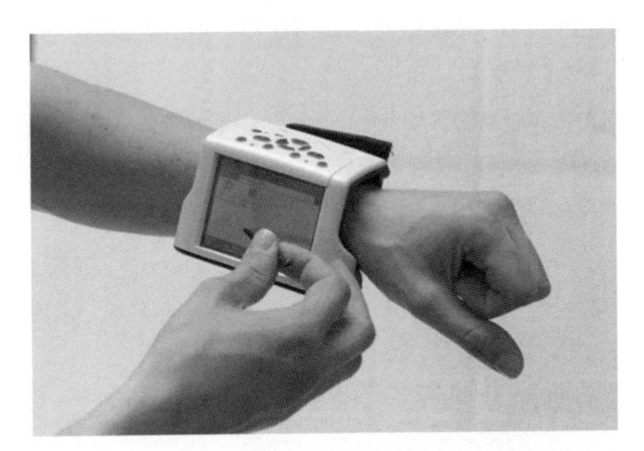

Figure 3.7. An example of wearable computer – the ZYPAD WL 1100 Wrist Worn PC. Some characteristics: CPU, PXA270 (400 MHz); O.S., Windows CE 6.0; wireless connectivity by Bluetooth and Wi-Fi (802.11 b/g); 320 × 240 screen resolution; RAM, 128 MB; flash memory, 128 MB; weight, 290 g.
Source: Courtesy of Eurotech Group.

This PC uses a 2.2 Ah, 3.6 V Li-ion battery with a power management system, allowing ~8 h of operation. This means an average current of 0.27 A (i.e. 1 W power consumption).

3.2.2. E-Book Readers

The first E-book reading devices were introduced in 2002, but met with little success. This was essentially due to two factors: the use of the PDF format, which is difficult to read on small screens, and the inadequate screen resolution. The turning point occurred in 2006, when Sony introduced its E-book reader (PRS-505, upgraded in 2007) with a more user-friendly screen that simulates the printed page. Furthermore, this device allows reading different file formats, for example PDF and audio (AAC/MP3). Sony's reader has a storage capability of 160 E-books and its Li-ion battery allows turning ~7500 pages before recharge. Memory stick or secure digital (SD) card can be used to have a larger memory.

In 2007, Amazon, the popular on-line bookseller, introduced its own reader – Kindle (Figure 3.8). Amazon will probably become the front runner in this rapidly growing market sector.

Figure 3.8. An E-book reader – Kindle by Amazon. Size: 19.1 × 13.5 × 1.8 cm; weight: 292 g. See text for other characteristics (See colour plate 1).

From the technical viewpoint, a book reader has features of both PCs and smartphones. Kindle's characteristics may be listed as:

- CPU: Intel PXA255
- Memory: 64-MB RAM, 180 MB available for storage, SD memory card expansion slot
- Operating system: Linux (2.6.10 kernel)
- Screen: 6″ with a resolution of 600×800 pixels, no backlighting
- Connectivity: USB 2.0, wireless modem, stereo headphone jack
- Battery: 3.7 V, 1.5 Ah Li-ion polymer; runtime: ~2 days with wireless connection *on*, ~7 days off
- Capacity: ~200 non-illustrated books
- File format: proprietary (AZW).

By supposing a 45-h runtime for the 1.5-Ah battery while the connection is *on*, a current of 33 mA is derived. This corresponds to a power consumption of 0.12 W only, which is made possible by Amazon's choice to use a black/white screen with no backlighting.

3.2.3. Cellular Phones and Smartphones

In Figure 3.9, the block diagram of a mobile phone is presented.

Owing to the large number of features, phones of this kind are also called smartphones: they are miniaturized computers taking pictures, playing music, making/receiving phone calls, acting as GPS, connecting to the Internet and receiving TV programs.

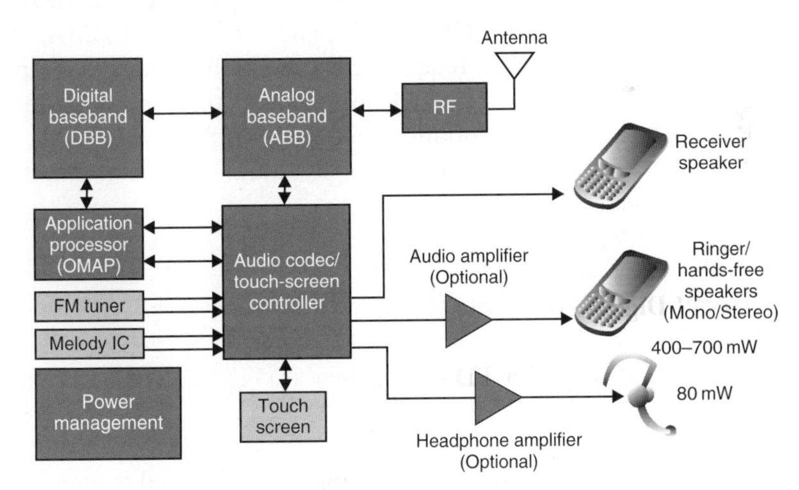

Figure 3.9. Block diagram of a mobile phone.
Source: Courtesy of Texas Instruments.

Table 3.1. Current consumption of a mobile phone in various operating modes.

	Standby Mode	Receive Mode	Transmit Mode
Percentage of time	80	10	10
Minutes per hour	48	6	6
Current drain during mode (mA)	10	62	325
Average current for mode (mA)	8	6.2	32.5

As a way of example, some of the technical characteristics of one of the most popular smartphones, the Blackberry 8830, are listed here.

- Size: $11.4 \times 6.6 \times 1.4$ cm; weight: 132 g
- Colour display with backlighting
- Several voice input/output options
- Several audio formats supported (e.g. MP3 and AAC)
- Memory: 64 MB (flash memory) expandable, support for SD card
- E-mail, web browser
- Wireless network: dual-band 900/1800 MHz GSM/GPRS; dual-band 900/1800 MHz CDMA2000
- Battery: 1.5 Ah Li-ion.

The battery allows a talk time of up to 300 min with the GSM/GPRS technology, and up to 200 min with the code division multiple access (CDMA) technology. This latter, mainly used in Asia and America, allows mobile phone users to share the same frequency channel.

By taking as a reference the above maximum talk times, the device has power consumption of ~1 W for GSM/GPRS, and ~1.5 W for CDMA.

In Table 3.1, typical current requirements of a mobile phone are reported. The battery used with this duty cycle will consume 46.7 mAh in 1 h. If the phone is a simple model, that is only working as a phone, an 850-mAh, 3.6-V Li-ion battery can be used. With this battery, the phone runtime will be 18.2 h, that is a full day of operation.

3.2.4. Personal Digital Assistants (PDA)

The block diagram of a PDA is shown in Figure 3.10 [2]. The core subsystems include:

- Processor/memory. The processor runs applications stored in memory. The operating system is stored in a non-volatile memory such as flash or ROM. Applications may be loaded in flash or DRAM.

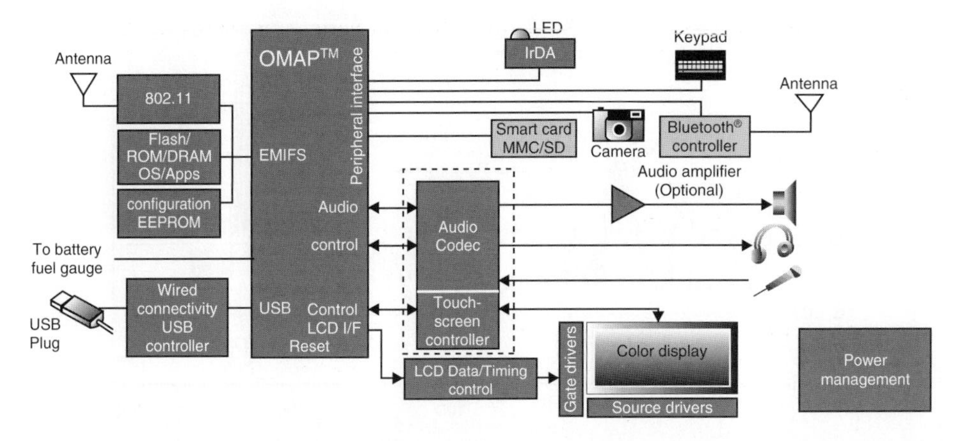

Figure 3.10. Block diagram of a PDA.
Source: From Ref. [2].

- User interface. It allows user to input/output data from the PDA using input commands via touch screen and output MP3 to the earplug.
- Connectivity. It allows PDA to connect to other computing machines.
- Audio codec. It performs compression/decompression of the digital files.
- Power conversion. It converts input power (battery or wall plug) to run various functional blocks.

The power consumption of a PDA in a typical application, audio record and playback mode, is shown in Figure 3.11. The initial conditions are display backlight *on*, PDA powered up, wireless *off*. The sequence is:

1. Select "Notes" (or equivalent) from main menu
2. Record message
3. Playback message

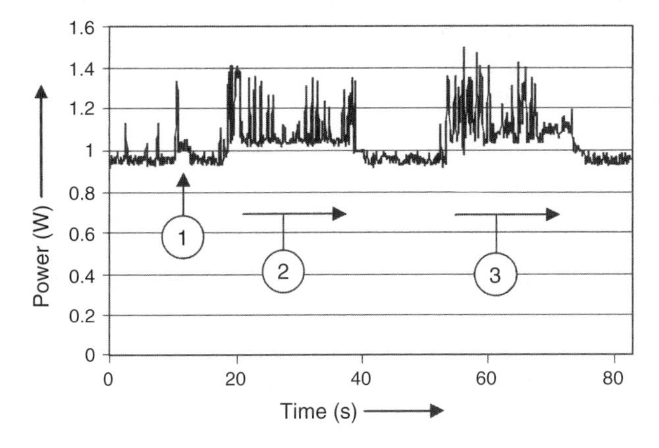

Figure 3.11. Power consumption of a PDA in audio record and playback mode.

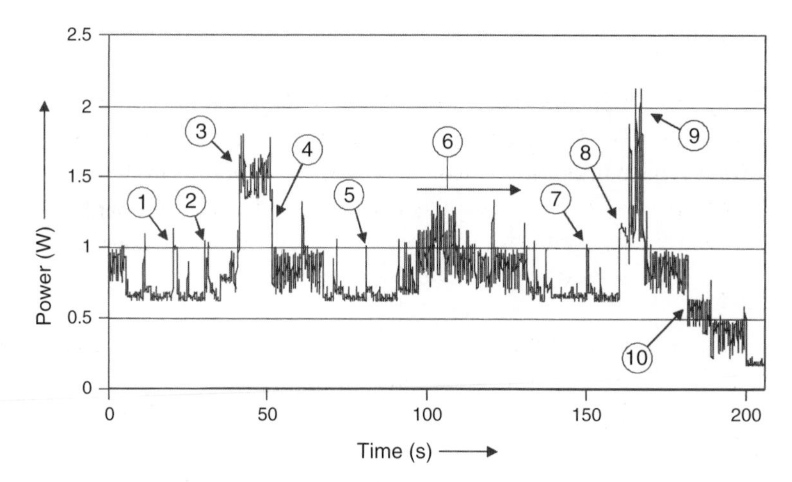

Figure 3.12. Power consumption for various PDA applications.

It appears that the record and playback functions increase the average power consumption by 20–30%.

The power consumed when a typical PDA performs various functions is shown in Figure 3.12. The initial conditions are: backlight *on*, PDA powered up, wireless *off*. The sequence is the following:

1. Select settings
2. Select display
3. Set brightness to maximum
4. Reduce to minimum
5. Select "Word" (or equivalent)
6. Use virtual keyboard
7. Select voice dialling
8. Attempt voice dialling
9. Announcement "phone is *off*"
10. Power down

As shown in the figure, the PDA of this test consumes between 0.3 and 2 W when used in various modes. For a Li-ion pack having a total energy of 5 Wh, the minimum consumption (backlight *off*) would give the battery a life of 5 Wh/0.3 W = 15 h. The maximum power consumption can result from a phone call while simultaneously using another PDA function with the backlight *on*. This would consume 2 W and, so, the battery would only last 2.5 h.

3.2.5. Mobile TV

TV programmes may be received on mobile phones, portable PCs or portable TV devices. The mobile TV service is delivered to subscribers via mobile telecom networks. South Korea is especially active in this sector, while British Telecom was the first company outside Korea to implement mobile TV.

There are some standards used for delivering mobile TV services. A non-exhaustive list is reported below.

- Digital multimedia broadcasting (DMB); terrestrial (T-DMB); satellite (S-DMB): developed in South Korea
- Digital video broadcasting (DVB); for handheld (DVB-H); satellite for handheld (DVB-SH); terrestrial (DVB-T): used in Europe
- MediaFLO (forward link only): a multicasting system developed by Qualcomm (USA); it is an alternative to the two technologies above.

DVB-H will probably become the European standard, also in view of compatibility with its terrestrial counterpart (DVB-T). This technology combines the standards of digital video with Internet protocol (IP): it divides the contents in data packets to be transferred to the handheld. However, fierce contenders will be T-DMB and S-DMB. China seems ready to exploit the DMB standard.

At present (April 2008), more than 90% of all mobile TV services rely on existing cellular networks. Indeed, there is still enough capacity in the 3G network and this is what the operators need to have a hot start.

A challenge for device manufacturers will be power consumption: battery life is seriously threatened by the upgraded mobile content and enhanced functions. To date, however, no technical reports on this matter have appeared.

In Figure 3.13, three portable devices are shown, which have mobile TV capability.

Samsung's cell phone integrates DVB-H and UMTS, supports videos in MPEG4 (and other formats), has a 2.2″ screen, and a 1.0-Ah Li-ion battery. Sony's portable PC has a 17″ screen and, among other features, a hybrid analog/digital TV tuner (DVB-T). The G1 player, by the Korean company iNavi, has a digital DMB tuner.

3.2.6. Digital Still Cameras (DSC)

The digital camera implements the image sensor (CCD or CMOS) to convert the light directly into series of pixel values that make up the image to

Figure 3.13. Mobile devices able to receive TV broadcastings. Top left: cell phone Samsung SGH-P910; top right: portable PC Sony VAIO VGN-AR31; Bottom: portable multimedia player iNavi G1.

be taken. The more pixels correspond to more detailed images. In Figure 3.14, the block diagram of a DSC is shown.

The core subsystems include [3]:

- Image sensor/front-end processor. In the digital still camera front end, CCD or CMOS image sensors convert light into electrons at the photosites. The analog signal from the image sensor is filtered, amplified and digitized by the high-speed A/D converter.

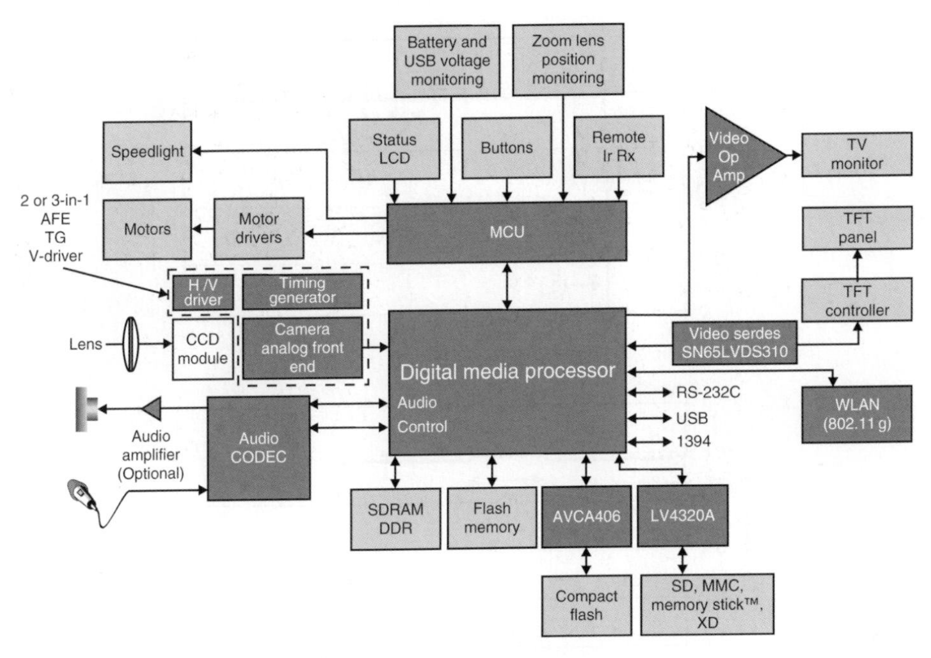

Figure 3.14. Block diagram of a DSC.
Source: From Ref. [3].

- Digital image processor. It provides processing power to handle various industry-standard imaging, audio and video algorithms. It also controls the timing relationship of the vertical/horizontal reference signals and the pixel clock.
- MCU. It controls the user interface. Performs general control such as system initialization, configuration, graphical user interface and user commands.
- Memory. It stores executing code and image data.
- Peripheral interface. It allows user to configure various features of the camera.
- Video interface. It allows processor to output NTSC/PAL composite, analog RGB, S-video and LCD outputs.
- Audio codec. It performs digital audio recording/playback under digital signal processor (DSP) control.
- LCD interface. It receives digital video to display camera image on the LCD.
- Power conversion. It converts input power from AC adapter or USB to charge the battery that runs various functional blocks.

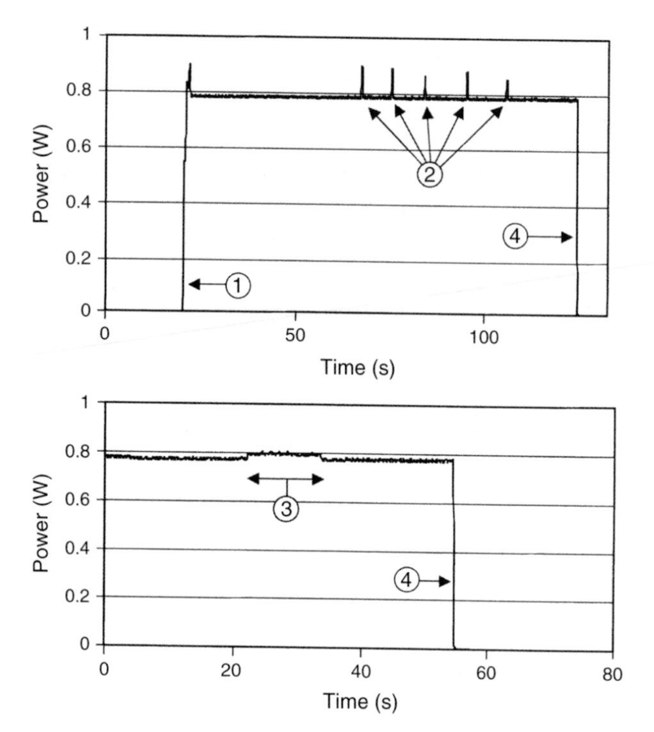

Figure 3.15. Power consumption of a digital still camera in playback mode. *Source*: From Ref. [1].

The power consumption of a digital camera when operating in playback mode is shown in Figure 3.15 [1]. The numbers correspond to:

1. Power *on* and "Play" mode selected
2. Select next picture
3. Play a short movie
4. Power *off*

In Figure 3.16, the power requested by the camera when operating in image capture mode is shown [1]. The action sequence is also described in the figure. It is interesting to note that capturing images, still or moving, means power consumption 1.5–2.5 times larger than viewing them in the display (see Figure 3.15). This is due to the process of converting light into electrons and forming a digital image. The larger power consumption when taking still pictures (~2 W) with respect to movies (~1.4 W) is due to the higher resolution of the first case.

With these powers, primary batteries, for example alkaline or primary Li, can be used in addition to Li-ion, which is anyway recommended if a long

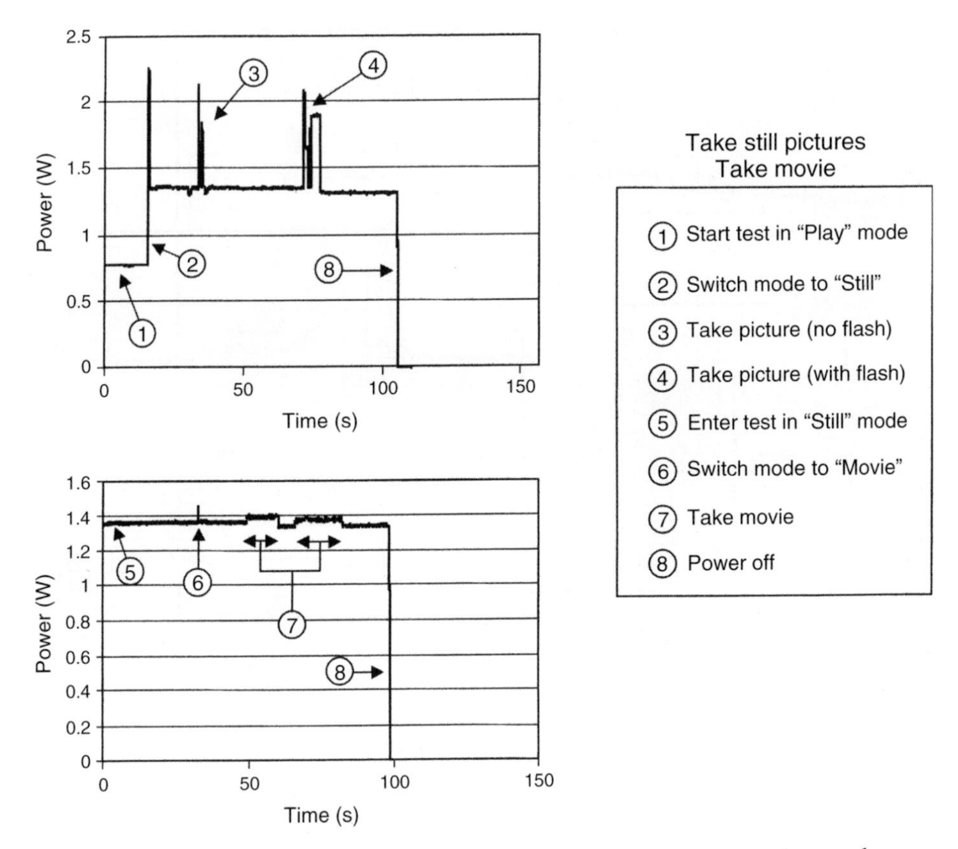

Figure 3.16. Power consumption of a digital still camera in image capture mode. *Source*: From Ref. [1].

runtime is also sought in high-resolution conditions. A prismatic 3.6-V battery of 600–700 mAh is adequate for a 5-megapixel (or more) DSC.

3.2.7. Digital Camcorders

A typical camcorder (analogue or digital) contains these basic parts:

- A camera section, consisting of an image sensor, lens and motors to handle the zoom, focus and aperture.
- A recording section.
- A viewfinder that receives the video image, thus allowing to see what is being shot. The viewfinders of modern camcorders have larger full-colour LCD screens.

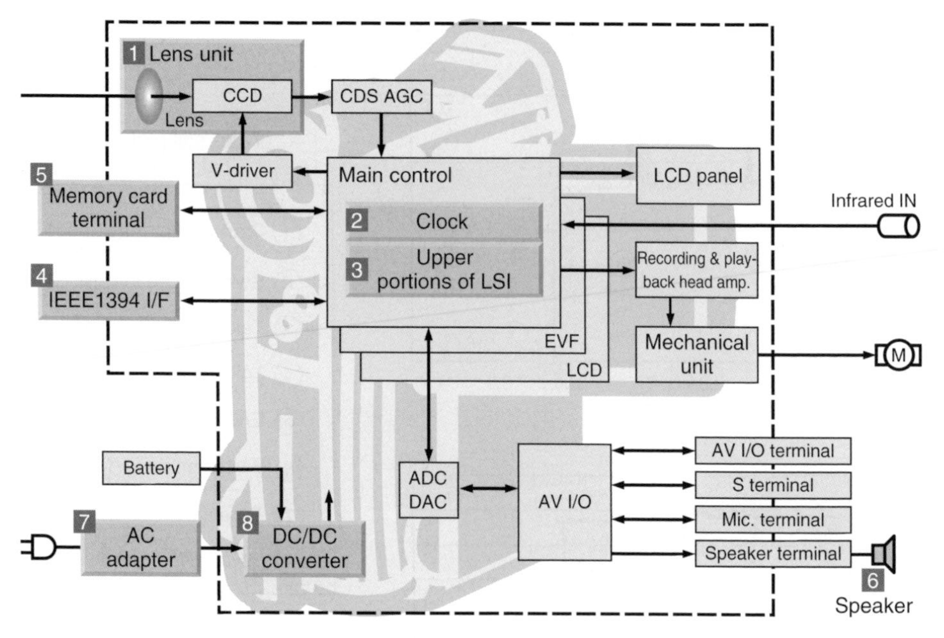

Figure 3.17. Block diagram of a digital camcorder.
Source: Courtesy of TDK.

In addition to the above elements, a digital camcorder has a component that takes the analogue information the camera gathers and translates it into bytes with an A/D converter. Instead of storing the video signal as a track of magnetic patterns, it records the picture as 1 and 0. A block diagram is shown in Figure 3.17.

The lens sends the light onto a small semiconductor image sensor, typically a charge-coupled device (CCD), which measures light with a small panel of 300 000 to 500 000 tiny light-sensitive diodes, the photosites. Each photosite measures the amount of light (photons) that hits a particular point, and translates this information into electrons. To create a colour image, a camcorder has to detect not only the total light levels, but also the levels of each colour.

Camcorders can use different storage mediums: (a) mini DV (compact cassettes); (b) standard Hi-8 mm tapes (used only by Sony); (c) DVD; and (d) memory cards, such as SD cards.

Of course, a digital video can also be downloaded to a computer.

3.2.8. Portable Players

Portable players include MP3, portable media player (PMP) and digital versatile disc (DVD). Their block diagrams are reported in Figures 3.18–3.20 [4].

mechanism moving the laser beam to follow the spiral tracks; (4) a video/audio decoder to obtain a composite digital signal from the bumps; and (5) a DAC to convert the digital input into an understandable output.

DVD players use lasers with a 650 nm wavelength capable of reading bumps of data 0.4 μm in size on spiral tracks 0.74 μm wide. The tracking system, therefore, must move the laser with a very high resolution.

All portable DVD players use Li-ion batteries with voltages of 7.2–9 V and capacities of 4–7 Ah, this giving the players several hours of operation.

3.2.9. Portable VoIP (Voice over Internet Protocol) Phones

VoIP is a technique turning analog audio signals into digital signals that can be transmitted using the Internet. This technique will likely replace the today's phone system, also in view of the fact that allows free calls by using appropriate software. The VoIP's heart is the audio codec (block diagram of Figure 3.21) that converts the audio signal into a compressed digital form for transmission and then back into an uncompressed audio signal for replay. Audio codecs have high sampling rates for the audio signal (e.g. 8000–64 000 times/s. With this frequency, the user never gets to know the missing of small audio pieces between samples.

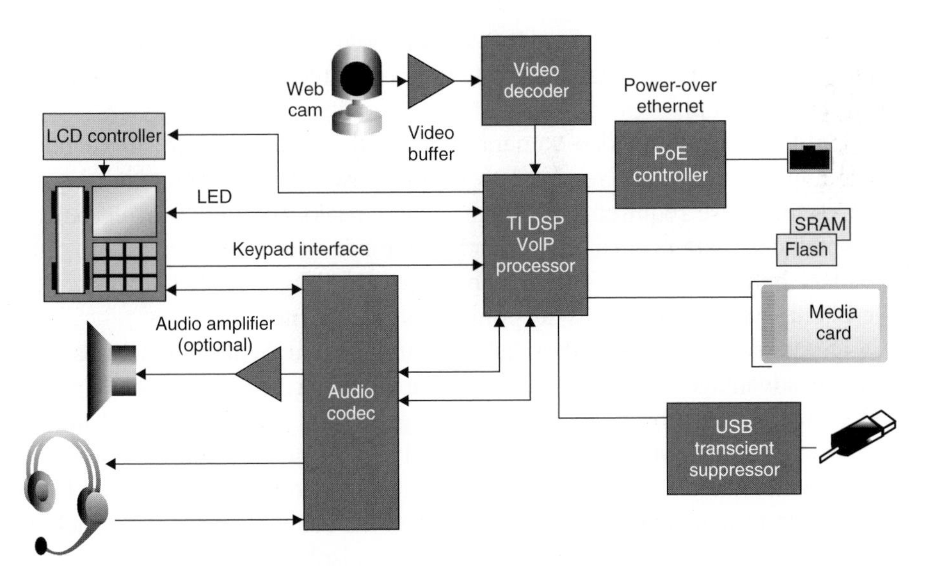

Figure 3.21. Block diagram of a portable VoIP phone.
Source: From Ref. [4].

Codecs work by using algorithms that help them sample, sort, compress and make packets of audio data. Indeed, VoIP telephony uses the packet switching technology instead of the circuit switching technology used by today's phone systems. In circuit switching, the connection is open and constant, while packet switching just opens brief connections to send small data portions (packets) from one system to another. With packet switching, several phone calls require the space occupied by only one in a circuit-switched network. Moreover, data compression may further reduce the size of a call.

In mobile VoIP, the Internet protocol is applied to mobile phones. In this case, the phone has to support high-speed IP communications, in most cases using the Wi-Fi technology (802.11 network). Session initiating protocol (SIP) is the standard used by most VoIP services and is now implemented on mobile phones as well.

Mobile telephony over IP can make use of special IP phones (see again the block diagram of Figure 3.21). The hardware and software needed to handle the IP call is embedded. These phones have an Ethernet connector, instead of the standard one, for connecting to a wireless network. The media card is a solid-state flash memory module.

Typical batteries for mobile VoIP phones are 3.6 V Li-ion batteries of 2.4–3.2 Ah.

3.2.10. Professional Audio/Video Equipment

Professionals working in the TV and movie business also use cordless audio/video equipment: stand-alone cameras, camcorders, lighting systems, audio recorders, etc.

Unlike consumer video cameras, the professional ones are more power demanding and are often required to supply power for auxiliary lighting. To comply with these requirements, large battery packs contained inside belts or even small cases are needed. These packs are made with primary or, more often, secondary batteries. Primary lithium batteries (Li/SO_2, $Li/SOCl_2$) can work in severe temperature conditions, as pointed out in Chapter 2.

Different types of secondary batteries are suitable. Sealed Pb-acid batteries have the advantage of a lower cost but are heavier; Ni–Cd have long been used owing to the ability of delivering high drains required by the lighting systems, high cycle number (>500) and good performance in outdoor conditions; Ni-MH battery packs are used to power cart-loaded audio recorders; Li-ion batteries are now increasingly used, and are often coupled with solar panels for their charge (e.g. when used on-board boats or in isolated areas).

Technical specifications of the packs include the following: voltage, 6–14 V for the camera, 12–30 V for the lighting systems; current, up to 5 A for the cameras, 4–10 A for the lighting systems; capacity, 4–9 Ah; operating temperature: –30/+50°C.

3.3. Medical Applications

As will be shown in the following, medicine has particularly benefited from advances in the electronic field. Measuring devices, diagnostic and therapeutic tools have been either modified or just created, and many of them have been made portable with the use of common or advanced batteries.

Portable medical devices with life-saving characteristics must have batteries (or packs) with a battery management unit (BMU) using redundant safety systems and protection circuitry. Packs with a high number of cells in series represent a particular concern due to the greater risk of cell imbalance. There are now new technologies to reduce this risk and some battery chemistries, for example Li-ion with Li–Fe phosphate as a positive electrode, are inherently safer, thus increasing the reliability of packs, including the ones to be used in large devices. On the other hand, the larger energy provided by the Li-ion systems allows using batteries of reduced dimensions in a number of small portable devices, for example the implantable ones.

To assess the suitability of a given battery for a specific device, real-word usage tests are necessary [5]. Tests in different conditions of temperature, current, shelf life, etc., must be carried out, not excluding possible misuses. All cells and packs must comply with the current regulations in terms of safety, especially as far as their materials are concerned.

Portable medical devices may be rather diverse in functions and sizes, ranging from small implantable devices to large apparatuses, for example for X-ray analysis or echography. Therefore, the circuit complexity, the power consumption and the battery characteristics may vary in a broad range.

3.3.1. Meters

A block diagram for a portable metering device is presented in Figure 3.22. The central part of the figure represents the core of the medical system and includes microcontroller, memory, display and battery with its management system [6]. Among the data transmission options on the right-hand side of the figure, the wireless transmission techniques are particularly relevant in remote sites. They will be described in Section 4.18.

In general, there are five system level blocks that are common to metering medical devices, such as blood glucose meter, digital blood pressure meter, blood gas meter, digital pulse/heart rate monitor or even a digital

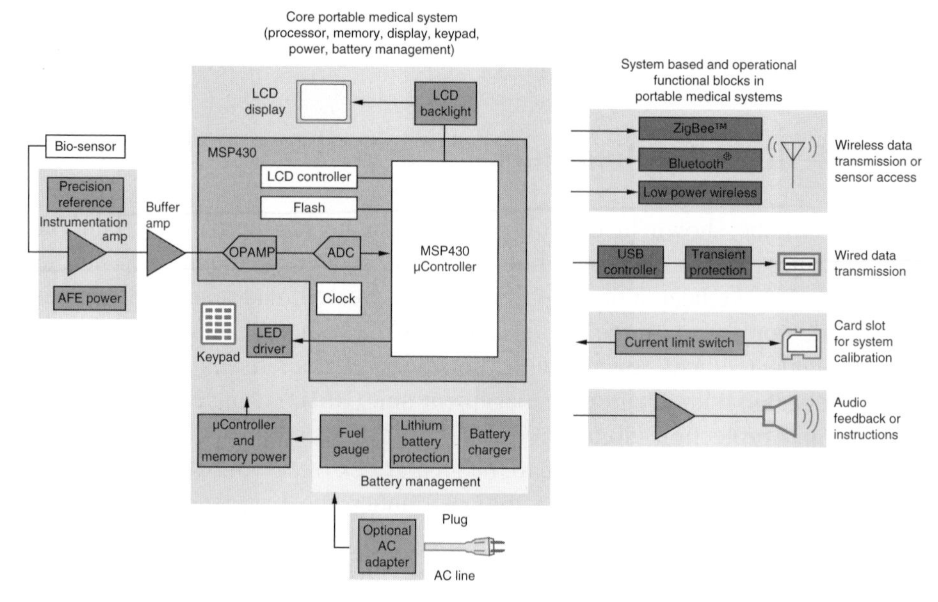

Figure 3.22. Block diagram of a portable medical metering device.
Source: From Ref. [6].

thermometer. These are (1) power/battery management (extensively treated in Section 3.5), (2) control and data processing, (3) amplification and A/D conversion of the sensor input, (4) some type of display and (5) the sensor element itself. Obviously, the actual utilization of these blocks depends on the meter type.

3.3.1.1. Glucose Meter

A typical block diagram is shown in Figure 3.23 for a modern glucose meter using just two devices – an ultra low-power microcontroller unit (MCU) and an amplifier with shutdown [7]. A test strip, wet with a small blood sample, generates a signal that is amplified and measured by the operational amplifier (TLV2763 in the figure). The amplifier's output is received by the A/D converter contained in the microcontroller unit (MSP430 in the figure). The MCU also has an integrated temperature sensor, as the chemical reaction of the test strip is temperature-sensitive. The measurements are often logged and later downloaded to a PC. Data logging is made possible by a flash memory also contained in the MCU: 8 kb of memory allow recording 1000 measurements.

To reduce power consumption, shutting off the analog circuit when inactive is very important. In this mode, the amplifier only draws current in the

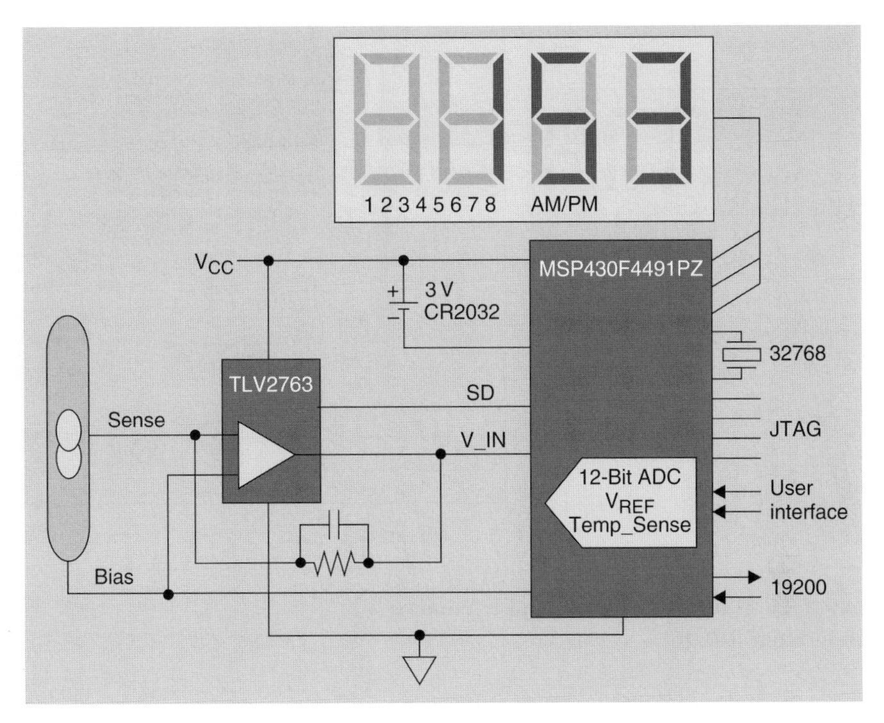

Figure 3.23. Block diagram of a glucose meter.
Source: From Ref. [7].

10 nA range. Regarding the microcontroller, it has to feature low standby power consumption and fast wake-up times. A high-speed clocked system in the MCU allows quick starts of the instrument.

The primary interface is a numeric LCD requiring a very small current when active (1–2 μA). Additional features include user input buttons, an alert buzzer and a serial communication link.

Modern meters have more or less the size of a cellular phone, response times of <15 s and require only microlitres of blood. They use primary batteries, either a couple of common AAA alkaline or a single Li–MnO_2 (the latter is preferred and indicated in the figure as CR2032).

3.3.1.2. Pulse Oximetry

A block diagram of this non-invasive device for measuring the oxygen content in the blood is shown in Figure 3.24 [6]. The microcontroller unit is connected to a probe, consisting of tweezers endowed with a pair of leads, in which the patient inserts his (her) finger. The probe includes two LEDs: one

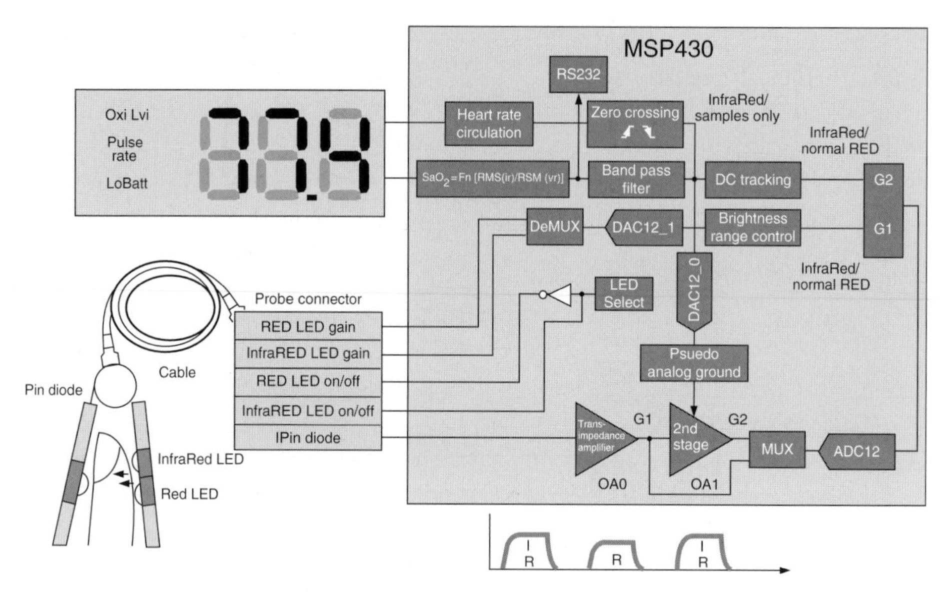

Figure 3.24. Block diagram of a pulse oximeter.
Source: From Ref. [6].

emits light in the visible red spectrum (660 nm) and the other in the infrared spectrum (940 nm). Deoxygenated haemoglobin allows more infrared light to pass through and absorbs more red light; highly oxygenated haemoglobin allows more red light to pass through and absorbs more infrared light. The percentage of O_2 in the blood is determined by measuring the intensity from each frequency of light, after its transmission through the body. The LCD display shows the O_2 level and the heart rate, and an alarm sounds if these decrease below preset values.

One-cell Li battery is preferred for this device, but two alkaline cells may also be used. A wearable pulse oximeter capable of wireless transmission within a sensor network will be described in Section 4.4.

3.3.1.3. Miscellaneous Meters

The blood pressure/heart monitor system is a well-known device; it employs a pressure cuff, a pump and a transducer and makes a measure in three phases: inflation, measurement and deflation [8]. The core subsystems include: microprocessor and memory (e.g. EEPROM), LCD to read the measurement, sensor interface allowing the processor to sense blood pressure, which is amplified and then digitized by the A/D converter. An alkaline battery is used, whose power is converted to run various functional blocks.

Fingerprint biometrics technology allows identification of a person by using his (her) fingerprints [9]. It is often used for restricted access purposes. The sensor captures an image of the finger and sends it to the DSP, which matches the captured image against stored fingerprints. If the match is successful, the DSP sends a signal to authorize access along with using some form of visual or audio signal to let the user and the system know that the user is verified. Should the match fail, a signal can be generated using the DSP to alert the users and the administrators.

A portable blood gas analyser allows measuring the contents of O_2, CO_2, CO and N_2. It requires a small blood sample and is based on the following subsystems: (a) an analog front end, in which chemical sensors inputs are treated and finally obtained in digital form; (b) microcontroller, which measures the gas levels and controls the interface with memory and peripherals; (c) LCD and (d) power management, which converts the battery input to run various functional blocks [10].

Using new technologies, different types of digital thermometers can be built. For instance, the ear thermometer measures with infrared sensors the heat of the eardrum, which reflects the temperature of the hypothalamus (the temperature-controlling system of the brain). Other digital thermometers, for example oral or underarm, use thermopiles or thermistors as sensors [6].

3.3.2. Therapeutic Devices

3.3.2.1. CPR (Cardio-Pulmonary Resuscitation) and AED (Automated External Defibrillator)

Fatal arrhythmias common in sudden cardiac arrest have a greater chance of being successfully terminated by electrical shock if cardio-pulmonary resuscitation (CPR) is performed first. A number of studies have confirmed that CPR can be life saving when provided either by laypersons or medical professionals.

CPR is simple to describe and remarkably difficult to perform, as humans generally do not have a good internal sense of timing to recognize 100 compressions or 8–12 ventilations per minute. A variety of new technologies has been developed over the past decade to assist in this goal. First, devices have been developed that detect CPR parameters during actual resuscitation, and these devices can trigger alarms or audio messages when incorrect CPR is detected, for example if the chest compression rate is too slow. Such devices assist the provider during human CPR by acting as real-time "CPR coaches", and are currently being marketed by a number of biomedical companies. Second,

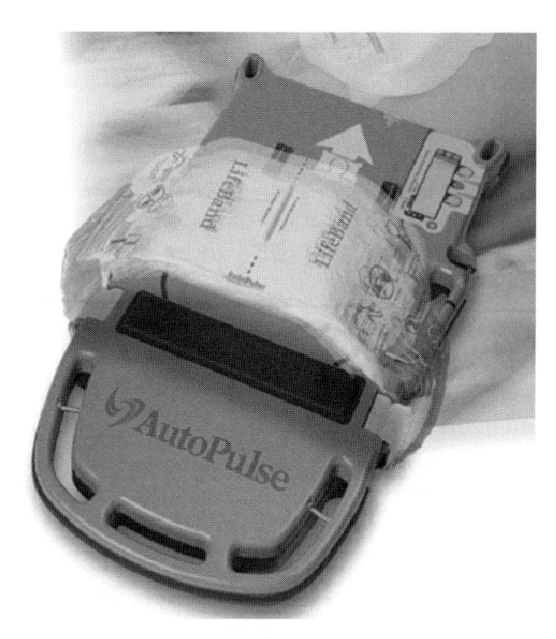

Figure 3.25. Motorized compression band for CPR.
Source: From Ref. [11].

devices are currently in use that deliver CPR mechanically, for instance via a battery-powered, motorized compression band wrapped around the patient's chest (see Figure 3.25) [11].

The automated external defibrillator (AED) is a highly sophisticated, microprocessor-based device that monitors, assesses and automatically treats patients with life-threatening heart rhythms. It captures ECG signals from the therapy electrodes, runs an ECG-analysis algorithm to identify abnormal rhythms, and then advises the operator about whether defibrillation is necessary. A basic defibrillator contains a high-voltage power supply, storage capacitor, optional inductor and patient's electrodes (see block diagram of Figure 3.26) [6]. It develops an electrical charge in the capacitor to a certain voltage, creating the potential for current flow.

The main front-end signals of the AED come from the ECG electrodes placed on the patient, which requires instrumentation amplifiers (see figure) to amplify their very small amplitude ($<10\,\text{mV}$). Another front-end signal is the microphone input for recording the audio from the scene of a cardiac arrest. Both ECG and microphone input are digitized and processed by the DSP contained in the dual core processor shown in the figure.

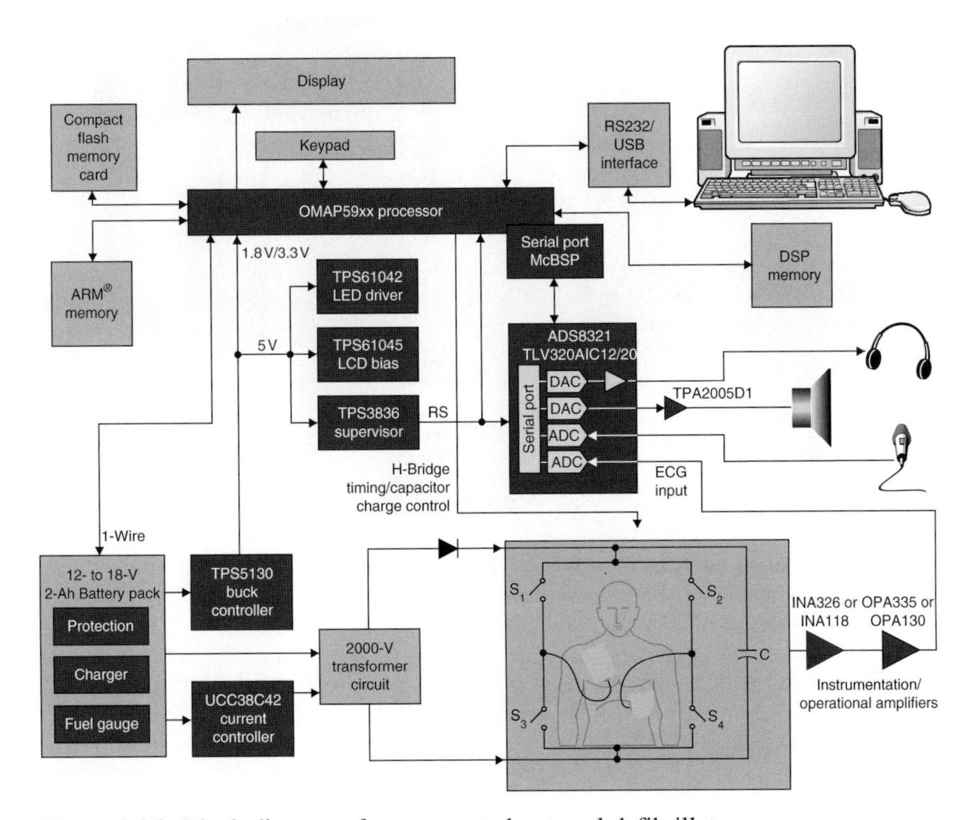

Figure 3.26. Block diagram of an automated external defibrillator.
Source: From Ref. [6].

Technical specifications of a typical AED include [12]:

- Base currents (*during standby periods*): a few microamperes
- Current during analysis phase: <500 mA for several minutes
- Stroke pulse (*defibrillation phase*): 10 A typically for the defibrillator for 5–10 s, while charging a capacitor, repeated every 15–20 s; about 2 A for the monitors
- Output voltage: 8–20 V (12 V typically)
- Operating temperature: from –20°C up to +60°C
- Runtime: up to several hundreds of discharges (even after 4 years of shelf life)
- Lifetime: up to 4 years.

AEDs are equipped with high-rate battery packs, made with spirally wound primary Li cells (e.g. Li/MnO_2 or Li/SO_2) or Li-ion cells.

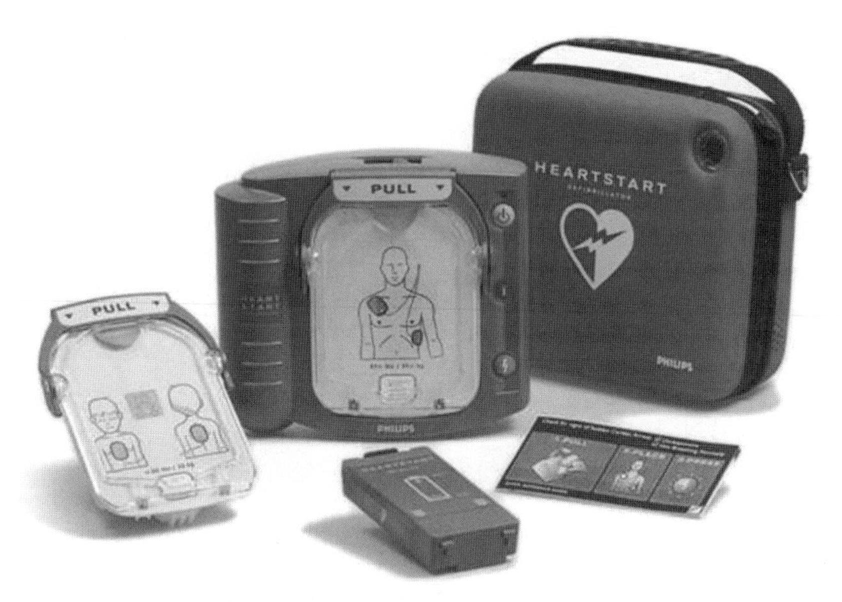

Figure 3.27. Components of the Philips AED.

Figure 3.27 shows the Philips AED. This defibrillator is equipped with a Li-ion long-life battery (in the figure front) that lasts 4 years with the defibrillator in standby mode.

3.3.2.2. Pacemakers and Other Portable Devices for Cardiac Rhythm Management

A cardiac pacemaker is prescribed when the cardiac rhythm is too slow. The device, implanted under the skin, detects the slow heart rate and sends impulses to stimulate the muscle. The pacemaker is connected to the heart with leads inserted in a vein. Several progresses have been made in this field since the early models available in the 1950s. Titanium casing has replaced the plastic materials used previously, thus shielding the device with respect to electromagnetic fields. A number of sources can emit such fields, for example power lines, radio frequencies (RFs), cellular phones, personal communication devices, microwave ovens, etc. If these external signals have frequencies in the 10–100 Hz range, that is the same range of the sense amplifier of the pacemaker, there could be sensing/pacing inhibition or asynchronous pacing. A further progress was made with the introduction of programmable pacemakers whose parameters could be set by external RF signals. Rate-responsive pacemakers are also available that detect body movements and accordingly increase or decrease the pacemaker's rate.

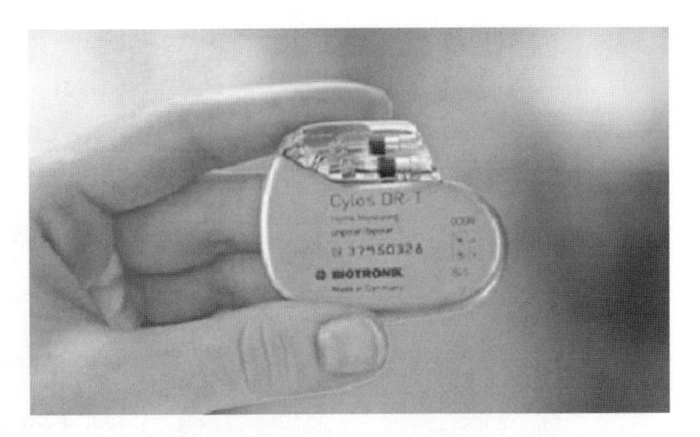

Figure 3.28. A typical pacemaker.
Source: Courtesy of Biotronik.

The pacemaker contains a timing device for setting the pacing rate, a circuitry that detects electrical signals from the heart, and a battery. The battery occupies more than half the total space. An example of pacemaker is presented in Figure 3.28.

Pacemakers may be of the single- or dual-chamber type. The former have a lead in the right atrium or in the right ventricle, while the latter have two leads, one for the atrium and the other for the ventricle, so that they can pace both. Furthermore, this latter type can coordinate the signals and contractions of the atria and the ventricles to help the heart beat more efficiently. One end of each wire is attached to the heart's muscle, while the other end is screwed to the pacemaker generator, as shown in Figure 3.29.

Since 1975, the most common pacemaker battery makes use of the Li/I_2 couple. This primary battery is requested to deliver currents in the microampere range and can last $\sim$10 years and up to 14 years [13]. Modern pacemakers may act as microcomputers. For instance, they can monitor the hearth activity so to change automatically the therapy delivered, and can store patient's data directly in the device memory for successive release via telemetry. These pacemakers need a medium-rate battery (mA range), and the Li/CF_x couple was found suitable.

The implantable cardioverter defibrillator (ICD) is designed to detect and treat episodes of ventricular fibrillation, ventricular tachycardia and bradycardia. This device contains a high-voltage capacitor that provides a stimulus to the heart when the regular pace is lost due to ventricular problems. In the ICD, batteries capable of delivering high current pulses in a few seconds (e.g. to charge the capacitor) are needed. Li/silver vanadium oxide (SVO) has proven especially suitable for this application [13].

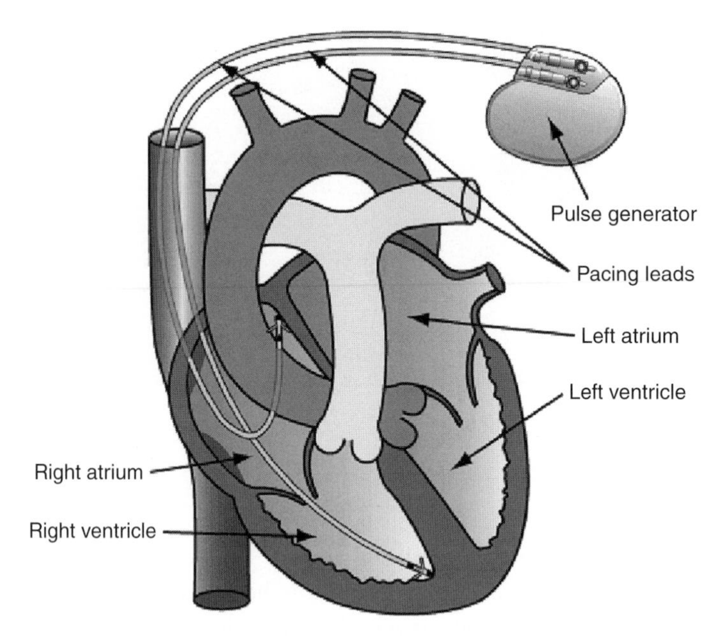

Figure 3.29. Dual-chamber pacemaker. The tip of one lead enters the right atrium, while the other enters the right ventricle.
Source: Courtesy of St. Jude Medical.

The left ventricular assist device (LVAD) is used to help this ventricle to pump blood more effectively. It is intended for patients who are no longer responsive to medical therapy and are not candidates for heart transplant. This device requires rechargeable batteries and the Li-ion have proved suitable for delivering $\sim$10 W with a 12-V pack. Current versions of this device are powered from external packs. In recent years, clinical trials have being going on with a LVAD in which the batteries are also implanted [14].

The total artificial heart (TAH) is a mechanical pump that completely replaces the patient's heart. Several experimental TAHs have been recently implanted. The design of this replacement heart contains two chambers, each capable of pumping over 7 l of blood per minute. This device utilizes an implantable Li-ion battery pack that is recharged through the skin.

3.3.2.3. Other Therapeutic Devices

A drug delivery system is a self-regulating therapeutic device, which eliminates the need of human intervention. A good example of systems of this kind is the intrathecal drug pump, or pain pump, of Figure 3.30 [15]. This system uses a small pump that is surgically placed under the abdomen's skin and

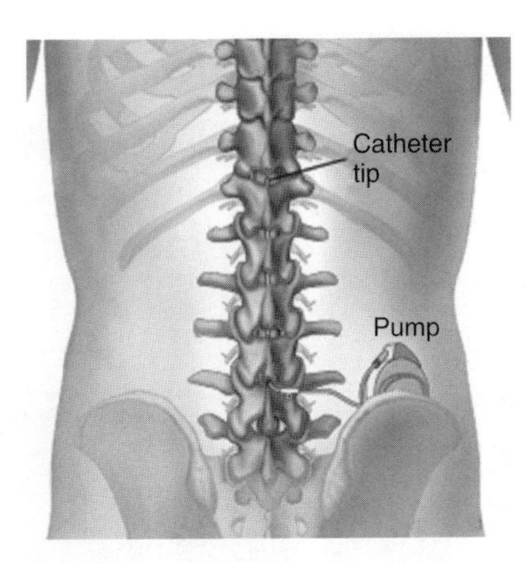

Figure 3.30. Intrathecal pain pump system.
Source: From Ref. [15].

delivers medication, for example morphine, through a catheter directly to the area around the spinal cord. This pain management option may be used if all other traditional methods have failed. Because the medication is delivered directly to the spinal cord, pain can be controlled with a much smaller dose than is needed with oral medication. The pump's reservoir holds the medication, and, when the reservoir is empty, the doctor or nurse refills the pump by inserting a needle through the skin and into the fill port on top of the reservoir. An external device allows programming the pump for a slow release of the medication over a period of time. The pump can also be programmed to release different amounts of medication at different times of the day. The pump stores the information about prescription(s) in its memory, and a doctor can easily review this information.

Devices of this type can be powered by either Li-ion or Li primary batteries.

Recently, portable breathing-assistance equipment for people suffering from chronic or acute respiratory failure has been marketed. A block diagram is shown in Figure 3.31 [6]. The respirator supports a patient with the appropriate O_2 level. One pressure sensor in front of the valve measures the breathe-in air and another one after the valve measures the breathe-out pressure. A microprocessor uses the data from the two pressure sensors and single flow sensor to calculate the output of the valve regulating the airflow. The medical staff can set the right airflow by a touch screen or key pad.

The Li-ion battery, with its management system, is preferred in this case too. Up to 11 h of autonomy can be obtained with one charge cycle. Another feature of this battery proved of importance for its adoption: its clear end-of-life indication enabling the respirator users to see easily how much power remains.

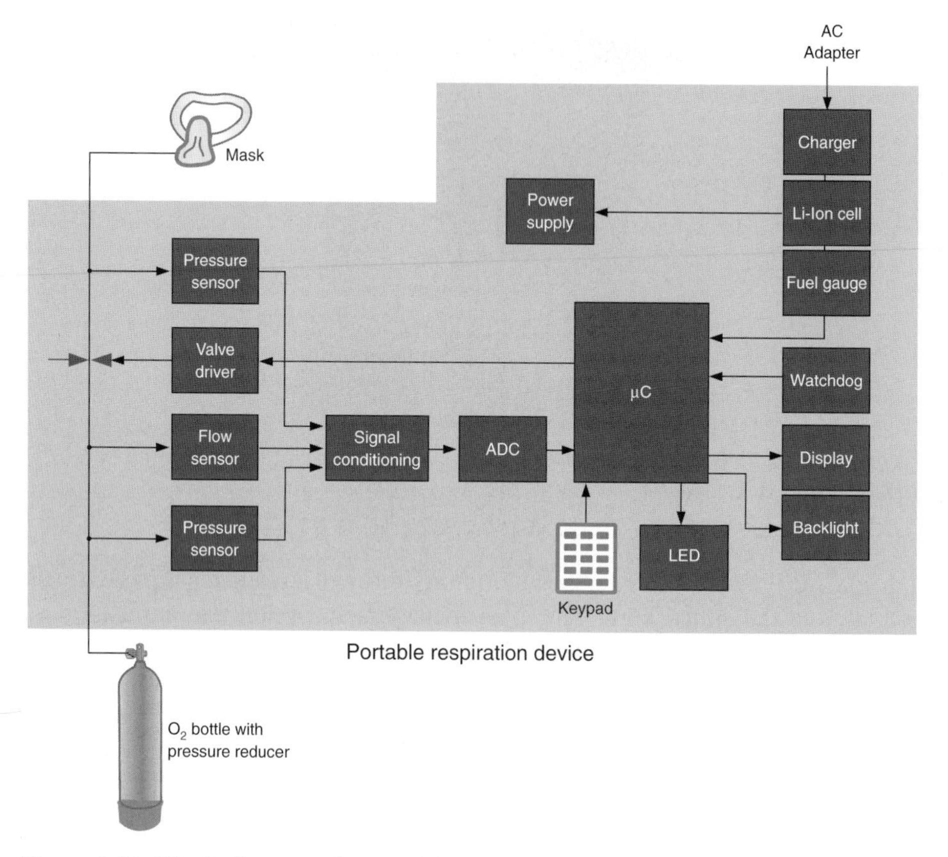

Figure 3.31. Block diagram of a portable respiration device.
Source: From Ref. [6].

Basic technical specifications for the battery include [12]:

- Current: ~0.5 A
- Voltage: from 4.8 V to, typically, 24 V
- Capacity: 2–9 Ah
- Charge time: 4–8 h.

An implantable microstimulator (Figure 3.32) is a tiny, programmable device that could help victims of stroke, Parkinson's disease, epilepsy, etc., by muscle and/or nerve stimulation [14]. The neurostimulator, which can be considered a bionic neuron, has been developed by Advanced Bionics Corp. and is powered by the small Li-ion polymer battery (2.9 × 13 mm) shown in the figure. The battery is endowed with a management system that control recharging and remote programming. Recharging is done wirelessly by an external electric

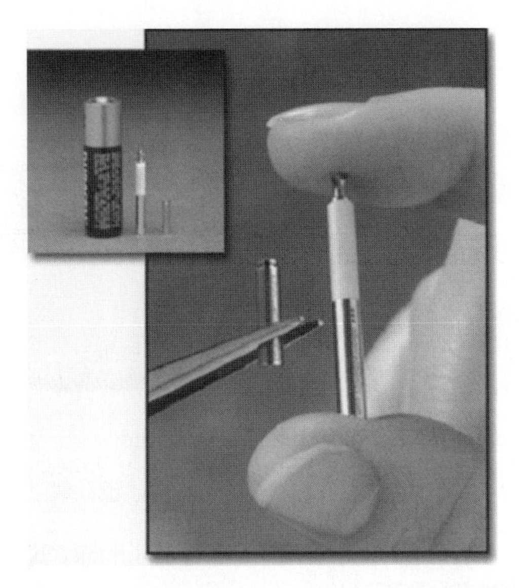

Figure 3.32. An implantable microstimulator (at right in the larger photograph) with its battery. The tiny battery is compared with an AA battery in the smaller photograph. *Source*: Courtesy of Advanced Bionics.

field. Owing to its very limited dimensions (3.2×27.5 mm, for a weight <1 g), the implant requires a minimally invasive operation.

A hearing aid is used to help moderately deaf people. The block diagram of a digital device (Figure 3.33) shows that its basic component are a clock, an analog to digital converter, a digital signal processor and an amplifier [16]. The use of a DSP allows software-controlled features, for example noise reduction, frequency shaping, reverberation reduction and possibility of a direct input from a TV or digital phone. A programmable DSP also means that the device's algorithms/features could be customized without changing the hardware.

The Zn-air primary battery is mostly used, but in the last few years the use of small rechargeable batteries has gained attention, especially in view of the fact that a Zn-air battery needs to be replaced every 5–15 days. A Li-ion battery ($C/LiCoO_2$) of the same dimensions of a typical hearing aid Zn-air battery has been tested at the regime of this application. This button cell of 10 mAh can sustain short-term 100-mA pulses superimposed to the 10 μA background current. It can be cycled more than 2000 times and proves capable of powering a hearing aid for 5 years on a daily use cycle [17].

The artificial ear, or cochlear, is an implantable device for profoundly deaf individuals. It is powered by a solid-state rechargeable Li microbattery, $Li/Lipon/LiCoO_2$, where Lipon stands for lithium phosphorus oxynitride. It

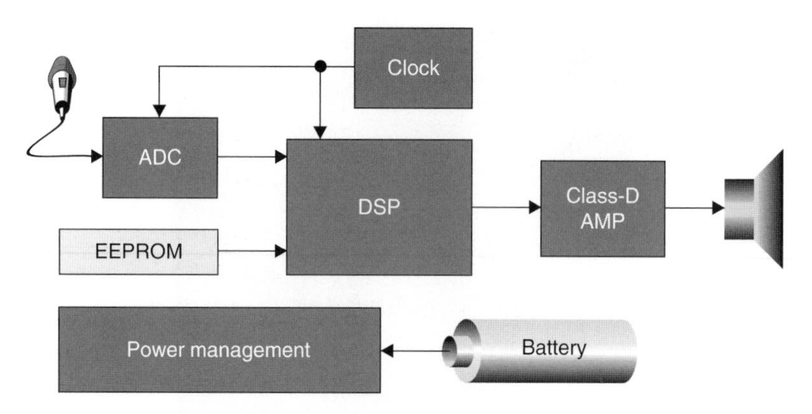

Figure 3.33. Block diagram of a DSP-based hearing aid.
Source: From Ref. [16].

Table 3.2. Portable medical devices and related batteries (see text for acronyms).

Medical Application	Battery
Pacemaker	Li/I_2, Li/CF_x
ICD	Li/SVO, Li/MnO_2
LVAD	Li-ion
TAH	Li-ion
Drug delivery system	Li-ion, primary Li
Portable respirator	Li-ion
Microstimulator	Polymer Li-ion
Hearing aid	Zn-air
Artificial ear	Li microbattery

is claimed to be cyclable for over 60 000 cycles and to be usable in a number of microelectronic applications. RF energy is used to recharge the battery remotely, thus enabling fully implantable product designs. The long cycle life provides permanent power and eliminates the need to replace the battery [14].

In Table 3.2, portable devices and their batteries are listed. As discussed above, some of these devices are implantable.

3.3.3. Diagnostic Devices

A gastrointestinal investigator is a battery-powered diagnostic capsule which, after ingestion, maps a portion of the gastrointestinal tract with a video camera. Given Imaging, a leader company in this area, has developed three

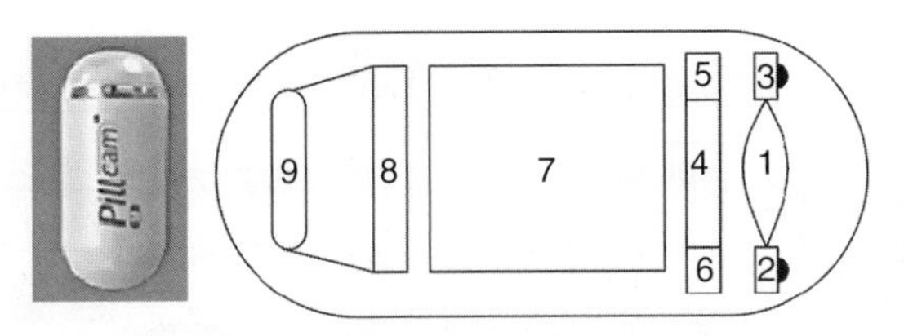

Figure 3.34. Video capsule visualizing the small intestine. (Left) Pillcam SB (11 × 26 mm); (right) capsule components: 1: lens; 2,3: illuminating LEDs; 4: CMOS sensor; 5: image compressor; 6: scanline controller; 7: battery; 8: RF transmitter; 9: antenna.
Source: Courtesy of Given Imaging.

different capsules for investigating the oesophagus, the colon and the small intestine (or small bowel, SB). This last capsule, shown in Figure 3.34, is briefly described here.

The capsule basically consists of light source, camera, processing electronics, data radio transmitter and Zn/Ag_2O primary batteries. The semiconductors of the CMOS sensor allow miniaturization and reduce energy consumption. The capsule takes about 8 h to move through the small intestine, taking two pictures per second with its single camera. The colour images are transmitted to a data recorder, powered by a second battery, carried by the patient. The images so transmitted help identifying such diseases as intestinal obstruction, ulcers, cancer and bleeding.

An ultrasonography (echography) can now be obtained with a portable ultrasound system (block diagram in Figure 3.35). The development of this system has been particularly pursued by General Electric (GE Healthcare) with their Vivid-i cardiovascular ultrasound system shown in Figure 3.36. It allows carrying out cardiac and vascular scanning with good image quality. The GE's imaging platform acquires the ultrasound data in a digital form and stores it in a raw format; the images can then be viewed and analysed without quality losses. This portable system, weighing 5 kg only, has wireless capability, so that the data can be easily transferred to other physicians for consultation. A rechargeable battery provides up to 1 h of scan operation.

Hospitals and other health care facilities use medical telemetry devices (block diagram in Figure 3.37) to monitor patients' vital signs and other important parameters, and transmit this information via radio to a remote location such as a nurses' station. Certain types of medical telemetry devices may be used in the home. The following is an example of how a patient can be monitored via telemetry.

A single patient with chronic cardiovascular disease monitors his heart via an X73-compatible intelligent Holter [18]. The patient is completely mobile,

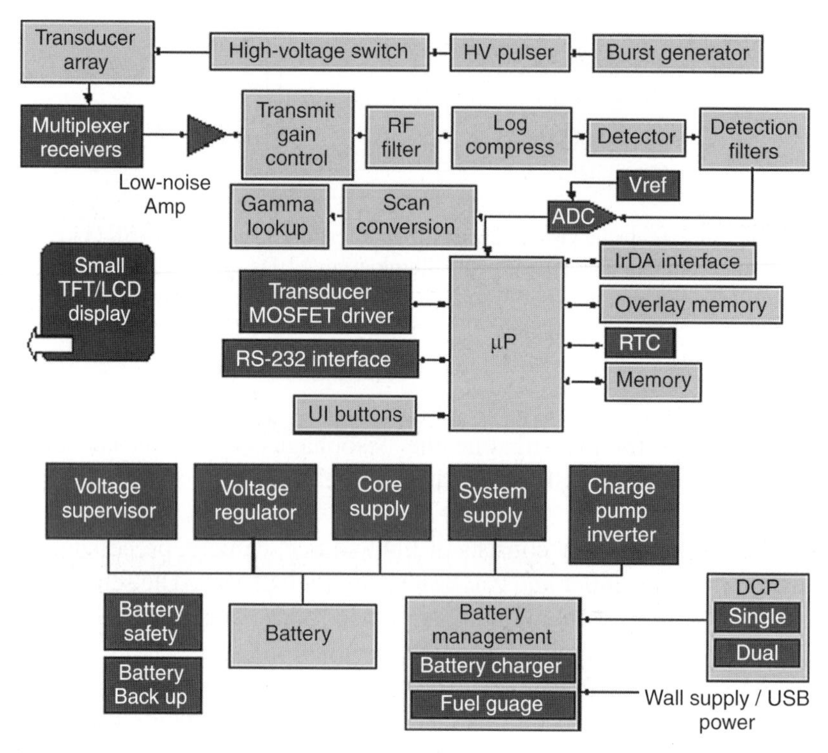

Figure 3.35. Block diagram of a portable ultrasound system.
Source: Courtesy of Intersil.

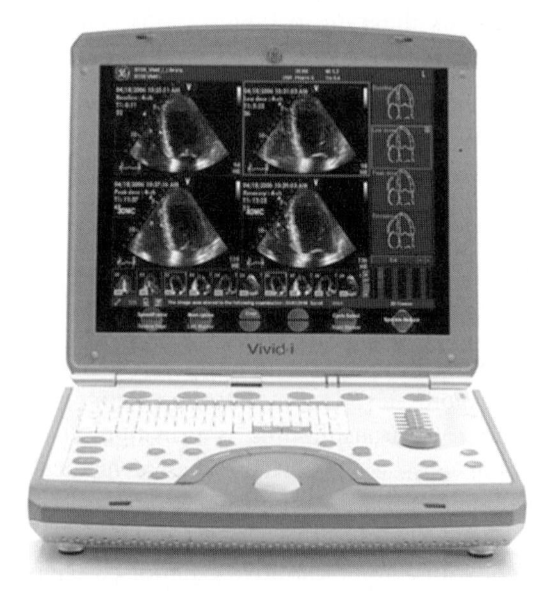

Figure 3.36. GE's portable cardiovascular system.
Source: Courtesy of General Electric.

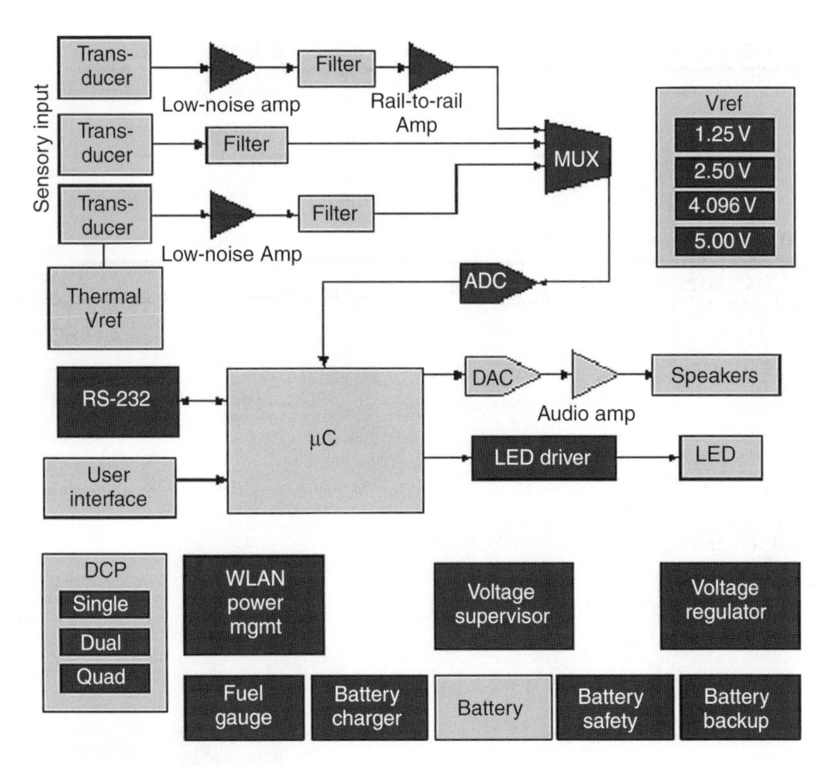

Figure 3.37. Block diagram of a medical telemetry device.
Source: Courtesy of Intersil.

and lives a normal life. Information is sent from the patient's Holter that detects, stores and transmits cardiac events through Bluetooth wireless technology. The Holter holds a cassette tape where the heart electrical impulses, detected by 5–7 electrodes, are recorded. AA-size alkaline batteries are normally used. The information acquired is sent to a mobile phone with an X73-installed application that acts as a gateway and transmits the data to a Telemedicine Server using a GPRS connection. The mobile application is prepared for interchanging X73-compatible data and configured for the specific service.

In the Telemedicine Server, the data can be analysed and reviewed by a health professional; he can see if the Holter is disconnected from the patient and can receive alarms from it; in that case, he will call the patient or a patient's relative and try to solve the problem. In Figure 3.38, the complete monitoring array is shown [18].

Another example of telemedicine is provided by the recently introduced mobile robotic doctor (Figure 3.39) [19]. The robot is remotely controlled by a doctor based at another location and acts much like him at the bedside when he cannot be present. It reads the patient's medical chart and checks his (her) vital signs on the bedside monitor. It eventually advises a nurse about adjusting medications before getting out of the room.

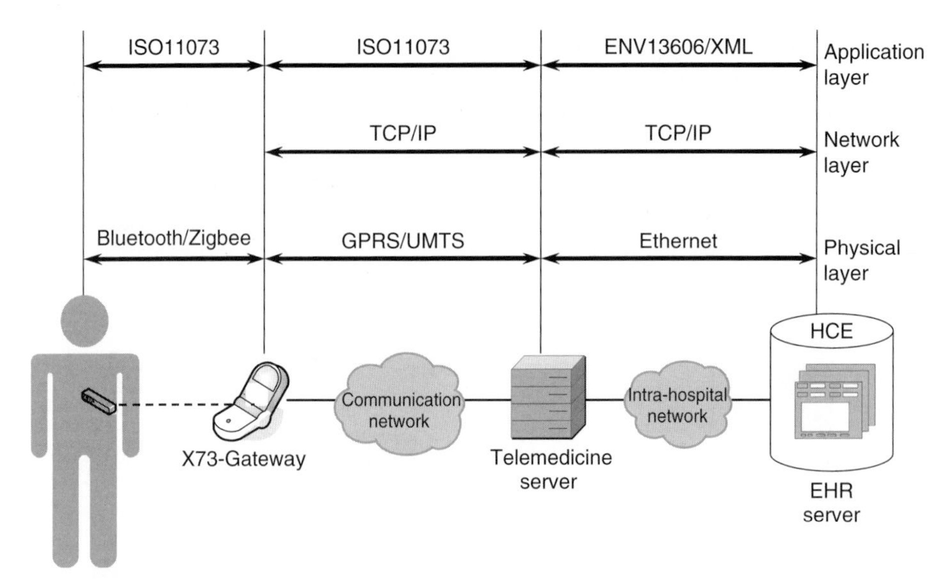

Figure 3.38. Mobile monitoring of a cardiac patient.
Source: From Ref. [18].

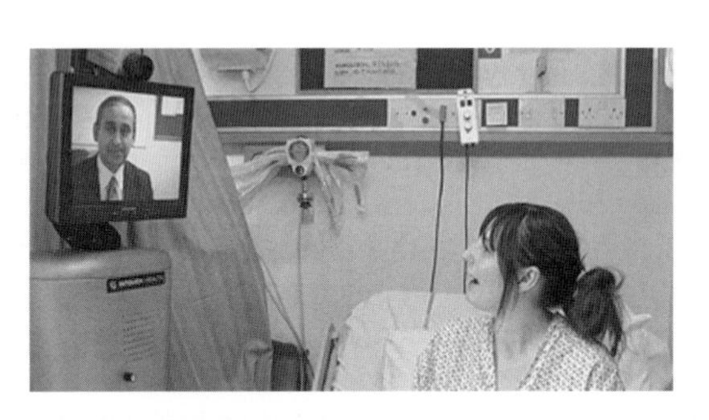

Figure 3.39. Mobile robotic doctor.
Source: From Ref. [19].

Basically, the robot is a broadband wireless video conferencing unit on wheels. It is driven by a joystick operated from a remote control station. The robot's head is actually a monitor which shows a live video image of the physician's face. Above the monitor, a video camera with 12× optical zoom is mounted. There is a speaker on the front torso of the robot and a separate microphone up near the camera. The rechargeable battery powering the electronics lasts up to 8 h depending on usage. The robot travels at a maximum speed of just more than 3 km/h in any direction and uses a holonomic drive system, which means it is capable of rotating while on the move in any direction. The drive system is powered by two lead-acid gel batteries lasting up to 8 h.

Portable X-ray equipment is now available. To give an example, a powerful instrument capable of delivering over 200 mA and 120 kV, by MinXray (USA), only weighs 43 kg, so being transportable anywhere with a mobile stand. Portable X-ray devices use Pb-acid batteries, of the sealed type, connected in series so to reach high-voltage values, for example 120–190 V.

Portable devices for electrocardiogram (ECG) and electroencephalogram (EEG) are other outstanding examples of portable diagnostic apparatuses made possible by the new technologies [6]. ECGs based on microsensors and electronics for computational and communication activity within wireless sensor networks are described in Section 4.4.

3.3.4. Miscellaneous Medical Devices

A list of devices not dealt with in the previous sections may include wheelchairs, surgical tools, dynamic prostheses, devices for pain relief, transdermal drug administration and emergency fluid warmers [12, 20].

Wheelchairs are mostly powered by lead-acid batteries, but Li-ion batteries are now taking over and can ensure a range above 50 km.

Dynamic prostheses are mainly used as substitutes for the semicircular canals in the ear, which act as rotation sensors to maintain balance. These prostheses, which use piezoelectric vibrating gyroscopes, are powered by Li primary batteries, which are mounted externally together with sensors and frequency converters [21].

Pain relief may be provided by transcutaneous electrical nerve stimulation: it is a non-invasive, drug-free treatment commonly used for short-term acute pain or long-term chronic pain, for example in knees, shoulders, and muscles. [22]. Pulses of some milliamperes for a few microseconds are provided with the help of 9-V alkaline primary batteries or 9-V rechargeable Ni-MH batteries.

Iontophoresis, or ion transfer, is an example of drug delivery through the skin: physically active ions penetrate the epidermis and mucous membrane of the body by using direct current. Iontophoresis devices, for example for physical rehabilitation, are available for home use. They may have the dimensions of cell phones, powered by alkaline batteries, or may be just patches applied to the patient's body. In this last case, the physician wet the patch with the required drug; the current is provided by an ultra-thin primary Li battery contained within the patch.

Cordless surgical tools look much like drills, with different bits for piercing, reaming and sawing. Their power sources may be primary ($Li/SOCl_2$) or secondary (Ni–Cd) batteries. Primary Li cells may be preferred in some surgical tools (e.g. planes and pointers) as these cells can stand the accidental exposures to the very high temperature (+140°C) of the autoclaves used to clean these tools [12].

Emergency blood and infusion liquids are stored at low temperatures (even <0°C) and, before using them for an intravenous infusion or transfusion, warming to 37°C is needed. In these life-critical missions, reliable control of the liquids temperature is essential. Li-ion batteries are preferred in this application because of their rate capability (typically 1 A must be delivered for 2.5 h), lightweight and high reliability [12].

3.4. Miscellaneous Applications

3.4.1. Hobby and Professional Power Tools

A number of cordless power tools, for hobby and professional purposes, are now available. Until a few years ago, Ni–Cd was the battery of choice for these applications; however, Ni-MH soon acquired a relevant share, and, in recent years, more and more tools have been equipped with Li-ion packs. Li-ion batteries are successful in these applications because even at high current rates their gravimetric energy density is 2–3 times greater than that of the other batteries.

Several manufacturers, for example Milwakee, Bosch, Dewalt, Makita, Black & Decker, Panasonic, Hitachi and Metabo, now offer tools powered by Li-ion batteries. All the battery packs rely on intelligent chargers and on electronic cell protection (see later). Milwakee has introduced the V18 Li-ion System, which includes a series of tools, such as percussion-drill, driver-drill, circular saw, impact wrench, reciprocating saw and rotary hammer. The same manufacturer has also introduced tools with 28 V Li-ion packs. These batteries have $LiMn_2O_4$ as a positive electrode, which well suits the power requirement of tools working in the ampere range. The 28 V pack is made up of 7 Li-ion cells (4 V, 3 Ah). In comparison with 18 V Ni–Cd batteries for the same tools, Li-ion batteries are reported to have 40–50% longer runtime and 20% lower weight. Furthermore, they can operate in a fairly wide temperature range. In particular, at sub-zero temperatures (down to –20°C), they do not experience the tremendous loss of power of Ni–Cd and Ni-MH batteries. A new high-rate charger by Milwakee allows charging Li-ion batteries in 60 min. The same charger can charge Ni–Cd batteries in 30–40 min.

36-V packs are also available, and the tool of Figure 3.40 shows an example. Bosch also commercializes a drill/driver with a 10.8 V Li-ion pack, which the manufacturer claims to have the same power of a standard tool, but with a 50% size reduction, this being made possible by the Li-ion technology. The special Bosch's charger cuts the charging time: the 36-V battery can be charged to 80% capacity in 25 min, while a traditional charger would require 40 min.

Fast charging of high-power Li-ion batteries implies a completely new design for the battery charger. A standard Li/$LiCoO_2$ battery may be charged in 3 h using a charger endowed with a 5-W power supply. To reduce significantly the charging time, much more power is needed: a 15-min charge would require a 60-W power supply. This poses a challenge to the designer, who has to (1) find solutions allowing efficient heat dissipation and (2) choose the charger's electronics that may both stand this heat and provide an accurate control of the battery, especially in terms of thermal management.

Ni–Cd and Ni-MH battery packs have capacities between 1.5 and 3.5 Ah and voltages between 7.2 and 24 V. Capacities and other features of Li-ion cells suitable for power tools are reported in Table 3.3 [23].

For high-power cells, the capacity is measured at a rate of 10C (or higher), while that of low-rate cells is measured at 1C. All cells of Table 3.3 show a reduced capacity loss at the 10C rate vs the 1C rate, thanks to a construction, based on thin electrodes, that minimizes internal impedance. Sanyo's cell uses Li–Mn oxide as a positive; Sony's cell has a $LiCoO_2$ positive with Co partially substituted by Mn and Ni; finally, Valence's cell has a Li–Fe phosphate positive. The gravimetric-specific energy has been reported to be 94 Wh/kg for Sony's cells and 90 Wh/kg for Valence's cell, as compared with 47 Wh/kg for Ni-MH and 35 Wh/kg for Ni–Cd. The Saphion chemistry can deliver high rate pulses, for example 30 A (2 kW/kg) for 30 s.

Figure 3.40. Drill/rotary hammer powered by a 36-V Li-ion pack.
Source: From Bosch's Catalog for Professional Power Tools.

Table 3.3. Characteristics of Li-ion batteries for power tools. See also Table 3.7.

Vendor/Model	Capacity at 1C or Less (Ah)	Capacity at 10C (Ah)	Impedance (mΩ) at 1 kHz	Charge Voltage (V)	Discharge Voltage (V)	Cycle Life
Sanyo UR18650W	1.6 at 1C	1.5	–	4.2	3.7	700 cycles to 75% initial capacity
Sony 18650VT	1.08 avg., 1.0 (rated capacity) at 1C	1.03 avg.	25	4.1	3.76	500 cycles to 90% initial capacity, 2-A discharge
Sony 26650VT	2.5 avg., 2.4 (rated capacity) at 0.2C	2.4 avg.	13	4.1	3.75	Same as above, but at 5-A discharge
Valence Tech. Saphion IFR18650p	1.1 at 1C (rated capacity)	1.05	<20	3.65	3.2 at 1C	>600 cycles to 70% capacity at 10 A

Source: From Ref. [23].

Dewalt (owned by Black & Decker) has launched power tools using 36-V Li-ion packs having nanosized Li–Fe phosphate as a positive. This material is claimed to confer incredibly high power to the pack, with current pulses as high as 100 C. Such a battery, commercialized by A123, is also regarded as a possible candidate for next generation of hybrid cars (see Chapter 5).

A list of the main cordless power tools, for hobby and professional use, include the following:

- Drills/Screwdrivers
- Rotary hammers
- Percussion drills
- Impact or torque wrenches
- Grinders
- Sanders
- Planers
- Circular, jig, band, or reciprocating saws

3.4.2. Portable Barcode Readers

Portable barcode scanners are battery-operated readers storing data in their memory for later batch uploading to a computer. A block diagram is shown in Figure 3.41. In addition to a bar code scanner, a portable reader usually has an LCD display to prompt the user what to do and a keyboard to enter variable data such as quantities. The scanner, of the laser or CCD type, may be separated from the unit (as shown in Figure 3.42) or may be built in. Particularly important parameters are battery life and ease of reading the display. Worth Data, a leading company in this field, has introduced voice prompt messages to supplement the display messages in a portable unit, thus avoiding lighting, language and message clarity problems. The Worth's unit actually announces when incorrect data are entered and when to change the batteries or upload data.

Li-ion cells are now the power sources of choice for these devices, with Ni-MH or Ni–Cd batteries still used in low-cost devices. Primary (alkaline) batteries may also be used.

General technical specifications for portable readers are [12]:

- Current: <0.2 A
- Battery capacity: 1.2–2.0 Ah
- Operating temperature: from 0 up to +40°C
- Runtime: operation throughout an 8-h shift.

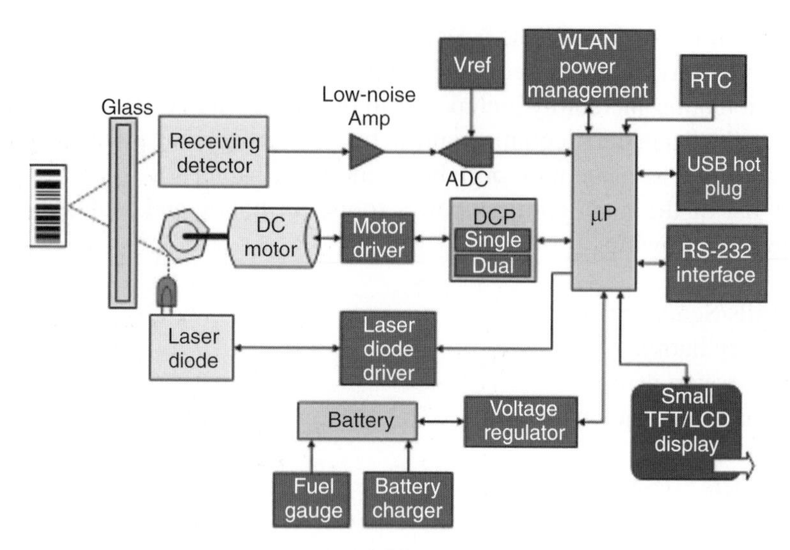

Figure 3.41. Block diagram of a barcode scanner.
Source: Courtesy of Intersil.

Figure 3.42. Portable barcode reader with separate scanner.
Source: Courtesy of Worth Data.

3.4.3. Portable Payment Terminals

Portable card readers for electronic payments are now available. An example is provided by the terminal of the Belgian specialist Banksys: this is a portable, battery-powered device with an embedded computer built around a dual-core system-on-chip with two 32-bit ARM processor cores, along with 32 MB each of flash and RAM memory. It targets on-the-road transactions, in-store promotions and the hospitality sector.

The terminal (see Figure 3.43) operates on Li-ion batteries that support, according to Banksys, up to 300 transactions with printed receipts between charges. It is available with a variety of wireless networking options, including 802.11 (Wi-Fi), and two varieties of tri-band GSM/GPRS. In the near future, it is expected that mobile phones may act as transaction terminals, thus avoiding cash payments even at the coffee shop or at the car parking.

3.4.4. Handheld GPS (Global Positioning Systems)

Portable GPS receivers (see block diagram of Figure 3.44) are based, as any other system of this type, on the communications with satellites, for example Galileo [12]. These satellites can locate anything having a GPS antenna

Figure 3.43. Portable card reader.
Source: Courtesy of Banksys.

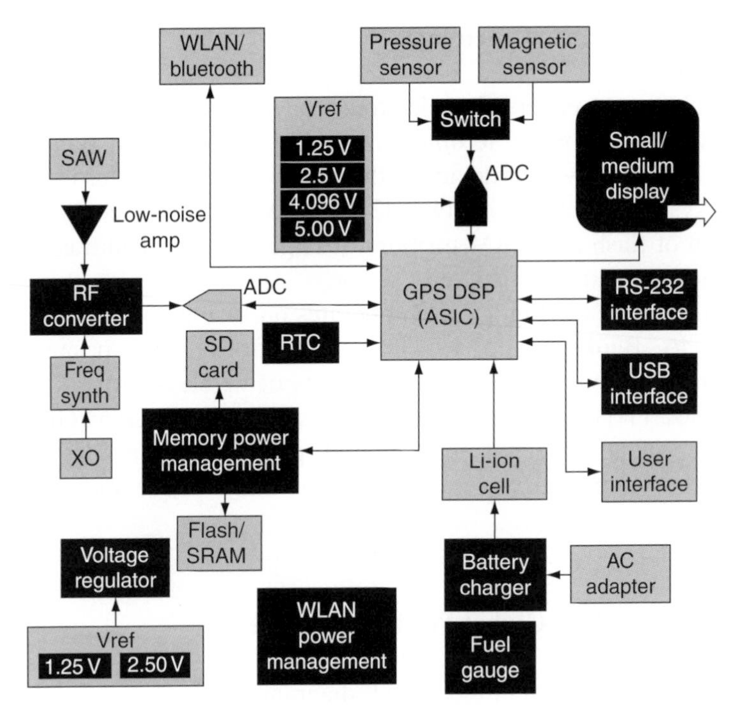

Figure 3.44. Block diagram of a portable GPS.
Source: Courtesy of Intersil.

and can help persons in their movements, thus acting as navigation aids. The main technical specifications of a portable GPS can be summarized as [12]:

- Base currents: <100 μA, sometimes a few milliamperes
- Pulses: 200 mA to 1 A (in case of backlighting) during several seconds
- Cut-off voltage: 5–10 V
- Operating temperature: –40/+80°C
- Runtime: from some hours to some years, or several hundreds of charge/discharge cycles.

A portable GPS receiver has two power sources: one for its internal clock and one as the main power supply. The former is often a 3 V Li coin cell, whereas the latter may be either a rechargeable (Ni–Cd, Ni-MH and Li-ion) or a primary battery (alkaline, Li), depending upon the recharging or replacement possibilities. The choice of the main battery also depends on the environmental operating conditions (temperature) and cost. Therefore, large primary lithium battery packs, made with R20 Li/SO_2, Li/MnO_2 or $Li/SOCl_2$ cells, are used in military receivers, whereas small size Li, alkaline or rechargeable cells are preferred in civilian applications.

A more detailed description of the GPS technology is reported in Section 4.9.2.1.

3.4.5. Fishing Aids

An interesting battery application is the portable fishfinder, basically consisting of a transducer and a display. The screen of the floating transducer shows fishes, depth, bottom structure and water temperature. Figure 3.45 (left), shows a high-resolution, wide-screen instrument, which can either be connected to mains or made portable with a battery-containing case. A wristwatch-like fishfinder is shown in Figure 3.45 (right), which can be worn thanks to its 130 g weight.

Figure 3.46 shows two lighted fishing floats especially useful for night fishing. They use primary Li cells. The penlight Li/CF$_x$ cell (25–50 mAh) is particularly suited for the torpedo-like float on the left.

3.5. Portable Device Power Management

In this section, emphasis will be put on the techniques adopted to reduce power consumption in portable devices, thus prolonging their runtimes. Approaches to be considered are (a) choice of the electronic components; (b) power-wise functioning of the device and (c) battery management.

Figure 3.47 represents the block diagram of an audio system applicable to several high-end portable devices, for example computers, DVD players and MP3 players. It outlines the presence of blocks connected to power

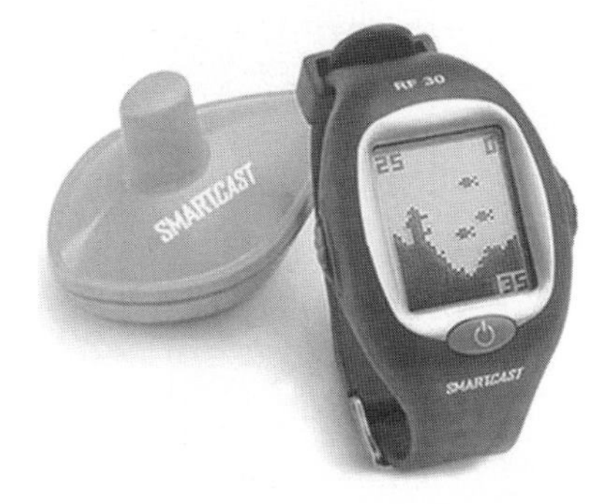

Figure 3.45. Left: wide-screen fishfinder (model 140 by Garmin: 4.7″ display; up to 180 m depth; powered by D size cells); right: wearable fishfinder (by Humminbird: 1.25″ display; up to 23 m range; powered by a 2450 Li/MnO$_2$ button cell).

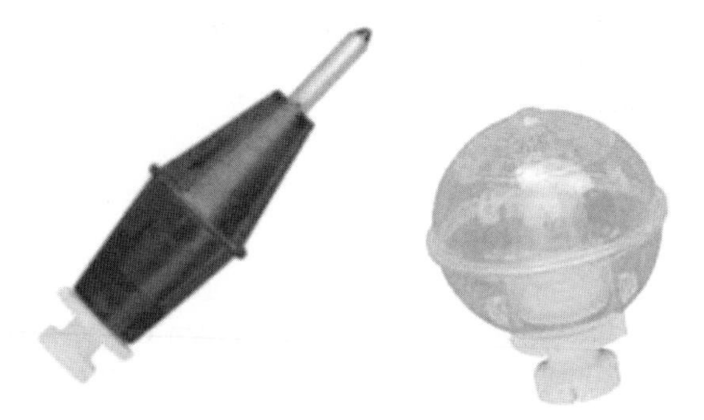

Figure 3.46. Lighted fishing floats.

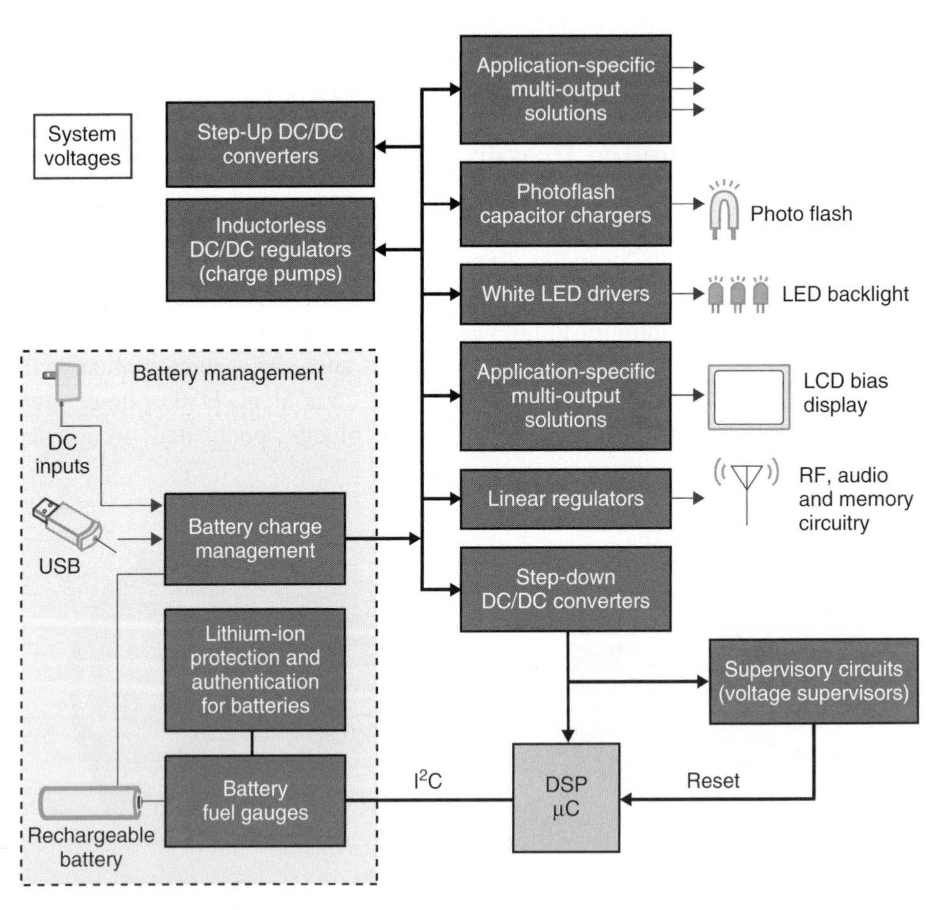

Figure 3.47. Block diagram of an audio system.
Source: From Ref. [4].

management: (1) voltage conversion and regulation; (2) core processor and peripherals and (3) battery management [4].

Power management is necessary for any device or system, including those used in industrial applications. However, this topic will be more extensively dealt within this chapter, as more information is available for portable electronic devices.

3.5.1. Power Management of the Device Components

3.5.1.1. Transistors

Low-power electronic circuits are the basis of such device components as: processors, voltage converters and regulators, LCD displays, application-specific integrated circuits (ASICs), field-programmable gate arrays (FPGAs), etc. The building block of integrated circuits (IC) is the transistor (plus capacitances, resistances, diodes, etc.): therefore, it is necessary to start from it to understand how power is consumed in a circuit and how consumption may be reduced.

The preferred transistor in modern electronics is the metal oxide semiconductor field effect transistor (MOSFET). A MOSFET is composed by the elements schematically shown in Figure 3.48. In this figure, the substrate (usually silicon) is positively doped (e.g. by B) and the channel between two n-type regions (drain, D, and source, S, doped with As or P) is filled by electrons when current flows. This is an N-channel MOSFET. With a negatively doped substrate, the conduction is due to positive holes and the MOSFET is a P-channel MOSFET.

The gate is commonly made of doped polycrystalline Si and is isolated from the substrate by an oxide layer (usually Si oxide). Critical dimensions in an MOSFET are the gate length (also called channel length) and the oxide thickness. The former is continuously decreasing in modern transistors and is now reaching values of 90–65 nm; the latter is about 2–10 nm [24].

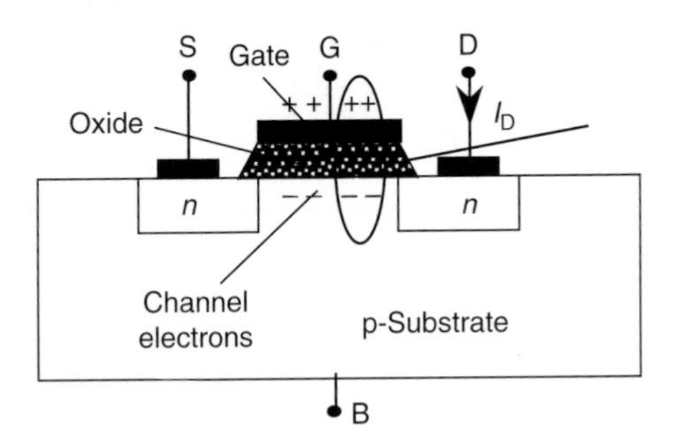

Figure 3.48. Elements of a MOSFET.
Source: From Ref. [24].

In the N-channel MOSFET, a positive gate voltage induces electron movement from source to drain (when the latter is positively biased vs source). For a gate voltage (V_G) equal to 0, no current flows in the channel (the MOSFET is *off*), while for a positive voltage above a given threshold (V_T) current starts to flow: the MOSFET is *on*. However, in practical transistors, these ideal conditions do not apply, as explained in the following, and this leads to power consumption greater than calculated.

Typical current (I_D) vs voltage (V_D) curves are shown in Figure 3.49, where the effect of the gate potential (V_G) is evidenced. The current rises linearly at low drain voltages and then reaches a saturation value, referred to as *on* current. The switching function of MOSFETs in digital circuit is obtained by operating them at zero-gate voltage or at the saturation conditions (where gate and drain are short-circuited). As the figure shows, higher gate voltages result in higher currents, as more electrons run through the channel. At the same time, it is desirable to have a high current in a short switching time.

As stated above, a switched-off MOSFET manifests a non-zero current (*off* current), whose influence becomes appreciable at low powers and increases as the gate length decreases. Indeed, for gate lengths <90 nm, the threshold voltage, V_T, may become small enough to allow the flow of a non-negligible current, called leakage current, in the channel from source to drain. A circuit may contain several million transistors, so the overall leakage current may become significant. It is unfortunate that, as the *on* current increases by lowering V_T, the *off* current also increases, as shown in Figure 3.50.

One method to approach the behaviour of an ideal MOSFET is the use of two threshold voltages: when the MOSFET is *on*, the threshold voltage is low for maximum *on* current and when it is *off*, the threshold voltage is changed to a

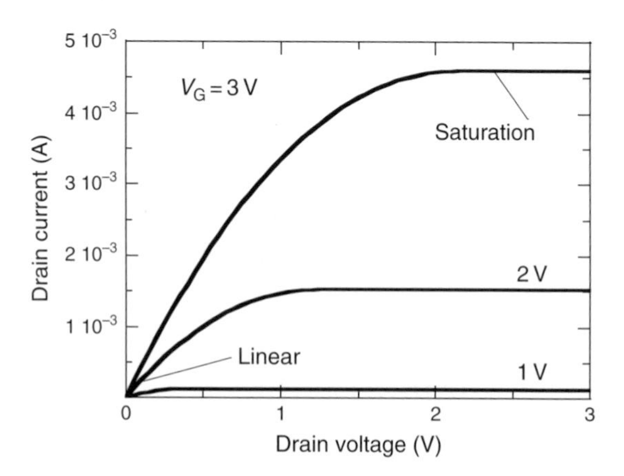

Figure 3.49. Drain current vs drain voltage in a MOSFET.
Source: From Ref. [24].

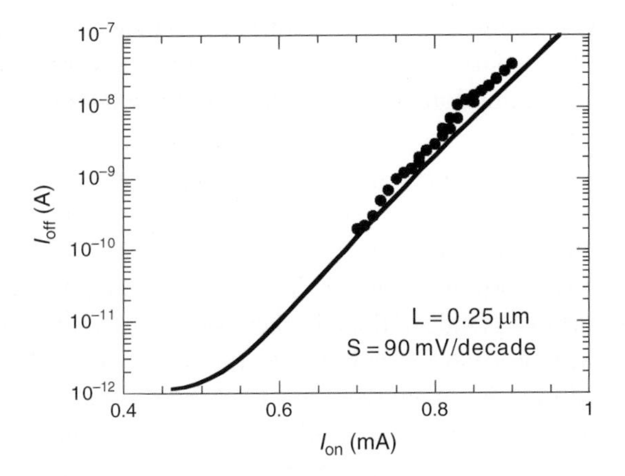

Figure 3.50. MOSFET *off* current vs *on* current.
Source: From Ref. [24].

higher value, reducing the *off* current. This can be done applying a voltage to the substrate (back bias), and adds complexity to the circuit.

By series connecting an N- and a P-channel MOSFET, a complementary MOSFET (C-MOSFET or CMOS for short) is obtained. When one of them is working, the other is *off*. A C-MOSFET consumes less power than a single N- or P-channel MOSFET, as the current flowing through the one that is *off* (leakage current) is very low. For this reason, C-MOSFETs are now preferred in low-power circuits.

In general, the power dissipated in a circuit, during a given operation time, by a component, for example a C-MOSFET (CMOS), is given by the relation

$$E = \int P(t)\ \mathrm{dt}$$

where E is the energy and P the power. The power dissipation may be represented as

$$P = P_{\mathrm{switch}} + P_{\mathrm{sc}} + P_{\mathrm{off}}$$

where P_{switch} is the switching power, P_{sc} the power corresponding to the simultaneous and short-time *on* state of the two MOSFETs, due to their non-ideal behaviour, and P_{off} is connected with the leakage current. P_{switch} accounts for the main power dissipation and is given by the equation

$$P_{\mathrm{switch}} = \alpha f C V^{2}_{\mathrm{DD}}$$

where α is the switching activity, f the frequency, C the capacitance and V_{DD} the voltage applied (the two drains, D, are connected in the C-MOSFET).

The energy consumption corresponding to the switching power is [24].

$$E = E_{\text{switch}} = CV^2_{\text{DD}}$$

In the CMOS, both the energy (E) and the power (P) dissipated are related to the square of the applied voltage. Therefore, reducing the voltage is of paramount importance, as indeed done in the last several years (see Figure 3.51). For instance, a microprocessor developed by Intel, Processor M, can work in the low-voltage and ultra low-voltage versions [25]. The former can work at 1.15 or 1.05 V, in maximum performance mode or battery optimized mode, respectively. The latter can work at 1.10 or 0.95 V, in the above modes. These values allow low-power consumptions and, at the same time, an excellent performance is provided in portable devices.

From Figure 3.51, it is clear that the threshold voltage V_T has been reduced with a lower rate vs V_{DD}. This is so because the *off* current would rapidly increase if V_T were reduced below a certain limit, as discussed above.

The leakage current in the *off* state is also called static current, hence static or leakage power is dissipated in these conditions. The *on* power is also called dynamic power [26].

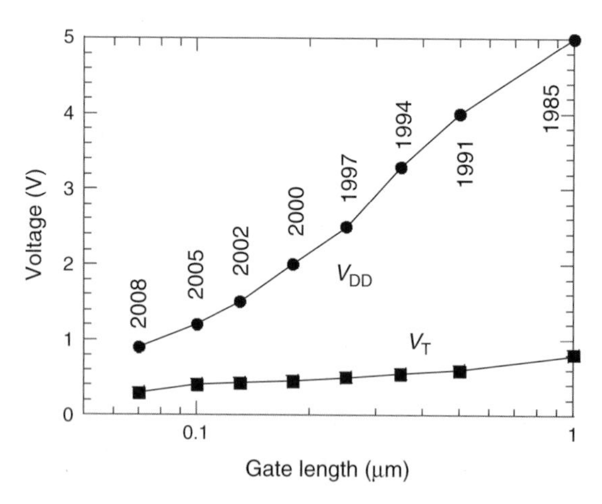

Figure 3.51. Trend of applied (V_{DD}) and threshold (V_T) voltages in C-MOSFETs, as a function of gate length, from 1985 to 2008.
Source: From Ref. [24].

3.5.1.2. Microprocessors and Microcontrollers

Let us now look at the power consumption of the device components mentioned at the beginning of the previous section.

A microprocessor can be defined as the integration of a number of useful functions into a single IC package (CPU). These functions are:

- The ability to execute a stored set of instructions
- The ability to access external memory chips to both read and write data from and to the memory
- The ability to access other I/O components.

In a microcontroller (μC for short), CPU, memory (both ROM and RAM) and some parallel digital I/O ports are contained in a single chip, as shown in Figure 3.52 [27]. In recent years, μCs have been developed around specifically designed CPUs, the processing modules of μCs, to correspond to specific functions. Figure 3.52 (right) illustrates a typical microcontroller device and the different subunits integrated onto the microcontroller microchip. The serial port allows transmitting (Tx) and receiving (Rx) functions. The three digital I/O ports allow interfacing of the μC to the environment (the numbers represent bits). The μC ports can be used to operate, for instance, LEDs, relays and switches. The timer module allows the μC to perform tasks for a given time period.

Microcontrollers and microprocessors form an important share of the power budget of portable equipment and, therefore, the power they use must

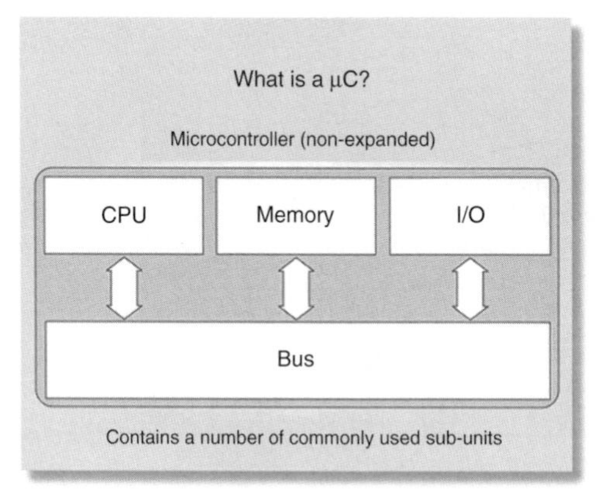

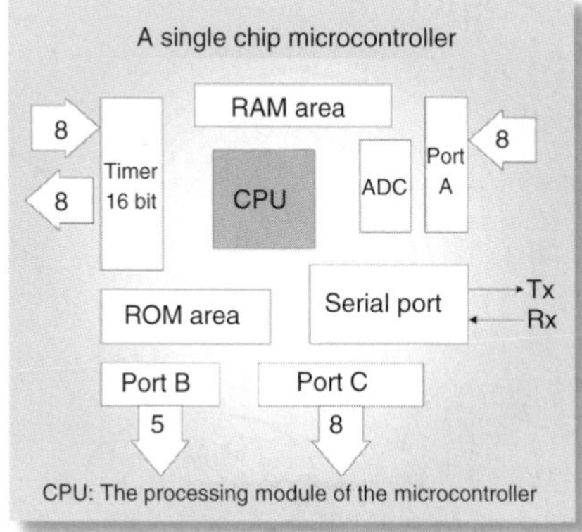

Figure 3.52. Left: non-expanded scheme of a microcontroller; right: expanded scheme showing sub-units. See text for details.
Source: From Ref. [27].

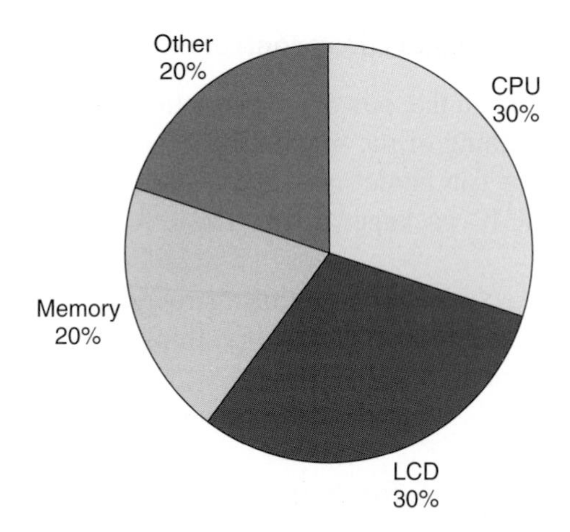

Figure 3.53. Power consumption of the main components of a webpad.
Source: From Ref. [28].

be minimized to extend battery life and device runtime. In Figure 3.53, the power consumption of the components of a webpad (a tablet PC), is shown [28]. In this example, the microprocessor power consumption can range from 720 mW to 1 W during normal operation, but for some low-power microprocessors can be as little as 250 mW.

Operating modes of a microprocessor include normal, run, sleep, suspend, standby, stop and idle operation. In common applications, for example in portable computing devices, often three modes – normal, sleep and idle – are present; the power consumption of an Intel microprocessor in the three modes is shown in Figure 3.54. In sleep mode, the power consumption would be $<90\,\mu W$.

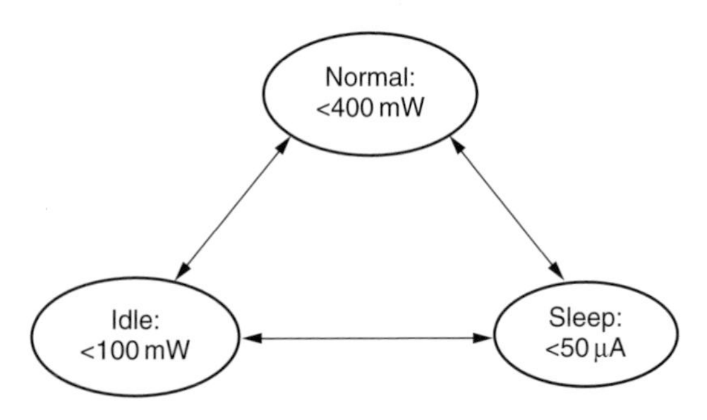

Figure 3.54. Power consumption of the microprocessor Intel StrongARM SA-1110 (206 MHz, 1.8 V internal voltage) in different modes.
Source: From Ref. [28].

Four strategies are suggested to reduce the power used by microcontrollers (microprocessors) [29]:

1. Reducing power dissipation of the transistor by shortening the gate length (see above), while ensuring at the same time a high operating speed.
2. Lowering the voltage supplied to the transistor (V_{DD} for a CMOS) by using, if necessary, a step-down voltage converter so to have a voltage <2 V.
3. Shutting off modules: microcontrollers include a large number of on-chip modules (such as controlling timers), but they need not run at the same time. The clock of modules not currently used may be shut off. This may be accomplished with the use of low-power complex programmable logic devices (CPLDs), which keep the device longer in idle or sleep mode.
4. Reducing the stabilization time of the oscillation, that is the signal created by the oscillator (a clock in a computer). For instance, while waiting for the oscillation to set, a computer wastes power before leaving the sleep mode and entering the active mode.

The problem of the oscillation stabilization may be alleviated using a double oscillator. The device will immediately start executing codes from a fast-starting oscillator, such as an RC one. Subsequently, the system will switch to the crystal oscillator. This is critical for applications where a device has frequent wake-up/sleep events.

The technique of power budgeting is important to predict power consumption and battery life. By examining the single operation modes, one estimates the amount of current used by them and, so, the total energy consumed. If too much power is requested, it is easy to determine where additional design consideration needs to be placed to reduce power consumption. Power budgeting also suggests size and type of the battery needed for a given application [30].

Graphics processors in portable computers deserve a particular mention, as these components can contribute significantly to the total system power consumption. The graphics subsystem performs 2D and 3D acceleration, renders images to the screen and provides hardware assistance for DVD playback [31]. Accelerating these processes can be very onerous for the battery; however an efficient graphics accelerator not only consumes little power, but also reduces the power requested by the LCD.

There are several power management techniques allowing to minimize power consumption while maximizing the performance of notebook graphics accelerators. Clock gating, for instance, is an activity-based, dynamic gate-level technique that automatically turns off the clocks of the idle parts of the graphics chip. For example, if the 3D engine is not in use, clock gating turns off the clock branches operating this block.

One of the most effective power management techniques is known as voltage and frequency throttling. This technology enables notebooks to lower the graphics processor core voltage as well as the core and memory clock speeds in situations when battery life is more important than performance. The option to adjust both the voltage and frequency settings of the graphics accelerator affords remarkable power saving. As power varies linearly with the clock speed (frequency) and varies with the square of the voltage setting (see the previous section), decreasing both the voltage and frequency results in cubic power savings.

Modern mobile graphics processors also reduce the LCD power consumption by reducing its refresh rate when the notebook is in battery mode. Furthermore, these power savings translate into space savings – a graphics component that consumes less power also produces less heat and does not require a heat sink or a fan to remove excess heat.

3.5.1.3. Voltage Regulators

Electronic devices usually include a variety of loads, for example LCD display, memories, microprocessor(s), USB, DSP, hard disk drive(s), etc. These loads require different operating voltages and currents. This means that the battery voltage must be regulated (converted) before powering the above components.

Commonly used DC-DC converters for battery-operated systems can be divided into three categories: (1) linear, for example low drop-out (LDO); (2) conventional switching (with inductor) and (3) inductorless switching.

Historically, linear regulators have been used to power core processors, as they are simple, cheap and require few external components [32, 33]. These regulators drop the difference in voltage between the DC power source for the processor and the battery across a pass transistor operating in the linear region – hence the name linear regulator. As the voltage required by digital circuits approaches 1 V, the power conversion efficiency of these regulators tends to decrease. For instance, in a system using a 3.6 V Li-ion battery and a 1.2 V processor, the regulator efficiency will be <40%, the remainder being lost as heat. Obviously, this heat would have a detrimental effect on the battery's life [32].

Switching converters have a higher power conversion (>80%), thus ensuring a longer battery life. They use low-loss components such as capacitors, inductors or transformers, as energy storage elements to transfer energy from input to output in discrete packets (with *on/off* switches). Buck converters (step-down) produce a lower regulated output voltage; boost converters (step-up) produce a higher output voltage; buck-boost converters can step down or up the voltage, and, finally, some switching converters may invert the input voltage. In Figures 3.55–3.58, different typologies of switching converters adopted in portable devices are presented.

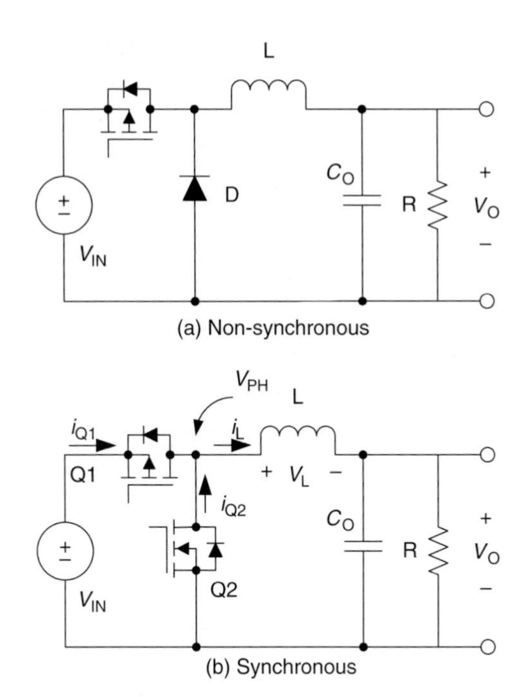

Figure 3.55. Schemes of buck converters.
Source: From Ref. [34].

The diode D of the non-synchronous buck converter is replaced by a MOSFET in the synchronous type for achieving higher efficiency at low-output voltage applications (Figure 3.55). Similarly, in a synchronous boost converter, the diode is replaced by a MOSFET (Q2 – see Figure 3.56) for improving power conversion efficiency vs the asynchronous one for limited output voltages. A buck-boost converter (Figure 3.57) is a combination of a buck converter followed by a boost converter, where an inductor is shared by both converters. For improving the power conversion efficiency, the synchronous type is normally used in portable power applications. The fly-back converter (Figure 3.58 (left)) is actually a buck-boost converter using a transformer to isolate between input and output. The synchronous type, suitable for high-voltage and low-power applications, is preferred. Finally, charge pump converters (inductorless converters) can step the voltage down or up and can double or invert it. Charge pump circuits (Figure 3.58 (right)) use capacitors as energy storage devices instead of inductors; in this way, any electromagnetic interference (especially important with cell phones) is alleviated [34].

In a system powered by a Li-ion battery, the selection of the most appropriate converter is illustrated in Figure 3.59. If the input voltage is always higher than the output, buck or LDO converters are normally the only solution. If the input voltage is lower than the output voltage, a boost converter can be employed, while a charge pump could be used for low-current and low-cost

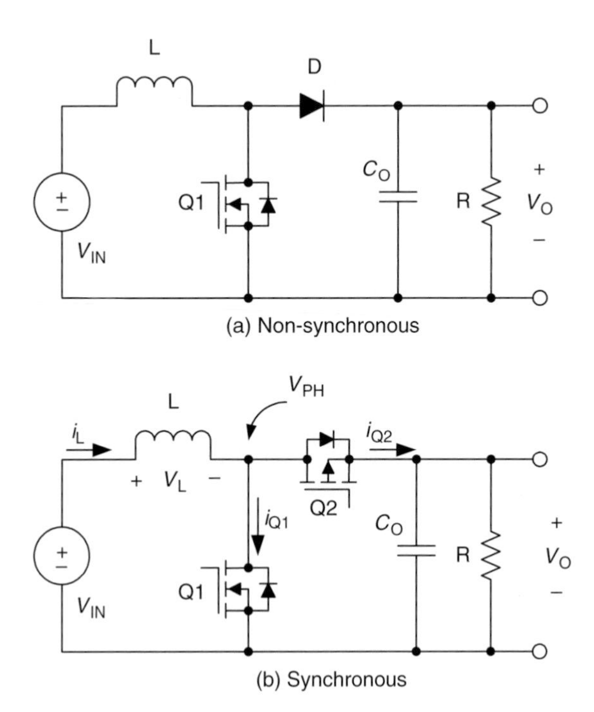

Figure 3.56. Schemes of boost converters.
Source: From Ref. [34].

applications. Buck-boost, charge pump and fly-back can be used when the battery voltage can be either higher or lower than the output voltage. However, fly-back is more suitable at higher output voltages [34].

The preference given to switching regulators based on a higher efficiency was entirely justified for devices with an output power rail of 3.3 V. (Power rail can be defined as the voltage source within the device from which its various functions can draw power.) However, the new generation of power rails, for example those of cellular phones, work at 1.5 V or less. Indeed, most of the large-scale ICs have these low operating voltages (Section 3.5.1). Examples are the baseband chipsets (1.375 V) and the video processing DSPs (1.2 V).

In devices having low-voltage ICs and powered by a 3.6 V Li-ion battery, designers have adopted a two-stage conversion solution: a high efficiency step-down converter drops the battery voltage down to 1.5 V; then, from this 1.5 V rail, another converter can further lower the voltage to supply the integrated circuits. A switching regulator could not be used to bring the voltage from 1.5 V to the 1.2 V of a DSP, as it would prove impossible activating the MOSFETs with such a low ΔV. A standard LDO would also fail, as it needs its dropout voltage normally exceed 0.7 V. Instead a very low dropout regulator (VLDO) could work: for a drop from 1.5 to 1.2 V, it shows 80% efficiency, and from 1.5

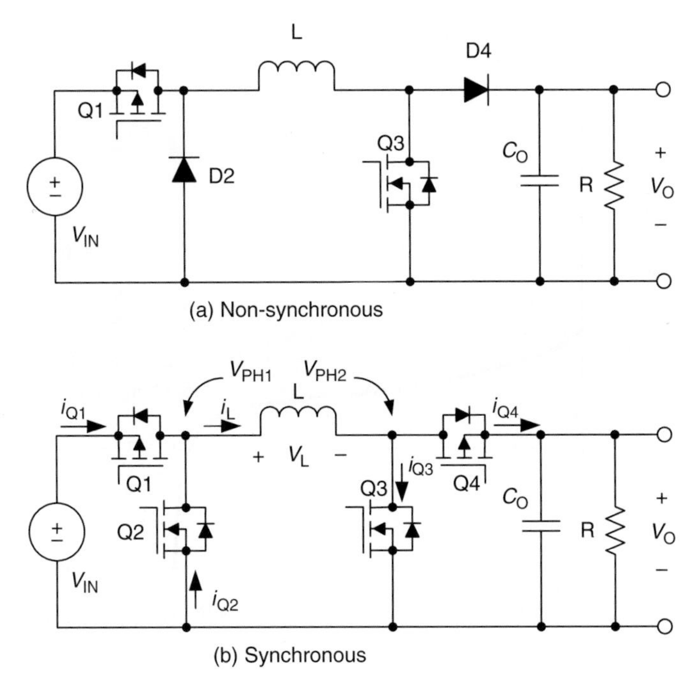

Figure 3.57. Schemes of buck-boost converters.
Source: From Ref. [34].

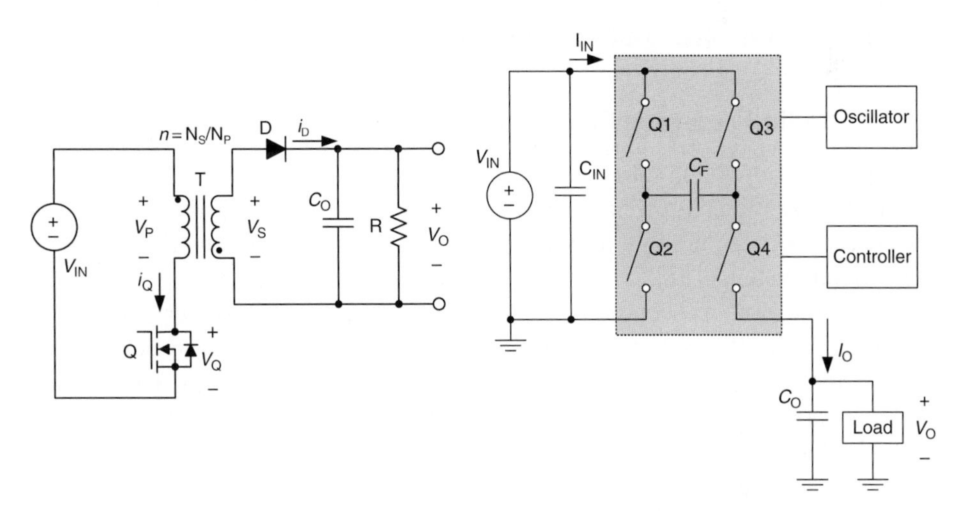

Figure 3.58. Scheme of a fly-back converter (left) and a basic charge pump converter (right).
Source: From Ref. [34].

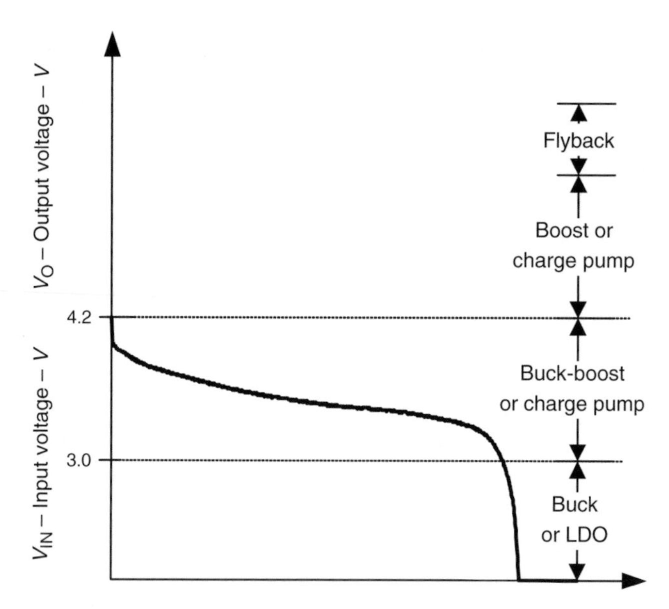

Figure 3.59. Discharge profile of a Li-ion battery and selection of converters. *Source*: From Ref. [34].

to 1.375 V the efficiency rises to 91%. Another advantage in the use of a VLDO is represented by its low noise level, this being especially important for digital ICs. However, a VLDO would pose problems of cost, size and stabilization [32].

Summarizing, the choice of an appropriate regulator has to be based on such factors as noise generation, space constraints, conversion efficiency and heat development. The last two factors are of direct relevance for battery life.

3.5.1.4. Radio-Frequency Communications

A very important subsystem in various portable systems is the RF communication section. Its circuits can range from relatively slow local data links using Bluetooth technology to more complex 802.11 Ethernet wireless LAN or 3G cellular systems [35].

For the core processor it is important to maximize power efficiency, whereas for the RF circuit reducing noise is mandatory. Therefore, the power that is applied to the RF circuit must be clean and noise-free. RF circuits typically operate at higher voltages (3 or 3.3 V) than the core processor in a portable system. Accordingly, the system does not usually have to step down the voltage from the battery as much as the core processor. At the same time, current consumption of RF circuits is usually much lower than the core processor and power efficiency is a less important issue. Given the need for a low noise supply and the reduced emphasis on efficiency, linear regulators offer a good option for RF applications. Switching buck converters are inherently noisy, while linear regulators provide much lower noise and input-to-output isolation.

3.5.1.5. Display

In the previous sections, it has been outlined that the display of portable devices is a power consuming subsystem if backlighting is operating.

A recent exciting change in these devices has been the move from monochrome to colour displays, which are being used to support a variety of new multimedia applications including location service, gaming and video capture and playback [35]. Of course, this has a deep impact on power requirements.

The most common display technology in use today is the colour TFT LCD. Typically, these displays are illuminated by white light from a backlighting source. To provide a pure white source needed for a colour-correct display, designers can employ a variety of methods. The most cost-efficient source of backlighting, and the most implemented today, is constituted by white LEDs.

White LEDs have a typical voltage of 3.5 V (20 mA). Therefore, when the Li-ion battery is fully charged (4.2 V) a linear step-down (buck) converter is needed, while at the end of discharge (<3.5 V) stepping up by, typically, boost converters is required. Supplying power to white LEDs must consider such factors as power efficiency, LED brightness consistency, footprint, design complexity and cost. Controlling the brightness of the LEDs is crucial to maximizing battery life. Most white LED drivers with inductive boost converters (and some with charge pumps) control the brightness of each LED by switching the converter off at a very high frequency using a pulse width modulation (PWM) control signal.

3.5.1.6. Port Power and Protection

Every portable device has some type of physical interface to input or output data – an example is the USB standard which defines a host/peripheral relationship. USB 2.0 specifies that 5 V must be supplied to the port. It is important to protect the system from any failure occurring at the port. To ensure system reliability and user safety, the system must provide protection to the internal power bus, for example port protection switches. Typically, they are comprised of P-channel MOSFETs with current limiting and over-temperature protection characteristics. Once a short circuit occurs, these devices protect the motherboard of the portable device by denying access.

A modification of the USB 2.0 standard is the USB-on-the-go (OTG) developed to support the direct interconnection of two portable devices. This new standard creates more power management challenges with respect to those of the traditional USB [35].

3.5.1.7. Accessory Lighting

Many devices now have colour LEDs to show that a function is enabled. Examples include a light indicating that the battery is being charged or the wireless connection is active.

White LEDs are often integrated into portable devices such as digital cameras to provide light to illuminate a subject. In many applications, the forward voltage of the diode, that is the voltage allowing current flow, may vary from 3.2 to 4.8 V. In extreme temperature conditions, this range widens (2.8–5.1 V). If the LED's voltage range is limited between 3.5 and 4.5 V, the charge pump technology is a good solution, as it is efficient, low cost and can supply a high current for a very short time. It is for instance well suited for applications like a photoflash where the system must supply 200–400 mA for 100 ms [35]. However, for a more extended voltage range, a buck-boost converter has to be used; in this case, pulses of 1 A may be needed [34].

3.5.1.8. Hard Disk Drives

Hard disk drive (HDD) storage technology is now largely used in consumer devices by offering the same advantages that it has offered computer users, that is the capability of storing economically large amounts of data, with fast transfer rates, and quick access to any specific piece of data [36].

In Table 3.4, devices using HDDs are listed together with their drive form factors. HDDs for small-size devices, for example cameras, MP3 players or cell phones, pose challenges in terms of form factors (≤1″) and power consumption. Indeed, given the size of these devices, the small batteries they use would soon be depleted if the HDD's power could not be

Table 3.4. Comparison of the hard disk form factor for several consumer devices.

Consumer Device/Size	3.5″	2.5″	1.8″	1″	<1″
Mobile phone				■	■
Audio/MP3			■	■	■
Digital video camera			■	■	
Digital still camera				■	■
Game consoles	■	■	■	■	■
Digital video recorders	■	■			
Laptop		■	■		

Source: From Ref. [36].

limited. Proper design and manufacturing of electronics that controls the drive's operation and manages its data path are of the utmost importance in this respect.

A great help in power reduction for HDDs <2.5″ is reported to come from new HDD preamplifiers [37]. With these preamplifiers, power consumption is low, for example 170 mW at 450 Mbps in write mode, this enabling manufacturers to offer longer battery lifetime between charges, or more compact products with the same battery life by using a smaller battery. Furthermore, these new products have power consumption of 250 μW in sleep mode, which further prolongs the battery runtime.

3.5.2. Thermal Management of the Device Components

Thermal management for high-end portable devices requires actions in three primary areas: power management, thermal management in the traditional sense and battery management from a thermal perspective [38].

The majority of power-related thermal problems can be linked to these causes: (1) use of high power consumption components; (2) use of components that do not provide power-conservation options. For instance, in a portable media player (PMP), if a buck regulator without a shutdown function powers the audio block, the entire audio block remains powered even when the player is in standby or operates in a mode that does not require audio output. The result is heat-generating wasted power, not only in the buck regulator, but also in the audio block components, e.g. the Codec and (3) use of components with low-power conversion efficiency. In the above example, if the audio block is powered by a non-synchronous buck regulator, the latter can lose as much as 30% of its power conversion efficiency during operation. A synchronous buck regulator converts this thermally harmful heat into useful output power.

Traditional thermal management problems involve non-optimized heat dissipation and inadequate heat-flow control. A cause of overheating may be the use of packaging, for example for a regulator, with a high thermal resistance. Another example is provided by a battery charger without a current limiter when the charger's temperature increases over a given value. One more example is given by the white LEDs for LCD displays (see Section 3.5.1.5): without a temperature sensor, overheating occurs and may damage the display or even other parts of the device. Thermal control requires the use of strategically placed temperature sensors throughout the system, followed by the action, if needed, of a thermal management microcontroller. In this way, the power reaching hot spots in the system will be shut down or reduced.

Thermal problems connected with the battery are dealt with in the following.

3.5.3. Battery Management

Any power management concerning the portable device's electronics has to be accompanied by an appropriate battery management for a satisfactory and long-lasting use of the device. The considerations that follow are also applicable to larger batteries for industrial and traction/automotive applications.

A battery management system (BMS) has three main objectives:

1. Optimizing the battery performance with respect to the requirements of the application powered by it
2. Prolonging the battery life
3. Avoiding damages to the battery and, at the same time, granting the host device's and the user's safety.

These objectives may be achieved if the BMS incorporate one or more of the following functions [39]:

- *Battery protection.* The operating conditions, for example temperature, current, charge/discharge voltage limits, must be carefully checked and controlled.
- *Charging technique.* This subject will be more extensively treated in Section 3.5.3.4.
- *State-of-charge (SOC) determination.* This may concern the battery or individual cells.
- *State-of-health (SOH) determination.* It could be defined as the fraction of nominal capacity that cannot be recovered as the battery ages.
- *Cell balancing.* In a multicell battery, voltage differences between cells may lead to severe problems: cells with a lower voltage will reach sooner the discharge end point and will be overcharged in the following charge. This may lead to a premature failure and to possible safety problems.
- *History.* Number of cycles, maximum and minimum voltages and temperatures, and maximum charging and discharging currents can be recorded for subsequent evaluation. These data can give an indication of the battery's SOH.
- *Authentication and identification.* With some BMS it is possible to record the cell type (chemistry and model), the manufacturer's name and the manufacturing date. Cell authentication makes it possible discarding replacement cells by non-qualified manufacturers (see below).
- *Communications.* Most BMS systems allow battery communications with its charger and the host device. In this way, BMS parameters may be modified, if needed.

3.5.3.1. The Concept of Smart Battery

For an intelligent battery management, the battery has to communicate with its host device and with its charger. This can be accomplished with the system management bus (SMbus) defined by Intel Corporation in 1995. This system, mainly used in personal computers for low-speed system management communications, is the basis for the smart battery system (SBS) [40].

Specifications of the SMBus can be found in the SBS Implementers Forum (www.sbs-forum.org). Version 1.0 of these specifications was issued on 15 February 1995 and version 2.0 on 3 August 2000. (The Forum is formed by these companies: Duracell, Energizer Power Systems, Fujitsu, Intel, Linear Technology, Maxim Integrated Products, Mitsubishi Electric Semiconductors, PowerSmart, Toshiba Battery, Unitrode, USAR Systems.)

The objective behind the SBS was to transfer charge control from the charger to the battery. With an efficient SMBus, the battery becomes a master, tells the charger, acting as a slave, about its chemistry, voltage and size, and thus dictates the algorithm. The battery controls such parameters as voltage, current, switching point, etc. [41].

The term *smart battery* may include rather different degrees of control. At the very least, the batteries should provide SOC indications. Smarter batteries can control overcharge and overdischarge; even smarter batteries provide more controls and information, this obviously adding to the battery cost.

Several companies are now producing more or less sophisticated circuits for smart batteries. They range from the single wire system to the two-wire system of the SMBus [42].

The single wire system uses only one wire for data communications, as shown in Figure 3.60 (left). In addition to the two battery terminals, another wire (thermistor) allows temperature sensing. This system stores the battery code and provides readings of T, V, I and SOC. It is relatively cheap and finds application in some transceiver radios, camcorders and portable computers. Most single wire systems have different shapes and cannot give standard measurements of SOH. Indeed, these measurements are only allowed when the hosting device is coupled to a designated battery pack. The original battery has always to be used, otherwise incorrect readings are obtained. Using the single wire system in a universal charger (i.e. a charger for all kinds of batteries) is not recommended.

The two-wire SMBus is the most complete system and represents the largest effort to create a standardized communications protocol and smart battery dataset (SBData). In this system, data and clock have separate wires, as shown in Figure 3.60 (right). It can be used in universal chargers, as each battery, regardless of its chemistry, would receive the correct amount of charge, as determined by its specific end-of-charge characteristics.

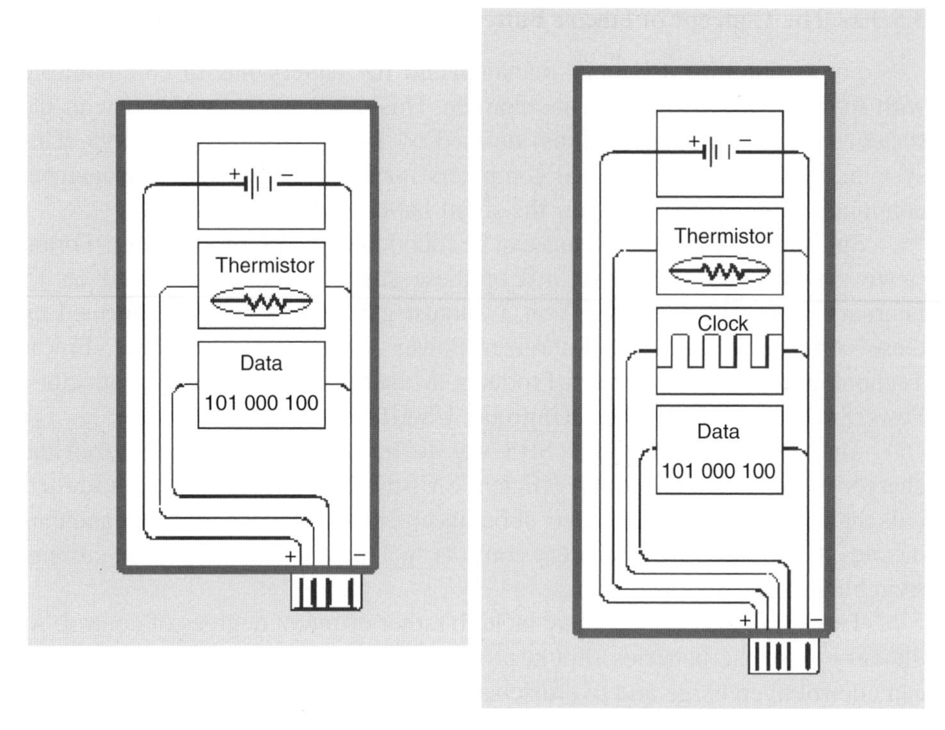

Figure 3.60. Single wire system of a smart battery (left); two-wire SMBus system: clock and data wires are separated (right).
Source: From Ref. [42].

An SMBus battery contains permanent and temporary data. The former are encoded by the manufacturer and include battery ID code, serial number and type, manufacturer's name and date of manufacture. Temporary data, that is cycle number, user pattern and maintenance requirements, is acquired during battery usage.

Full SBdata implementation includes (1) execution of all data value functions; (2) meeting the accuracy and precision requirements of all data functions; and (3) maintaining the proper SMBus timing and data transfer protocols [40]. The data values can be divided into the following categories:

- Historical and identification
- Measurements
- Capacity information
- Time remaining
- Alarms and broadcasts
- Mode, status and errors.

Carrying out measurements at a sampling rate allowing a sufficient accuracy is crucial. The rate needs to be higher when a high current is flowing through the battery.

The SMBus is divided into levels 1, 2 and 3. The first one is not used anymore, as it did not permit chemistry independent charging. Level 2 SMBus allows battery charging within the host (e.g. a laptop computer); the charging circuit may be contained in the battery pack. Level 3 is enclosed in full-featured external chargers. Of course, chargers with level 3 SMBus are sophisticated and expensive. To cut the cost, some chargers with this bus may be not fully SBS compliant. However, in very demanding applications, such as biomedical instruments and precision data collection devices, full compliance is necessary.

In Figure 3.61, a smart-battery compliant charger (level 2) is shown. The circuit contains both a switching regulator (with a power efficiency of 90%) and a linear regulator (the latter operating at low currents only). The high switching frequency (250 kHz) of the controller IC (MAX1647 in this case) permits the use of a small inductor (33 μH).

The block diagram of an IC for full protection of Li-ion battery packs is shown in Figure 3.62. This IC also integrates an I^2C-compatible bus (a two-wire serial bus) to extract battery parameters such as battery voltage, individual cell voltages and control output status. Other parameters, such as current protection thresholds and delays, can also be programmed to increase the flexibility of the BMS. This IC provides safety protection for overload, short circuit in charge and short circuit in discharge. In overload and short circuit conditions, the IC turns off the field-effect transistor (FET) drive autonomously, depending on the internal configuration setting. The communications interface allows the host to observe and control the IC status, enable cell balancing, enter different power modes, set current protection levels and set the blanking delay times.

Knowing the true state of charge of a battery is not a simple task. Measurements only based on voltage or capacity may prove incorrect, if such factors as temperature, ageing, self-discharge and charge/discharge rate are not taken into account [43].

Real-time impedance measurements have proven very useful for determining a true SOC, as they eliminate differences in impedance created by battery aging [43]. Texas Instruments has exploited this concept in an IC, which can be coupled with that of Figure 3.62 devoted to battery protection. This SOC determination technology is shown in Figure 3.63. Its IC complies with the smart battery specifications SBS 1.1 and its impedance-based technique provides SOC estimates with an accuracy better than 1% over the battery lifetime.

As anticipated above, battery authentication is of prime importance to avoid performance and safety problems with low-quality replacement batteries. To cut prices, some vendors may remove from the battery the capacity

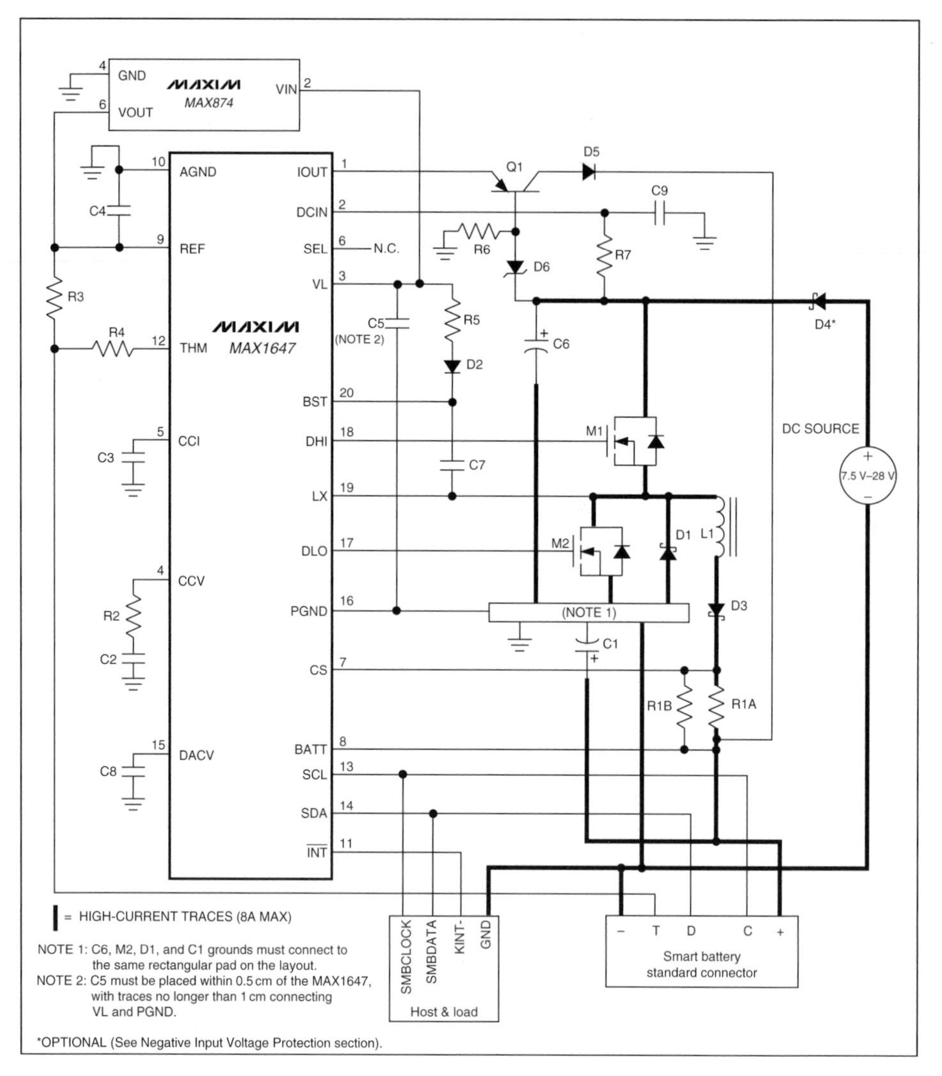

Figure 3.61. Smart-battery compliant charger based on the MAX1647 integrated circuit.
Source: Courtesy of Maxim.

monitoring, the thermistor or even the protection circuit for over-voltage, over-temperature and over-current protection [44].

A very secure way to authenticate a battery is the so-called random challenge-response authentication shown in Figure 3.64.

When the responder, an identification component associated with the battery pack, receives the challenge data, it combines it with a plain-text version of the secret stored in a private secure memory. It performs the authentication

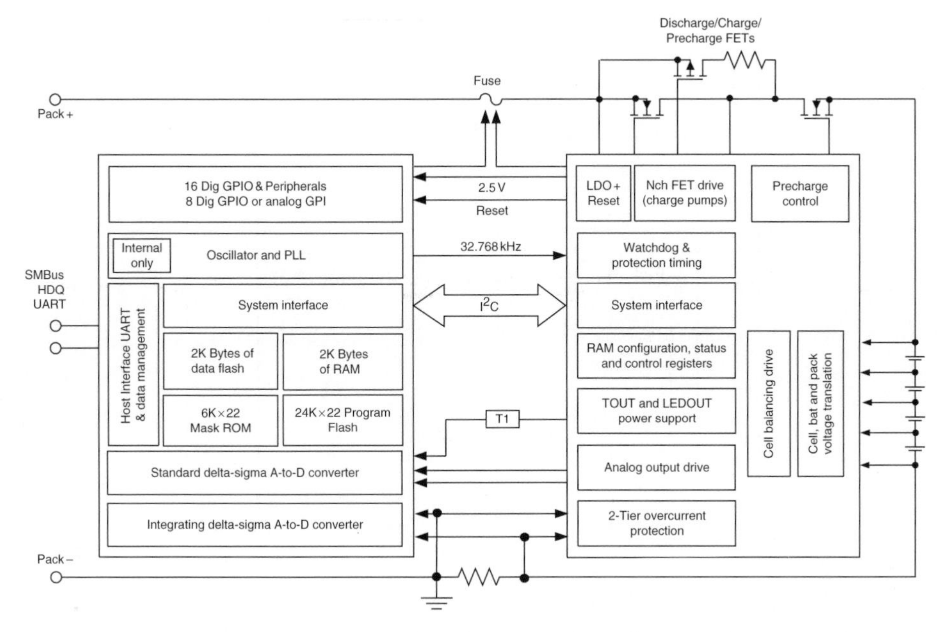

Figure 3.62. IC (Texas Instruments bq29330) for the full protection of multiple-cell Li-ion battery pack.

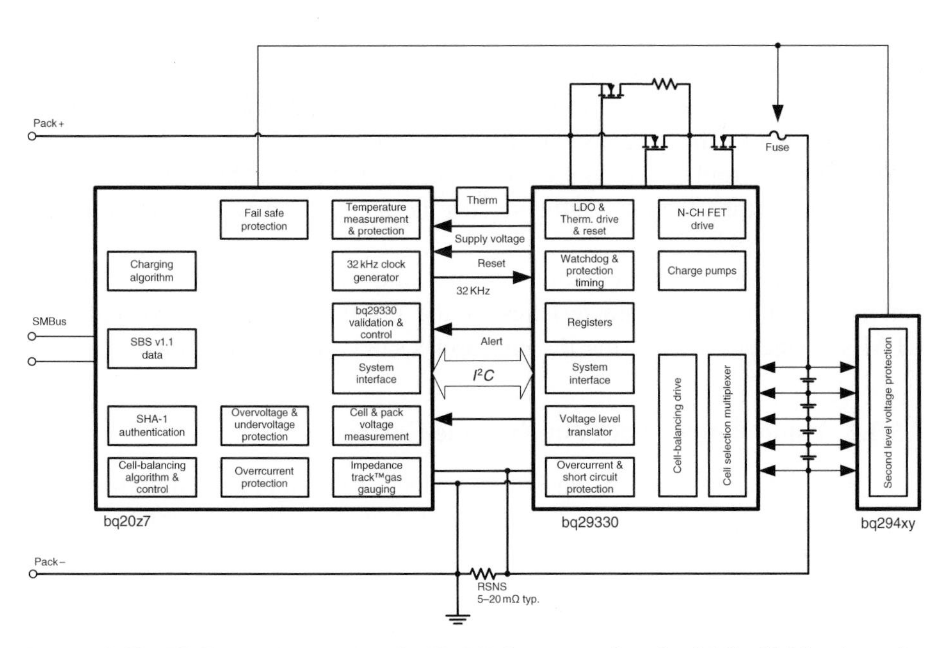

Figure 3.63. IC (Texas Instruments bq20z70) for measuring the SOC of Li-ion batteries in conjunction with bq29330.

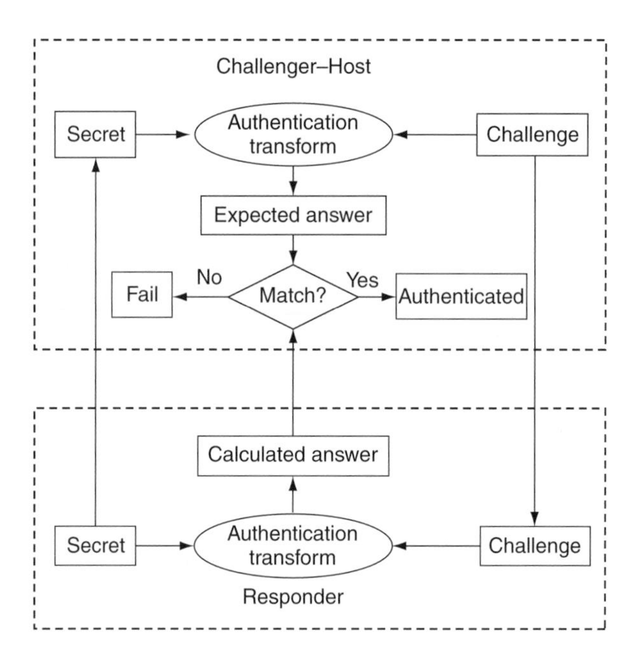

Figure 3.64. Challenge and response-based authentication scheme.
Source: From Ref. [44].

transform to calculate the response. On the other side, the host performs the same transform using the same challenge data and the plain-text version of the secret or decrypted secret from the device. The host compares the value that it computes against the response obtained from the identification device. If the calculated data from the authentication component matches the expected answer from the host, then the host authenticates the battery and allows the system to start operation [44].

An authentication technique of this kind ensures that the replacement battery has the same characteristics of the original one, thus protecting the OEM's business and ensuring the end-user's safety and satisfaction.

3.5.3.2. Using Battery Packs in Extreme Environments

It is not uncommon using some devices, for example radios, telemetry monitors, test equipment, weather station instruments, etc., at temperatures as low as −40°C or as high as +80°C.

The elements of a typical battery pack are shown in Figure 3.65 [45]. The printed circuit board allows advanced functions described in previous sections, thanks to the presence of fuel-gauge (to know the remaining pack capacity), protection circuitry, thermal sensors used to monitor internal pack temperature,

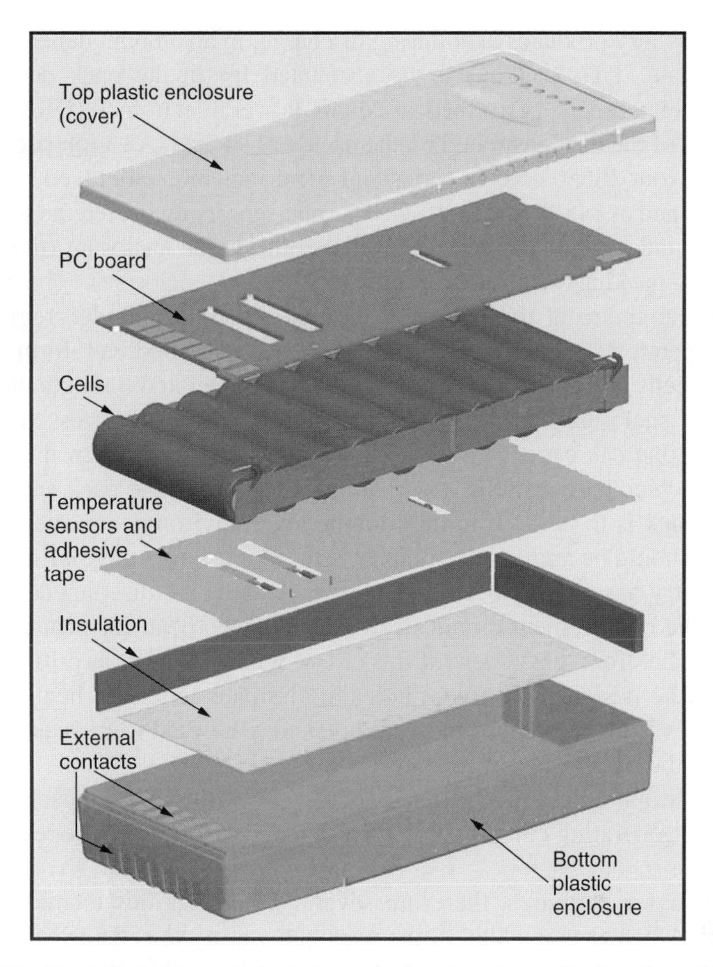

Figure 3.65. Exploded view of a battery pack showing, in particular, the cells, the PC board and the temperature sensors.
Source: From Ref. [45].

and a bus communicating with the host device. All the elements of Figure 3.65 can be customized when designing a battery pack for high- or low-temperature operation. However, the cells are the elements more sensitive to extreme temperatures and a judicious choice has to be made to avoid pack failures.

Li-ion cells, now largely used in packs for high-end portable devices, normally operate between $-20°C$ and $+60°C$, a range recently extended to $-30°C$ to $+80°C$. In addition, each single cell has a separator that melts and becomes insulating above a temperature threshold, and a vent to relieve internal pressure [46]. For temperatures outside the above range, primary Li cells can be used. As already reported, some types of $Li/SOCl_2$, Li/MnO_2, Li/SO_2 and Li/CF_x can work even at $-55°C$ or at 125–150°C (or more for special $Li/SOCl_2$ cells) [46].

Any battery produces heat during discharge, in an amount depending on the discharge rate. This heat has to be accounted for in the pack design. In an interesting experiment performed at Micro Power Electronics (USA) [45], the distribution of the heat generated by the pack was studied. A Li-ion pack made by assembling four 18650 cells in series and six strings in parallel (see Figure 3.65) was discharged at 145 W and at 45°C. Thermal sensors registered the temperature at the core and edge of the pack, as well as outside the enclosure (the pack was wrapped in packing material to simulate a plastic enclosure). The thermistor placed in the centre of the pack and the two others at the edge registered the same temperature: 64–65°C. This means that the cylindrical form factor of the 18650 cells can uniformly distribute the heat generated throughout the pack. The plastic enclosure had a lower temperature of 54°C. This test also demonstrates that one can expect up to 20°C rise in temperature when the pack is in operation, which may result in a 9°C temperature rise of the pack enclosure.

If a pack is to be used in high-temperature environments, specific design principles should be applied to dissipate heat. The use of a thermal sensor in the pack circuitry has already been mentioned: the cells are disconnected on overheating. The management circuit itself may produce heat, for instance through the FETs. Therefore, placement of this circuit within the pack is critical. Finally, packs may be designed with vent holes to dissipate generated heat or possible vented gases from cells; multiple vent holes may be used to increase airflow in and out of the pack.

In addition, the designer has to consider the position of the pack in relation to any heat-generating components, such as high-performance processors, in the host device.

At low temperatures, there are several design options to maximize performance. A heater embedded in the pack can warm the cells prior to use. The embedded heater can be powered from the cells within the pack or by an external source, such as a charger or another battery pack.

The host device may be designed to pulse discharge cells before primary discharge: this self-warms the cells. This technique is applicable, for instance, using super capacitors embedded within a pack, which may provide immediate energy to the host device while cells warm up.

3.5.3.3. Radio Frequency Interferences

Another important aspect the designer has to take into account is the one connected with radio frequency interferences (RFI) in battery-powered devices using RF signals, for example cell phones. RFI is any undesirable RF signal that interferes with the integrity of electronic and electrical systems. Wires, circuit components and printed circuit boards can transmit RFI. Advanced techniques must be used to protect batteries from RFI; their design has to be such to conform to both RF and battery management parameters [47].

Smaller and lighter mobile communication devices have battery packs usually residing close to their hosts' circuitry. Multiple-cell packs with their complex electronics cause an even greater concern as far as RFI are concerned. The electronics for battery management can demodulate the RF energy that the phone transmits and can generate noise. This noise, in turn, can interfere with the phone's power-management system and eventually find its way into the phone's receiver system, thereby degrading the receiver performance. The aluminium case often used in lighter Li-ion batteries can bypass the battery's electronic circuit and can be a medium to send RFI to the phone receiver [47].

If the designer overlooks battery-circuit RF immunity and susceptibility, protective ICs and passive support circuitry may operate at undesired voltage references, altering functional voltage- and current-protection parameters. Static battery data, such as chemistry and charge/discharge terminations, are less prone to RFI, because these variables involve hard-coded information. However, loss of battery information may result from the lack of immunity of battery's EPROM to RFI. Dynamic battery data, for example SOC measurements and monitoring of charge and discharge cycles, are more prone to RFI and low-frequency noise interference.

3.5.3.4. Battery Charging

From the electronics viewpoint, the battery chargers can be divided into linear- and switching-regulator chargers.

For some devices, for example digital cameras, preferred battery charging is done by using a cradle charger in which the device, or just the battery, is placed. With this method, the heat dissipated by the charger has much less importance with respect to that of an embedded charger. In these cases, a linear-regulator charger is used, which has relatively large dimensions and an efficiency hardly exceeding 60% (see also Section 3.5.1.3).

In some larger devices, for example notebooks, the battery charger is part of the system and, therefore, heat generation has to be limited. A switching charger is remarkably smaller than a linear one and can be embedded in the device. This is made possible by the use of a switching type regulator (or switch-mode power supply (SMPS)). The main drawback of switching regulators is the noise level generated by their switching action (10 000 times more than in linear regulators). However, there are means to cope with this drawback, for example connecting special capacitors from the AC input terminal to ground [48].

The electronic industry produces a number of ICs for the correct charging of batteries according to their chemistry. Original equipment manufacturers make their choice as a function of the IC performance and price. Therefore, they might want just a single chemistry charger (Ni–Cd/Ni-MH or Li-ion or Pb-acid), or a multi-chemistry, smart-battery-compliant charger/controller.

Multi-chemistry charging is now available in several ICs. Examples of chemistry-independent chargers are shown in Figures 3.61 and 3.66. The latter charger is based on bq2000, an IC by Texas Instruments, whose details are given here as a way to illustrate how similar ICs work. Bq2000 is a programmable IC that can charge Ni–Cd, Ni-MH or Li-ion batteries. The battery chemistry is detected by monitoring the voltage profile in the early charge stages. Afterwards, the charge is brought to completion with an appropriate algorithm. A programmable timer is also available to allow a further charge control. Fast charge is not allowed until voltage and battery temperature fit in a given range. If V is low, the IC provides a trickle charge to raise it to an acceptable level. If T is too high (typically above 45°C), charge is only started after cooling down; if T is low (typically below 10°C), a trickle charge is applied. The power efficiency is more than 90% and power dissipation is low. Other features of chargers/controllers of this type include: sleep mode for low power consumption, battery removal and insertion detection, continuous temperature monitoring and fault detection, user-programmable charge current and voltage.

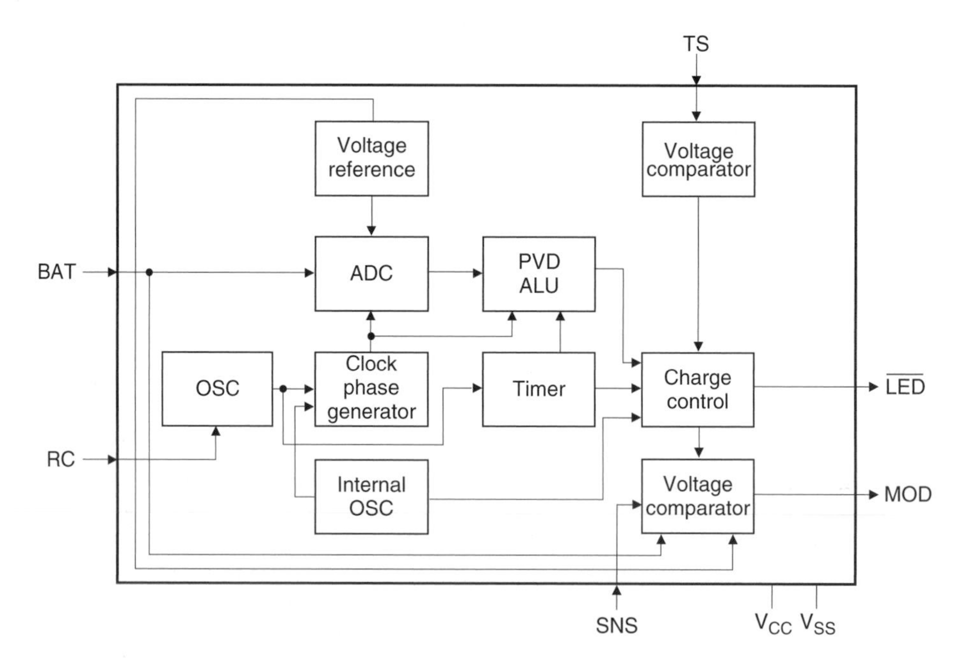

Figure 3.66. Bq 2000 block diagram. The functions of the various blocks are indicated. *Source*: Courtesy of Texas Instruments.

In multi-chemistry chargers, Ni–Cd and Ni-MH batteries are charged at constant current and charge termination is based on voltage, temperature or time values. In the first case, the peak voltage typical of the end of charge of these batteries is used. If a top-up charge is requested, this is made at a lower current and termination is time-dependent. Finally, a trickle charge, usually at a rate between C/20 and C/100, is applied to the battery to compensate for self-discharge.

For Li-ion batteries, smart chargers normally apply the constant current (at the 1C rate), constant voltage (within 1%) method (Figure 3.67). Charge is typically stopped when the current in the constant-voltage step falls below C/30 or when the timer tells that time is out. Li-ion batteries have low self-discharge rates, so trickle charge is not applied. Depending upon the power requirements of the application, battery packs can consist of up to four Li-ion cells in a variety of configurations (see Table 3.5). These batteries can be charged with different power-source types: AC adapter, USB port, or car adapter [48].

As stated above, and further remarked in Table 3.5, chargers may be linear or switching type. With reference to Li-ion batteries, more details are given below.

A linear charger is preferred when the input voltage of the charger is only moderately higher than the OCV of fully charged cells and the 1C fast-charging current is not much higher than 1 A. For example, there is typically one Li-ion battery cell in an MP3 player, with capacity ranging from 700 to 1500 mAh and an OCV of 4.2 V. Because the power source for the MP3 player is either an AC/DC adapter or a USB port with a well-regulated 5-V output, a battery charger with a linear topology offers the simplest and most cost-effective solution. Because of the high power dissipation of linear chargers, they are only suitable for small-capacity devices (<~1300 mAh) [48].

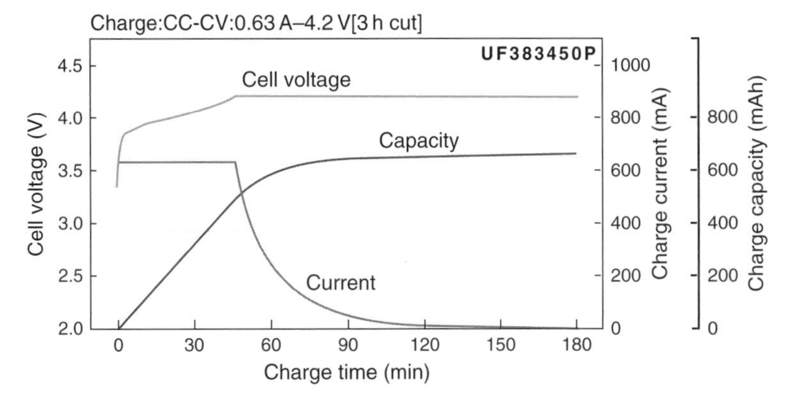

Figure 3.67. CC-CV charging of a Li-ion cell.
Source: Courtesy of Sanyo.

Table 3.5. Capacity, configuration and charging characteristics of different Li-ion battery packs.

Application	Battery Capacity (mAh)	Cell Configuration	Power Source	Charger Normal Input Range (V)	Suggested Charging Topology
Bluetooth headset	90–150	1 cell	AC Adapter, USB	5	Linear
Wireless handset	70–200	1 cell	AC Adapter, USB	5	Linear
Mobile phone, PDA	700–1500	1 cell	AC Adapter, USB	5	Linear
MP3 player	700–1500	1 cell	AC Adapter, USB	5	Linear
PMP (HDD)	1200–2400	1 cell	AC Adapter, USB, Car Adapter	5, 9–16	Linear, Switching (Buck)
P-DVD/P-TV	1800–3600	2 cells (2 in series, multiple parallel)	AC Adapter, Car Adapter	9–16	Switching (Buck)
Notebook PC	1800–3600	3 to 4 cells (3 to 4 in series, multiple parallel)	AC Adapter, Car Adapter	9–16	Switching (SEPIC)

Source: From Ref. [36].

The buck, or step-down, switching charger is a better choice when the 1C charging current is higher than 1 A or when the input voltage is much higher than the open-circuit voltage of the fully charged cells. For example, in a hard-drive-based PMP, a one-cell Li-ion battery with OCV of 4.2 V and a capacity of 1200–2400 mAh is typically used. PMPs are now frequently charged by car kits, which output a voltage between 9 and 16 V. The high voltage differential (at least 4.8 V) between the input voltage and the battery voltage makes the linear topology very inefficient for this application. This inefficiency, coupled with the high fast-charging current (1.2–2.4 A, i.e. 1 C) can create heat-dissipation problems. To avoid this, the buck technique needs to be adopted. Technical details on a high-efficiency switching buck battery charger can be found in Ref. [49].

In certain applications where three or even four Li-ion/Li-polymer cells are connected in series, the input voltage to the charger might not always be higher than the battery voltage. For instance, a laptop PC uses a three-cell Li-ion battery pack, with a 12.6-V (4.2 V $\times$ 3) OCV at full charge and 1800–3600 mAh capacity. The input power source can be either an AC/DC adapter with an output voltage of 16 V or a car kit with an output voltage between 9 and 16 V. As such, the input voltage can be either lower or higher than the battery voltage. Obviously, neither a linear nor a buck charger is capable of charging a battery pack under this circumstance. This gives way to a switching charger with the same structure of a buck-boost regulator (see also Figure 3.59).

A recently introduced method for charging Li-ion batteries, that is constant-current pulse charging, combines the benefits of a linear charger and a switch-mode charger. It limits the charging current by employing a current-limited AC wall adapter. The current from the converter is switched to the battery for constant-current charging. As the battery voltage rises to the voltage limit, the current source is switched *on* and *off*, thereby supplying the required average current to the battery without exceeding the voltage limit.

Power dissipation is low, because the switch is either *on* or *off*, as for a switch-mode charger. Yet the circuit is simple, as in a linear charger, because no output filter is required. It can dissipate more power while in the current-limit mode (depending on the AC wall adapter used), but that has little effect on the battery or its load if the maximum safe temperature is not exceeded. The circuit of this charger is smaller and less complex than a switching charger: a complete charger can be made with only two capacitors and one resistor in addition to the IC (e.g. 1679 by Maxim) and external MOSFET [50].

According to Sanyo, a producer of this type of charger, full charge can be reached in 90 min vs 2–3 h for a conventional CC-CV charger. Pulse width modulation may also be used: as the battery reaches full charge, the period between pulses increases. The objective of using a pulse current is to increase the rate of electrode reactions. When a sufficiently high charging current flows continuously through a cell, the electrodes becomes polarized. Consequently, the cell voltage increases and tends prematurely to the upper limit. If the current is stopped for a while, some relaxation may occur, that is the electrodes can have some degree of depolarization and the ionic flux at the electrode/electrolyte interface can be faster.

Reconsidering the power sources of Table 3.5, the most flexible source is certainly the AC wall adapter. However, the USB port of a notebook, desktop computer, or computer peripheral offers an interesting alternative. Indeed, charging a cell phone, MP3 player, or digital camera from a notebook computer is highly convenient and can be performed anywhere; moreover, it allows users to perform two functions at the same time, such as downloading music or updating files, while recharging their battery.

A typical notebook's USB port provides a 5-V source with up to 500 mA of current. The ability of a USB port to supply a charge for another portable device is highly dependent on how much power the portable system draws to run other functions, while the battery is being charged. When the system requests power to run subsystems or applications, the current available to charge the battery will be reduced because the USB port can only provide a finite amount of current, and this will extend the charging time [51].

3.6. Trends in Battery Selection for Portable Devices

In Table 3.6, some common portable applications and the type of batteries they use are listed. The number of primary and secondary, aqueous or non-aqueous, batteries is still very high (see Chapter 2). This is justified by the different power requirements and by considerations based on battery price and performance in a given application.

For low-end applications, for example radios, lights and watches, primary batteries are mostly used and will be used again in the future. This will also be the case for very specific applications, where primary Li cells with long life, resistance to temperature extremes and capability of delivering high power may be selected. In contrast, high-end consumer applications will increasingly be based on secondary batteries. Among them, Li-ion is becoming the system of choice: these batteries have the positive features already mentioned and their price is steadily decreasing. Therefore, the focus of this section will be on these batteries.

In Figure 3.68, the specific energies and energy densities of different rechargeable batteries are shown. The values for Li-ion batteries (up to 240 Wh/kg and 630 Wh/L) include the latest developments, especially based on the shift from a traditional positive electrode, $LiCoO_2$, to one also containing Ni and Al ($LiCo_xNi_yAl_zO_2$ – see also Table 2.6) [52–57].

Table 3.6. Applications in common portable devices, typical currents, and use of primary and/or secondary batteries. Preferred battery types are in bold.

Application	Specific Application	Typical Current (mA)	Primary Battery	Secondary Battery
Light	Flashlight	100–700	**Alk**, Zn-C	
	Lantern-type	300–800	**Alk**, Zn-C	SLA
Watch (LCD)		10–25	**Silver**, alk, button Li	
Toy	Radio-controlled	600–1500	Alk	
	Electronic game	20–250	**Alk**, button Li	
	Video game	20–250	**Alk**, button Li	
Audio	Walkman	200–300	**Alk**, Zn-C	
	CD player	100–350	Alk	Ni-MH, **Li-ion**
	MP3 player	100–350		Ni-MH, **Li-ion**
	Handy radio	70–200	Alk, Zn-C	
Calculator (LCD)		<1	Button Li, silver, alk	
Telecommunications	Cellular phone	60–400		Ni-MH, **Li-ion**
	Smart phone/ PDA	250–550		Li-ion
	Cordless phone			Ni–Cd, Ni-MH
Computer	Palmtop	400–800		Li-ion
	Notebook	1000–2800		Li-ion
Digital camera		220–450	Several primaries	Li-ion

(*Continued*)

Table 3.6. (*Continued*)

Application	Specific Application	Typical Current (mA)	Primary Battery	Secondary Battery
Camcorder		700–1000		Ni-MH, **Li-ion**
DVD player		2000–3000		Ni-MH, **Li-ion**
Handheld GPS		<100 cont.; <1000 pulse		Ni-MH, **Li-ion**
Medical instruments		500–10 000	Primary Li	SLA, Ni–Cd, **Li-ion**
Power tools		10 000–15 000		Ni–Cd, Ni-MH, **Li-ion**

The evolution in terms of capacity and energy for a Li-ion cell is reported in Table 2.7 for the most common size, that is the 18650 cell (18 mm diameter, 65 mm height).

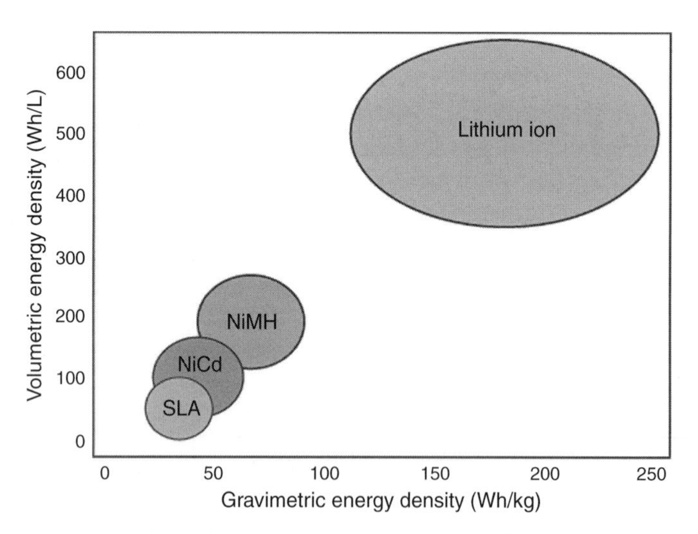

Figure 3.68. Specific energies and energy densities of the four systems mostly used in portable applications. For the Li-ion battery, the improved values brought about by the latest developments have been taken into account (see text).
Source: From Ref. [57].

The new Li-ion battery developed by Sony cell, Nexelion, in which both electrodes have been modified with respect to the conventional C/$LiCoO_2$ couple, has already been introduced in Chapter 2.

Some Li-ion batteries use polymeric electrolytes. Li-ion polymer batteries have typical specific energies in the range 130–175 Wh/kg and energy densities in the range of 300–400 Wh/L. The use of a polymeric electrolyte allows fabricating cells with a thin form factor.

Other positive electrodes proposed for Li-ion cells are the Mn spinel $LiMn_2O_4$ and the phosphate $LiFePO_4$, which have been introduced when dealing with power tools (see Table 3.3). $LiFePO_4$ having nanosized particles is especially capable of sustaining very high currents. The high-power versions of these Li-ion batteries pay a penalty in terms of capacity and energy vs the low-rate ones: a high-rate 18650 cell has a typical capacity of 1.6 Ah and an energy density of 250 Wh/L (compare with Table 2.7).

In Table 3.7, the most common types of Li-ion batteries commercialized today are presented together with their applications.

In the above table, the excellent power capability of some cells is emphasized. High-power Li-ion batteries are phasing out Ni–Cd batteries that have long dominated this field. Some of the Li-ion cells of Table 3.7 are claimed to have up to 50% more power, longer run times, faster charge times, and longer lifetimes than Ni–Cd [56]. The success of Li-ion packs is also due to the need to replace Cd-containing products. Indeed, Ni–Cd batteries do not comply with the ROHS (Reduction of Hazardous Substances) and WEEE (Waste from Electrical and Electronic Equipment) legislations [57].

The possibility of fast charging, already mentioned in the section on power tools, is particularly intriguing and several producers claim outstanding advances over the 3-h charge time of traditional $LiCoO_2$-based batteries. Toshiba's 600-mAh battery can be recharged to 80% in 1 h. Valence's IFR18650p (1.1 Ah) can be charged to 95% of its capacity in 30 min. A123 claims for its ANR26650-M1 cell (2.3 Ah) a recharge time of 5 min only, as it can accept a current of 10 A. This battery, with a cathode formed by nano-sized $LiFePO_4$, would provide a power of 2.7 kW/kg on continuous discharge at 30C and 3.5 kW/kg on long pulse discharge at 80C. This is made possible by the very low internal impedance (8 mΩ) and by a new type of cell construction based on a dual plate tubular design optimized to deliver maximum power with high efficiency. This new design uses an "all laser welded" construction.

In addition to power tools and various consumer electronics, high-power Li-ion batteries are considered ideal for some portable medical devices. The Li-ion system allows connecting a lower number of cells in series to have the high voltages often required. Indeed, Ni-MH systems can be configured with a maximum of 10 cells in series, this resulting in a voltage of 12.5 V, but this value can be doubled, and the pack weight reduced, by just connecting

Table 3.7. Rate capabilities and applications of commercial Li-ion batteries with different positive electrodes.

Positive Electrode	V Limit on Charge	Charge/ Discharge C Rates	Applications	Notes
$LiCoO_2$	4.2	1C limit	Cell phones, cameras, laptops	
NCA	4.2	C limit	Laptops	$LiCo_xNi_yAl_zO_2$, 620 Wh/L, enhanced safety
NCM	4.1	~5C cont. 30C pulse	Power tools, medical equipment	$LiCo_xNi_yMn_zO_2$, compromise between high rate and high capacity
Mn spinel	4.2	10C cont. 40C pulse	As above	High rate, fast charging, lower capacity and energy
Fe phosphate	3.6	30C cont. 80C pulse	As above	Nanosized $LiFePO_4$, very high rate, fast charge, long cycling

NCA, Ni–Co–Al; NCM, Ni–Co–Mn; Mn spinel, based on $LiMn_2O_4$ (Mg-doped); Fe phosphate, based on doped $LiFePO_4$ (nanoparticles). NCM, spinel and phosphate are particularly optimized for high-power applications.
Source: From Ref. [56].

7 Li-ion cells. Such products as left-ventricular assist devices, ventilators, and automated external defibrillators (see Chapter 2) enable patients to survive potentially fatal events, and now most of these devices are powered by Li-ion batteries [5].

Temperature issues are of the utmost importance in these applications. Many medical devices are expected to operate at temperatures from –20 to 60°C. In addition, the heat generated by the operating battery must be strictly monitored, as previously described. The same applies when the battery is being charged. Excess heat from the battery can affect the runtime of diagnostic equipment, such as portable ultrasounds, or even cause the failure of live-saving equipment, such as automated external defibrillators.

The self-discharge rate may also be a determining factor, as the need of recharging after storage may prove very detrimental in case of emergency [5].

In conclusion of this section on battery choice, it can be useful summarizing the types of batteries used in the most common consumer applications. The devices selected are: camcorders, cameras, cordless phones, PDAs and cellular phones.

- Camcorders. Ni-MH and, mainly, Li-ion are used. Ni-MH is preferentially used as a 6 V battery with capacities of 2.1–4 Ah. For Li-ion batteries: 7.2 or 7.4 V, with capacities in the range of 0.7–5.5 Ah.
- Cameras. In this case, several primary batteries can be used in addition to Li-ion: 3 V Li/MnO_2, 6 V and 12 V alkaline, 1.5 V Li and 1.4 V Zn/Ag_2O.
- Cordless phones. 3.6 V Ni-MH with capacities from 0.3 to 1.8 Ah. 2.4, 3.6 and 4.8 V Ni–Cd with capacities from 0.3 to 0.6 Ah.
- PDAs. 3.7 V Li-ion of 0.8–1.8 Ah.
- Cell phones. 2.4–7.2 V Ni-MH with capacities from 0.6 to 1.4 Ah. 3.6–7.2 V Li-ion with capacities from 0.55 to 1.8 Ah.

References

1. B. Haskell, *Portable Electronics Product Design and Development*, McGraw-Hill, New York, 2004.
2. Texas Instruments, "System Block Diagrams, Personal Digital Assistant (PDA)".
3. Texas Instruments, "System Block Diagrams, Digital Still Camera".
4. Texas Instruments, "Audio Solutions Guide", Report Slyy013, 1Q 2006.
5. R.S. Tichy, "Reliable Power Sources for Medical Devices", *Portable Design*, July 2006.
6. Texas Instruments, "Medical Applications Guide", Report Slyb108b, 2Q 2007.
7. Texas Instruments, "Information for Medical Applications", Report Slyb108a, 2Q 2004.
8. Texas Instruments, "System Block Diagrams, Blood Pressure Monitor".
9. Texas Instruments, "System Block Diagrams, Fingerprint Biometrics".
10. Texas Instruments, "System Block Diagrams, Portable Blood Gas Analyzer".
11. B.S. Abella, "The Importance of Cardiopulmonary Resuscitation in Improving Patient Outcomes from Cardiac Arrest", *Emergency Medicine & Critical Care Review*, 2004, p. 36.
12. M. Grimm, in *Industrial Applications of Batteries. From Cars to Aerospace and Energy Storage*, M. Broussely and G. Pistoia, Eds., Elsevier, Amsterdam, 2007.
13. E.S. Takeuchi, R.A. Leising, D.M. Spillman, R. Rubino, H. Gan, K.J. Takeuchi and A.C Marschilok, in *Lithium Batteries – Science and Technology*, G.A. Nazri and G. Pistoia, Eds., Kluwer Academic Pub., Boston, 2004.
14. G. Pistoia, *Batteries for Portable Devices*, Elsevier, Amsterdam, 2005.
15. Mayfield Clinic, "Intrathecal Drug Pump", www.mayfieldclinic.com/PE-PUMP
16. Texas Instruments, "Digital Hearing Aids"
17. S. Passerini, B.B. Owens and F. Coustier, *J. Power Sources* 89 (2000) 29.
18. "Mobile Wellness Monitoring of a Single Cardiac Patient", Report WGIV/n06–19, www.tc251wgiv.nhs.uk/pages/docs/wgiv_n06_19
19. "Telemedicine – Robo-Doc Makes Hospital Rounds", *Middle East Health Magazine*, 2005.
20. R. Latham, R. Linford and W. Schlindwein, *J. Power Sources* 172 (2004) 7.
21. W.S. Gong and D.M. Merfeld, *Ann. Biomed. Eng.* 28 (2000) 572.
22. C.F. Holmes, *Interface* 12 (2003) 26.

23. D. Morrison, "Li-ion Cells Build Better Batteries for Power Tools", *Power Electronics Technology*, February 2006.
24. D.K. Schroder, "Low Power Silicon Devices", in *The Encyclopedia of Materials: Science and Technology*, K.H.J. Buschow, R.W. Cahn, M.C. Flemings, B. Ilschner, E.J. Kramer, and S. Mahajan, Eds., Elsevier, Amsterdam, 2001.
25. Mobile Intel Pentium III Processor-M Datasheet.
26. R. Pelt and D.L. Martin, "Low-Power Software-Defined Radio Design Using FPGAs", Altera Corporation Report, 2005.
27. Microelectronics Industrial Centre. Northumbria University, Newcastle, UK, 2002.
28. Xilinx Report, "Decrease Processor Power Consumption using a CoolRunner CPLD", May 2001.
29. M. Sugai, K. Nishimura, K. Takamatsu and T. Fujinaga, *Hitachi Rev.* 48 (1999) 313.
30. G. Kavaiya, "Microcontrollers Offer Various Methods to Maximize Battery Life", *Portable Design*, August 2003.
31. T. Vora, "Managing Graphics Power Consumption in Portable Computing Devices", *ECN*, November 2001.
32. T. Armstrong, "The Question of Linear *vs.* Switching Regulators in Portable Products", *Portable Design*, August 2006.
33. M. May, "Understanding Low Drop Out (LDO) Regulators", *Texas Instruments' 2006 Portable Power Design Seminar.*
34. L. Zhao and J. Qian, "DC-DC Power Conversions and System Design Considerations for Battery Operated Systems", *Texas Instruments' 2006 Portable Power Design Seminar.*
35. D. Brown, "Optimizing Power Management for Portable Multimedia Devices", *Advanced Analogic Technologies Report*, 2006.
36. M. Kendall and D. Furness, "Increasing Storage for Portable Consumer Electronics", *EDN Asia*, 2004.
37. STMicroelectronics Report, "HDD Preamplifiers Lower Power Demand in Portable Devices", August 2006.
38. Q. Deng and H. Lee, "Thermal-Management Challenges and Solutions for Portable Device Designs", *Portable Design*, 2006.
39. MPower, "Battery Management Systems", 2005.
40. D. Friel, "The Importance of Full Smart Battery Data (SBData) Specification Implementation", PowerSmart Inc. *Report*, 1998.
41. W. van Schalkwijk, in *Advances in Lithium-ion Batteries*, W. van Schalkwijk and B. Scrosati, Eds., Kluwer Academic Publishers, Boston, 2002.
42. I. Buchmann, *Batteries in a Portable World,* Cadex Electronics Inc., Richmond, B.C., Canada, 2001.
43. Y. Barsukov and B. Krafthofer, "Predicting Runtime for Battery Fuel Gauges", *Power Management DesignLine*, August 2005.
44. J. Qian, "Authentication Distinguishes Authentic Battery Pack and Peripheral from Third Party", *Power Management DesignLine*, October 2005.
45. J. VanZwol, "Designing Battery Packs for Thermal Extremes", *Power Electronics Technology*, July 2006.
46. G. Pistoia, in *Industrial Applications of Batteries. From Cars to Aerospace and Energy Storage*, M. Broussely and G. Pistoia, Eds., Chapter 1, Elsevier, Amsterdam, 2007.
47. M. Krishnamurthi and A.K. Shamsuri, "RF-Interference-Design Considerations for Portable-Device Batteries", *EDN*, March 2006.
48. Q. Deng and H. Lee, "Select the Right Li-Ion and Li-Polymer Battery Charger", *Portable Design*, July 2006.

49. J. Qian, "How to Charge Li-Ion Batteries for Portable Devices More Efficiently", *Power Management DesignLine*, August 2005.
50. Maxim, "Portable Devices Need High-Performance Battery Chargers", Application Note 721, December 2000.
51. D. Brown, "Battery Charging Options for Portable Products", *Power Management Design Line*, April 2006.
52. D. Morrison, "New Materials Extend Li-ion Performance", *Power Electronics Technology*, January 2006.
53. M. Conner, "New Battery Technologies Hold Promise, Peril for Portable-System Designers", *EDN*, December 2005.
54. R.S. Tichy, "Create an Ideal Battery Pack for Mobile Medical Devices" *EDN*, September 2005.
55. Sony Corp. Info, "Sony's New Nexelion Hybrid Lithium Ion Batteries", February 2005.
56. V. Biancomano, "Lithium-Ion Batteries Make a Grab for Power", *Power Management Design Line*, May 2006.
57. R.S. Tichy, "Reliable Power Sources for Medical Devices", *Portable Design*, July 2006.

Chapter 4

INDUSTRIAL APPLICATIONS (EXCEPT ROAD VEHICLES)

4.1. Introduction

This chapter includes all applications that have a prevailing industrial interest. Therefore, even though objects allowing portability are sometimes mentioned, for example some sensors or mini-robots, a further subdivision was avoided.

A non-exhaustive list of the applications described in this chapter includes:

- Aerospace applications: satellites, launchers, planetary exploration missions (probes, rovers), aircraft
- Stationary applications: telecommunications, uninterruptible power supply (UPS), load levelling, solar/wind energy storage, power quality
- Metering (electricity, heat, gas, water, flow)
- Emergency, alarm systems
- Gas/oil prospecting
- Oceanographic applications
- Seismic instrumentation
- Meteorological instruments
- Wireless connectivity
- Munitions and missiles
- Radio-frequency identification
- Unmanned vehicles
- Remote monitoring
- Navigations aids
- Tracking systems
- Transmitters
- Agriculture
- Electro-mechanical systems
- Robotics
- Video-surveillance
- Access control devices
- Sensor networks
- Data loggers

4.2. Meters

Industries, offices and individual users are connected to utilities, such as water, gas and electric power, whose consumptions have to be measured timely, precisely and economically. Traditional mechanical and electromechanical meters are rugged and require no maintenance, but have a limited accuracy and require on-site reading by personnel sent by the service provider. These meters may be upgraded by adding electronic devices that can convert the measured value into an electronic signal, which is then transmitted to a central computer (see later). In recent times, fully electronic meters have been installed, most of them enabling automatic meter reading (AMR) and data transfer.

Apart from utility meters, several other meters (e.g. for flow, distance and leak measurements) are commonly used. All of them rely on batteries as primary or backup power sources. More precisely, batteries for electronic meters are requested to provide the following functions [1]:

- powering the microprocessor, LCD, and, if needed, the prepayment unit of a portable and/or remote meter
- powering remote reading modules, accessories (such as valves) and auxiliary devices (such as reading repeaters)
- acting as a backup, in case of mains outage, for memory, real-time clock and display.

As a function of the operations performed, these basic specifications apply to meter batteries:

- a few microamperes in sleep mode
- a few milliampere pulses during awakening and listening
- up to several A pulses during data transmissions
- wide-operating temperature ranges: typically −40/+70°C
- runtimes of 5–10 years or more
- ability to stand adverse environmental conditions (in several instances the meters are located outdoor).

Primary Li batteries comply with these requirements and are widely used. Alkaline batteries may be an alternative when cost is an issue and Li-ion batteries when recharging avoids frequent replacements.

4.2.1. Power Meters

Electronic power meters allow numerous functions such as multi-rate tariffs, remote reading, load switching, and energy management thanks to time-of-use registration, etc. These new meters contain a mains-powered memory where power consumption data is stored. Of course, a power outage can lead to loss of information, but the presence of primary Li cells for memory backup avoids this inconvenience. In this case, the pulse currents do not exceed few mAs.

A $Li/SOCl_2$ cell is ideally suited for this application, in view of its high operating voltage, full hermeticity, ability to operate over a wide temperature range, and low self-discharge rate that allows very long runtimes (10–20 years).

4.2.2. Gas Meters

In the early 1990s, a British company introduced an ultrasonic gas meter. Since then, more than one million units have been installed in Great Britain [2]. Other countries have adopted this technology, even though cost issues contrast its widespread use.

As detailed in Section 4.2.5, either the transit-time or the Doppler technology can be used to measure the flow rate of liquids and gases. Transit-time is especially used for natural gas pipes. Tests on $Li/SOCl_2$ and Li/CF_x batteries have shown the better performance of the former, which grants a longer runtime before reaching the limit voltage of 2.5 V [2]. If high power is requested to operate valves, alkaline cells can be selected, as they are cheaper, easy to replace, and can be located outside the meters [1].

4.2.3. Water Meters

Mechanical meters of domestic water consumption can be fitted with an electronic device collecting and sending data to a central computer. The wireless data transmission is normally done by GSM. In this case, current pulses of up to 2 A may be needed. Cylindrical $Li/SOCl_2$ cells are mostly used, because of the high cutoff voltages (2.5–3.2 V) and the high-energy density needed. These cells may power a control unit and even a valve switch in prepaid meters [1].

Instead of high-rate $Li/SOCl_2$ cells, which have a limited capacity, low-rate cells coupled to a hybrid-layer capacitor (HLC) managing high pulses (e.g. to actuate valves) can be used [3]. On a smaller scale, Li/MnO_2 cells (2/3A size) can also been used.

$Li/SOCl_2$-HLC hybrids can power a device (AMR) able to collect and transmit data related to a network of up to 10 water or gas meters – see Figure 4.1 [3]. The hybrid battery has a runtime of at least 10 years.

For industrial applications, fully electronic meters are applied, which measure the speed of propagation of ultrasounds in clean or dirty waters, using

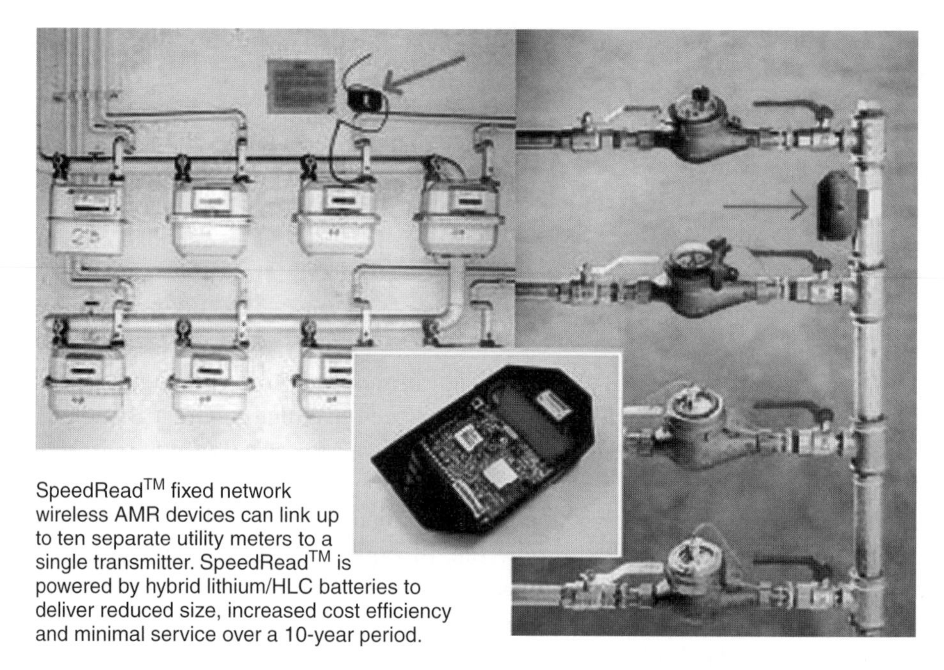

Figure 4.1. Electronic device for automatic meter reading (AMR) in a network of water and gas meters.
Source: From Ref. [3].

the transit-time or Doppler technology, respectively. The functioning of ultrasonic meters is detailed in Section 4.2.5.

4.2.4. Heat Meters

These meters are commonly used for measuring the consumption of hot water in residential and industrial buildings. Basically, they measure both the flow rate and the temperature of water. Flow can be measured either mechanically (with an impeller) or by ultrasound transmission (transit-time technique – see later). Measuring the temperature involves the use of two sensors, one for the forward and the other for the return flow. An electronic device is needed for the calculation of heat consumption over a given lapse of time. Remote data transmission is possible, for example via RF, RS232 or optical interface. Water temperature in the conduits often exceeds 70°C to prevent germ proliferation: this poses serious challenges to the meter batteries, which are expected to work at temperatures above 45°C and up to 80°C [1].

These devices have to sustain pulses of up to 50 mA, with a voltage not lower than 2.5–3.0 V. Their runtimes usually are in the range 6–12 years, but some manufacturers claim 16 years. Cylindrical primary Li batteries are once more preferred (Li/CF_x, Li/MnO_2 and Li/$SOCl_2$). Modern meters may have several functions (such as remote data transmission): in this case, large Li/$SOCl_2$ batteries (R20 size) are preferred for a longer runtime.

Heat cost allocators are electronic devices measuring the amount of hot water consumed in single apartments or rooms. They are directly mounted on the radiators and may work with different principles. With the evaporation method, the radiator's heat is transferred, via a conductive aluminium plate, to a measuring ampoule. The amount of liquid evaporated is proportional to the heat consumed. Another method measures the temperature difference between the radiator surface and the room by using two sensors. Some allocators only require a limited current: in this case, coin cells of low capacity may be used.

4.2.5. Flow Meters

Ultrasonic flow meters are now widely used for measuring flow rates in liquids and in some gases. There are basically two methods for these measurements: transit time and Doppler. The former is used for "clean" fluids, that is those pure or containing less than ~5% of gas or solid particles; the latter is instead used for "dirty" liquids, that is those with significant impurity levels. In some high-end devices, the Doppler method can also be used to measure the flow in clean liquids; for instance, by measuring the turbulence just downstream of a 90° elbow along the piping.

In Figure 4.2, an ultrasonic meter and the basis of the two methods are shown. In the Doppler method, the transmitted frequency (>20 kHz) is reflected by the particles present in the fluid; in the transit-time method, the speed of wave propagation is measured along the flow and in the opposite direction: the difference in the transit times is used to calculate the flow. The transit-time transducers can be located in different positions, as shown in the figure, but in all cases the sonic beam follows a single path. The transducers are connected to a converter that can output a 4–20 mA DC signal.

When these methods are applied to gases, the transit time is ideal for natural gas pipes, whereas the Doppler is useful for gases with sound reflecting particles.

Both primary and rechargeable batteries can be used to power these meters. In the first case, alkaline batteries (e.g. four of the AA size) can provide more than 30 h of operation; in the latter, 12-V gel lead-acid batteries are used (24–36 h).

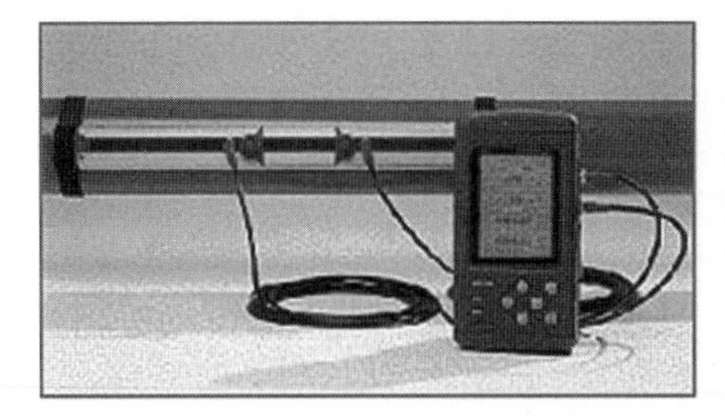

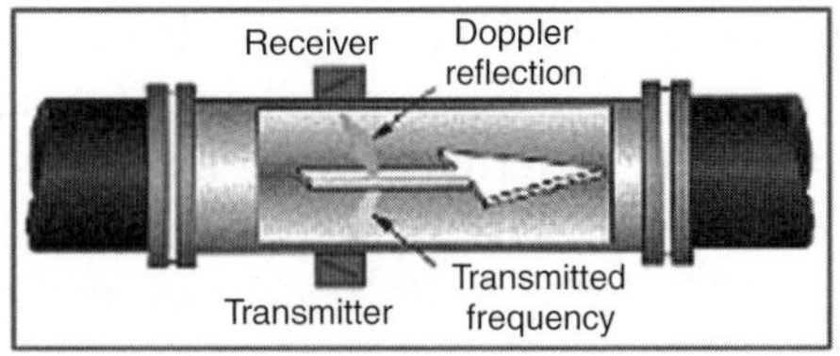

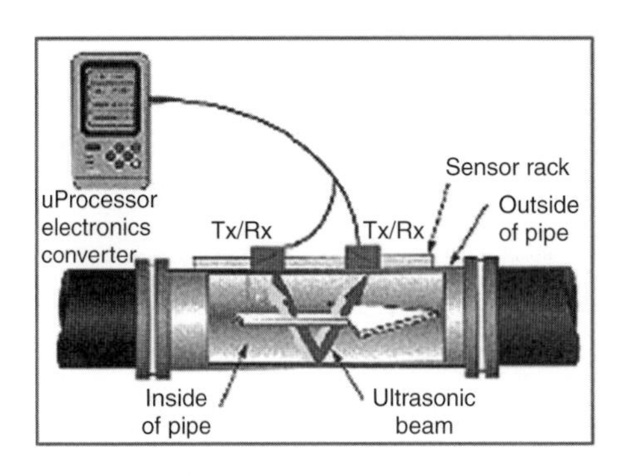

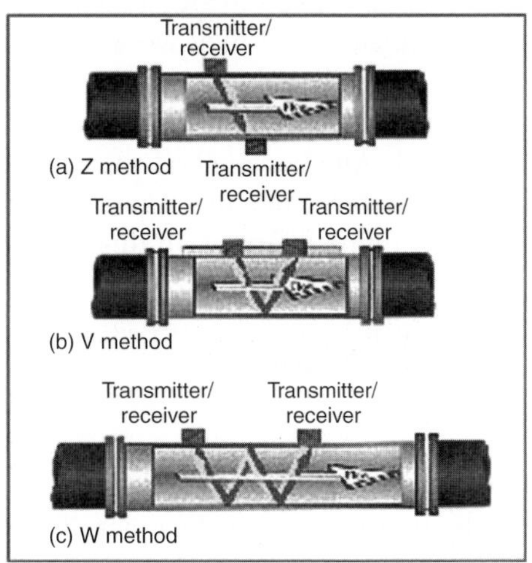

Figure 4.2. Ultrasonic flow meter. Top left: clamp-on meter with rail-mounted transducers; top right: principle of the Doppler method: the reflection of gaseous or solid particles in the liquid is used; bottom left: principle of the transit-time method: the travel time of ultrasound waves is measured in the direction of the flow and in the opposite direction; bottom right: possible positions of the transducers in the transit-time method. *Source*: From Ref. [5].

4.2.6. Other Meters

4.2.6.1. Parking Meters

Upgrading from mechanical to electronic battery-operated parking meters brings about some advantages, for example lower purchase and maintenance costs, higher accuracy (due to the quartz clock), no moving parts or exposed wires, easy reprogramming for time and rate changes.

These meters – see Figure 4.3a – usually run on a 9-V prismatic battery, either alkaline or lithium, the latter lasting 3–5 times longer. As average currents of only 40–60 μA are drawn, a Li battery can grant a runtime of $\sim$1 year. Of course, the runtime is a function of such factors as temperature, display backlighting, and use of cards or coins. The allowed temperature range is -30 to $+80$°C.

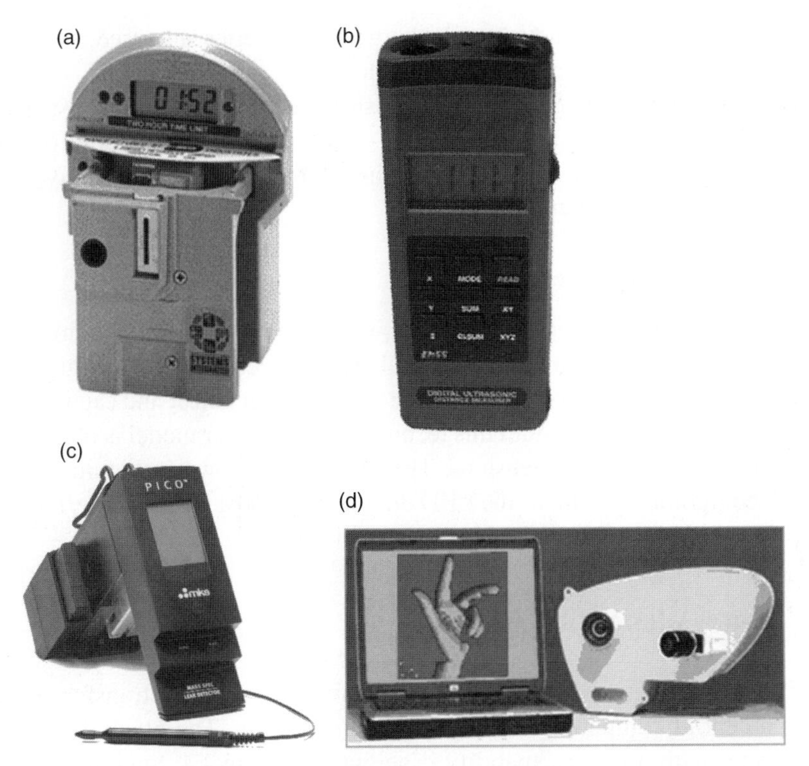

Figure 4.3. Examples of battery powered meters. (a) Parking meter; (b) distance meter with laser pointer; (c) sniffing He leak detector; (d) 3D metrology apparatus.

4.2.6.2. Distance Meters

Electronic distance meters (Figure 4.3b) use ultrasonic waves and allow calculations of length, area and volume. For increased accuracy, these handy devices have a laser pointer allowing to precisely identify the point from which the distance is measured. The power supply is a 9-V battery.

4.2.6.3. Gas Leak Detectors

Sophisticated portable leak detectors (Figure 4.3c) are now available, which are based on the detection of Helium by mass spectrometer-based designs. These meters can reveal holes or imperfections in products or systems. Measurements are usually performed by the vacuum or sniffing techniques; in both cases, trace amounts of He are sought with sensitivities of up to 10^{-11} atm cm^3/s. Typical applications include air-conditioning systems, chemical processes, power plants and the high-purity gas industry. Rechargeable batteries are used, with Li-ion batteries granting operation times of 2.5–3 h.

Simpler detectors of reduced dimensions use ultrasound waves: they are most sensitive to sounds around 40 kHz – well above the threshold of sounds audible by human ears – this excluding interferences by audible noises. Their power source is a 9-V battery.

4.2.6.4. 3-D Portable Metrology

Measuring the dimensions of objects can be done with portable instruments based on a high-speed camera and digital light processor (DLP). In this way, a series of patterns can be projected onto the surface of an object and captured in less than a second by a camera. With this technique, a computer model is obtained with fine details and accurate dimensions. The instrument shown in Figure 4.3d has a LED-based lighting system; as the LED does not require a high voltage, the system may be powered by a 12- or 20-V battery pack [4].

4.2.7. Meters with AMR Capability

Electronic meters of the AMR type can automatically transfer collected data to a central computer for analysing and/or billing. The block diagram of a utility meter with transfer capability is shown in Figure 4.4. Such meter is used for real-time acquisition of data concerning consumption, time-of-use and peak

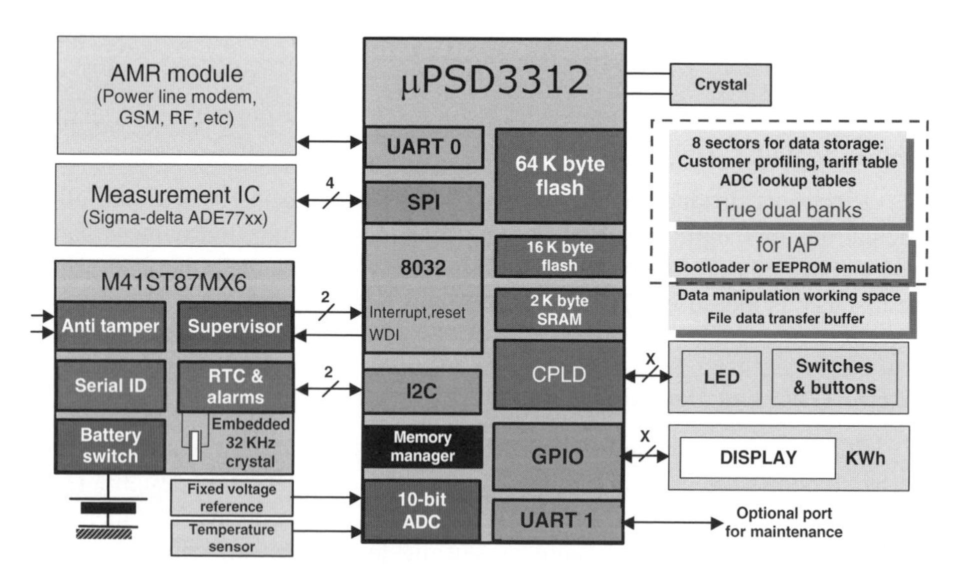

Figure 4.4. Block diagram of a meter with AMR capability.
Source: Courtesy of ST System Solutions.

demand of water, gas or power. Data transmission, especially if remote, is mainly based on radio frequencies (RFs), but a power line carrier or a wired telephone line can also be used. RF can be of the two- or one-way type: in the former, the transmitter also acts as a receiver and vice versa; in the latter, the meter can only transmit and the receiver can only read data. Common RF systems are (a) handheld, (b) mobile and (c) fixed network.

In a handheld system, the meter sends its data to a small computer carried by a reader walking by the meter location ("walk-by meter"). Mobile reading is done with the receiving computer installed in a vehicle ("drive-by meter"). In this case, the aid of a global positioning system (GPS) is often required for mapping and orientation. In the fixed-network AMR, an infrastructure receives data from the meter for further transmission to a central computer. There are basically two types of network configurations: star and mesh. The star network has a collector or repeater for receiving the meter's data: the collector receives and store data for processing; the repeater transmits data without storing it. The repeater is sometimes converted to a telephone or Internet protocol (IP) network. In the mesh network, the meters act as repeaters passing the data from one another until the collector is reached.

Internet-enabled AMR (especially for utilities) is schematically shown in Figure 4.5. The router enables the IP network to get in touch with local area networks. Wired (PSTN) and wireless (GSM) networks can also communicate with the Internet. Utility companies can have some advantages from Internet-enabled metering. In addition to data on consumption and demand, these meters can inform on such issues as leaking or tampering, can be remotely controlled/commanded, and can allow networking of appliances within the home, with the meter as a gateway.

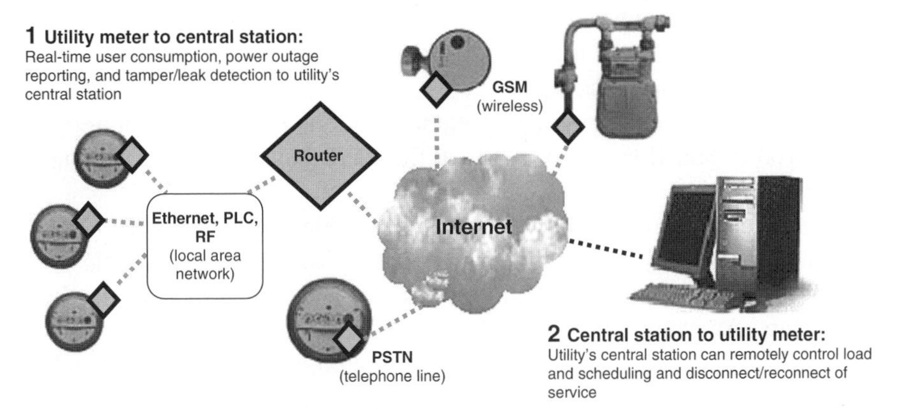

Figure 4.5. Automatic meter transmission using the Internet. PSTN, public switched telephone network; PLC, power line carrier.
Source: Courtesy of eDevice.

As mentioned in Section 4.2.3, the primary bobbin-type Li/$SOCl_2$ cell works fairly well in these applications. For instance, the transmitter units of a utility company's star fixed network use these batteries. They can last up to 20 years in utility products and can stand extreme temperatures (−40 to +85°C). However, the Li/$SOCl_2$ cell has a couple of limitations: it can hardly sustain the high current pulses needed in some transmission, and shows passivation phenomena on storage at high temperature. In these cases, the Li/$SOCl_2$-HLC hybrid is more conveniently used (see Section 4.2.3 [6]).

4.3. Data Loggers

Data loggers are processor-based devices recording data in connection with sensors or instruments that may be either built-in or external. These devices are often battery-powered to ensure portability and remote location use.

The basic features of a data logger may be summarized as follows:

- large memory to store data even for months while unattended; the memory may be battery-backed, sram, flash or eeprom
- ability to operate in remote locations and in harsh environmental conditions
- high-power efficiency to ensure a long battery life
- high reliability: unattended data loggers have to store data as long as their battery works – no failure has to occur.

Data loggers of last generation have wireless communication capabilities. They can also publish web pages and exchange files over networks using the IP. More generally, communications may occur through GSM, GPRS, SMS, Ethernet and low-power radio. Portable data loggers with a built-in GPRS modem can send e-mails with their data to the user.

Typical applications include weather station, hydrographical recording (water level, depth, flow, etc.), road traffic, environment, vehicle testing, gas pressure and factory processes.

Examples of portable data loggers are presented in Figure 4.6.

Data loggers may be networked with a LAN (Ethernet) or a wireless LAN (WLAN), and the data can be sent, via a router and an Internet connection, to remote users. An example of networking and remote transmission can be found in Figure 4.7: data loggers for temperature and humidity are connected and their data sent to a computer or a cellular phone. The type of remote transmission obviously depends on the distance between the logger(s) and the receiver. For large distances, as is the case of multinational companies, the worldwide Internet is used.

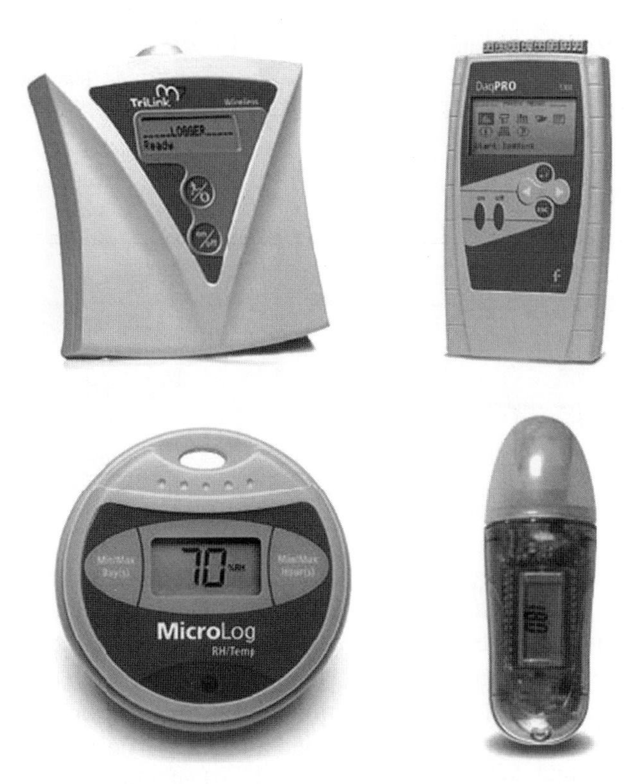

Figure 4.6. Examples of portable data loggers and their characteristics: 4-sensor, fast sampling rate, Bluetooth wireless communication (top left); 8-channel, graphic display and built-in analysis function, can work without connection to a computer (top right); compact, for temperature and humidity, can be connected (together with up to 200 loggers) to a PC via the cradle technology (RF transmission) (bottom left); small temperature logger (weight, 45 g), can work in the range of −40 to 80°C with a coin Li/MnO_2 battery lasting up to two years (bottom right).
Source: Courtesy of Fourier.

Several batteries can be used to power portable loggers, depending on their characteristics: primary Li, Li-ion, Ni–Cd or Pb-acid.

4.4. Sensors and Sensor Networks

Portable electric or electronic sensors encompass a wide range of types and applications. By the type, they can be broadly divided into chemical/biological or physical. The former class includes devices that can sense specific elements or molecules; the latter class includes a number of devices able to detect such physical parameters as temperature, pressure, sound, electricity, magnetism, motion, optical radiation, etc.

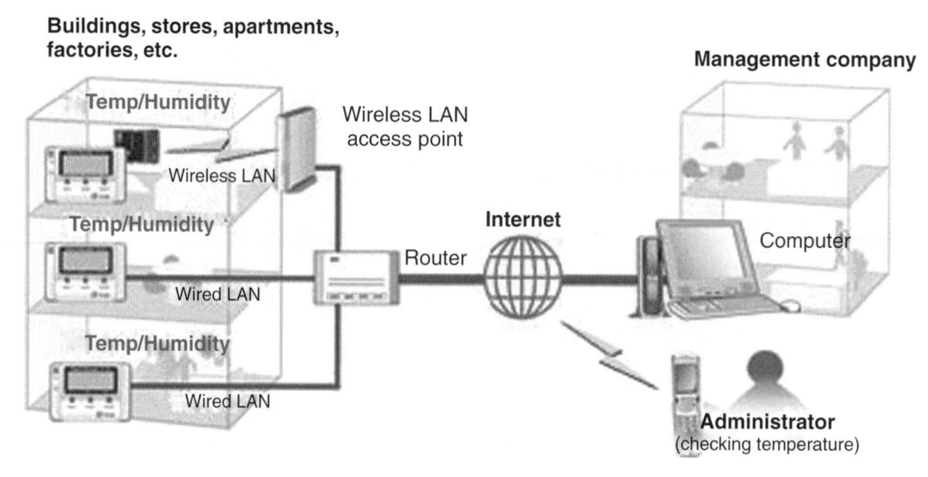

Figure 4.7. An example of data logger networking.

Fields of applications are medical, automotive, industrial processes, aerospatial, environmental, biological, etc. Table 4.1 summarizes measurement types and related measurable quantities.

Sensors may have a passive function, that is they can only provide indications of a given quantity, or may command an actuator doing some mechanical action, for example open/close valves or electrical circuits. Sensors may be produced on a microscale with the same technology used for MEMS (see later); these microsensors often have a higher sensitivity than their macro analogues. Examples of sensors of reduced dimensions are given in Figures 4.8 (sensors for physical quantities) and Figure 4.9 (sensors for medical care). The wearable

Table 4.1. Examples of parameters measurable by sensors.

Measurement Type	Measurable Quantities
Thermal	Temperature, heat, heat flow, entropy, heat capacity, etc.
Radiation	Gamma rays, X-rays, ultraviolet, visible and IR light, microwaves, radio waves, etc.
Mechanical	Displacement, velocity, acceleration, force, pressure, mass flow, acoustic wavelength and amplitude, etc.
Magnetic	Magnetic field, flux, magnetic moment, magnetization, magnetic permeability, etc.
Chemical	Humidity, pH, concentrations, toxic and flammable materials, pollutants, etc.
Biological	Sugars, proteins, hormones, antigens, etc.

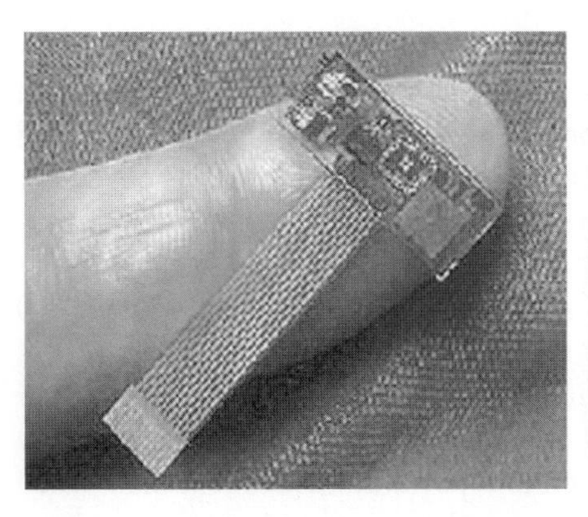

Figure 4.8. Examples of physical microsensors. Left: position sensor (*source*: From Ref. [7]); right: fluid monitor for *in situ* density and viscosity measurements (*source*: courtesy of Sandia National Laboratories).

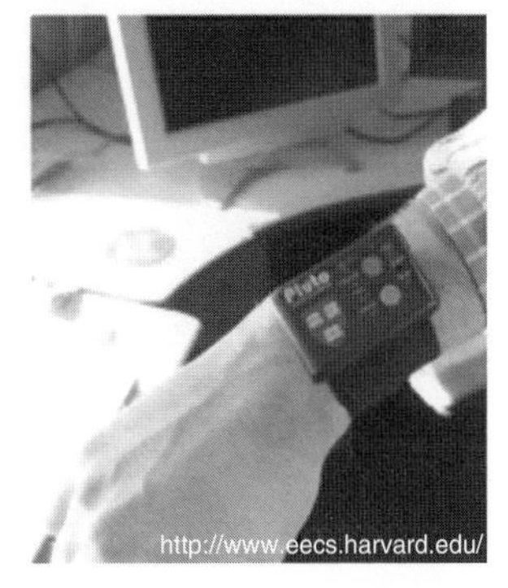

Figure 4.9. Examples of wireless vital sign sensors. Wearable pulse oximeter sensor (top left); wearable two-lead EKG (top right); wearable sensor for monitoring physical activity and motor functions (bottom left); same sensor with case and strapping (bottom right). *Source*: From Ref. [8].

pulse oximeter and the two-lead EKG of Figure 4.9 are part of a wireless sensor network (WSN) [8, 9]. They collect vital signs, such as heart rate, O_2 saturation and EKG, and relay them over a short-range ($\sim$100 m) WSN to receiving devices, for example PDAs, laptops or terminals on ambulances. Their power sources are two AA batteries with a lifetime of up to several months if programmed appropriately.

Within the network, these sensors are located in nodes (see Figures 4.10 and 4.11), which can be programmed to sample, transmit, filter or process data [8, 10]. The sensor for motor functions of Figure 4.9 is powered by a very small Li-ion battery embedded in the chip. The wearable oximeter can be compared with that of Figure 3.24, which is a portable instrument also based on a microcontroller, but is neither miniaturized nor inserted in a network.

In general, sensor networks are dense wireless networks of small, low-cost sensors, which collect data in a given area and transmit it to a receiving device. The nodes of a WSN have various energy and computational constraints, because of their inexpensive nature, and ad hoc method of deployment. Considerable research has been devoted at overcoming these limitations (see later) [11]. A node is formed by these basic components: sensor(s), mote, power source, and packaging (see Figure 4.12 [12]). The mote is the brain of the node, hosting a microprocessor (microcontroller), memory, ADC and a radio-frequency transceiver or other communication device. The block diagram of a sensor node is shown in Figure 4.13 [13]. Using small microcontrollers increases energy efficiency, allowing them to operate at a power of $\sim$1 mW, while running at a frequency of $\sim$10 MHz. As the standby power can be $\sim$1 μW, if the microcontroller is active for a limited time, its average power

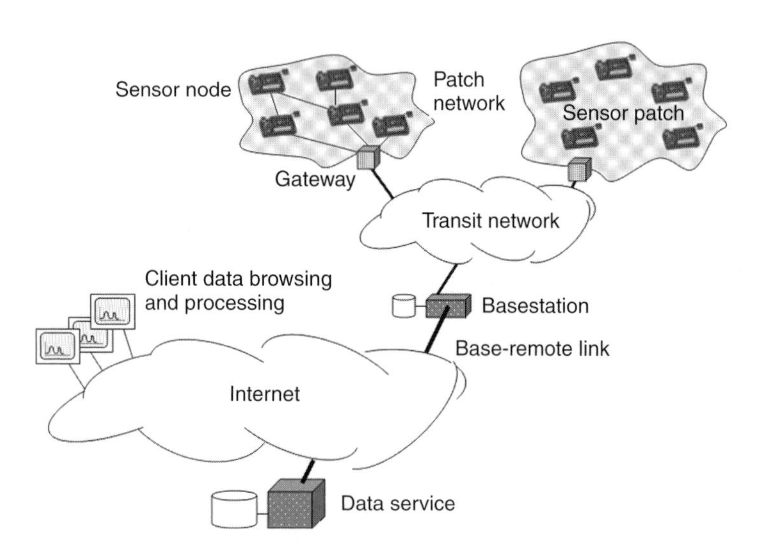

Figure 4.10. Architecture of a sensor network for habitat monitoring – see text. *Source*: From Ref. [10].

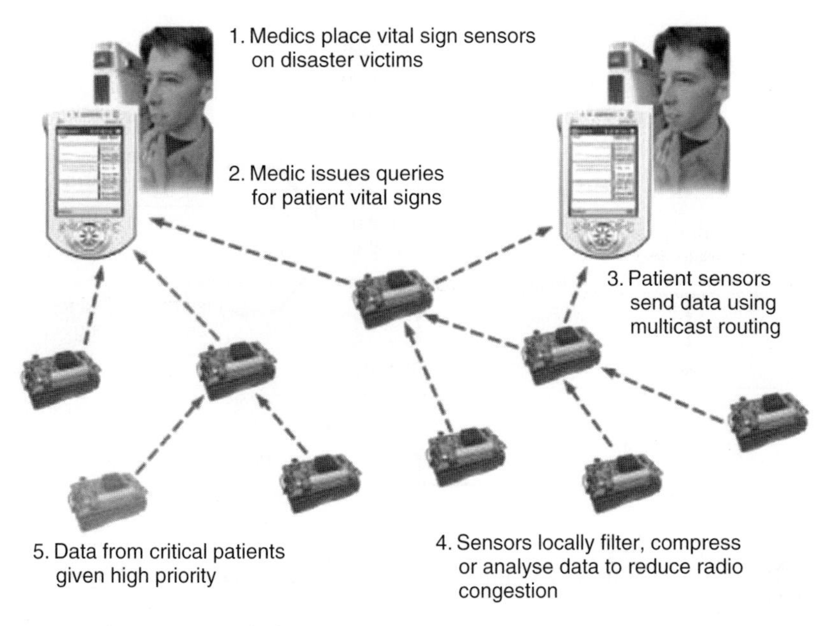

Figure 4.11. Sensor network for emergency response.
Source: From Ref. [8].

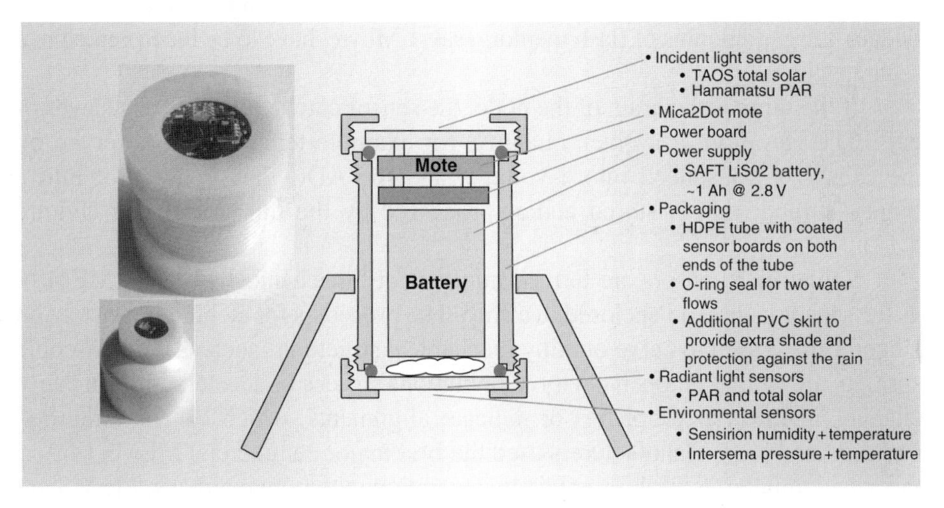

Figure 4.12. A wireless sensor node for environmental monitoring – see text.
Source: From Ref. [12].

consumption is just a few microwatts [12]. A 1-cm^3 battery can have a capacity of 1 Ah, so centimeter-scale devices can run almost indefinitely in many environments. However, low-power microprocessors have limited storage ability, typically <10 kbytes of RAM for data and <100 kbytes of ROM for program storage. This

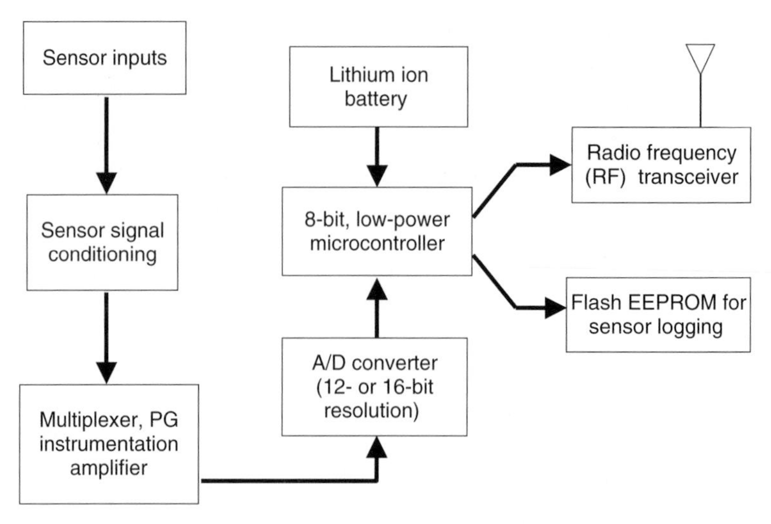

Figure 4.13. Block diagram of a wireless sensing node with data logging and RF communication capabilities. The microcontroller manages power used by the sensors and responds to commands from the central station.
Source: From Ref. [13].

limited amount of memory consumes most of the chip area and much of the power budget. Larger amounts of flash memory, say 1 Mbyte, have to be incorporated in a separate chip.

If the mote is the brain of the node, the sensor can be regarded as its eyes or ears. The variation of a given quantity, for example temperature, observed by the sensor is transduced into a voltage, and the ADC then converts it into a binary number that is stored and/or processed by the microcontroller (Figure 4.13).

Some microsensors are termed micro-electromechanical systems (MEMS), as the sensor is directly enclosed in a chip. The processes for etching transistors on silicon can be used to carve out tiny mechanical structures, such as a microscopic springboard within an open cavity. Gravitational forces or acceleration can cause changes in material properties or delicate alignments, which can be amplified and digitized [12]. Manufacturers used the first major commercial MEMS sensor, the accelerometer, to replace previous-generation automotive airbags. MEMS are also used, for instance, in heating and cooling systems to improve energy savings. In addition to sensors, MEMS can have other mechanical elements, the actuators, which can be thought of as the device's arms, as they can do such actions as operating switches or valves.

More details on MEMS can be found in Section 4.14.

The sensor nodes perform general-purpose computing and networking in addition to application-specific sensing. They may be deployed in dense patches (see Figure 4.10) and transmit data through the sensor network to the gateway. This transmission is one of the most energy-consuming operations, thus many WSNs have the capability of processing data within the network. The gateway is responsible for transmitting data from the sensor patch through a local transit network to the remote base station that provides WLAN connectivity and data logging. The base station connects to database replicas across the Internet. Finally, the data are accessed by users again through the Internet [10].

The deployment of sensors in a network for emergency care is shown in Figure 4.11. These sensors have the ability to filter, compress or analyse data so to reduce radio transmissions. Data from critical patients (see light-grey device on the left-low corner) are sent with high priority to the receiving device(s) [8].

The uses of a WSN can be roughly differentiated into:

- monitoring space
- monitoring persons, animals, plants and things
- monitoring the interactions of the above subjects with each other and the surrounding space.

The first category includes environmental and habitat monitoring, precision agriculture, indoor climate control, surveillance and intelligent alarms. The second includes structural monitoring, ecophysiology, condition-based equipment maintenance, medical diagnostics and urban terrain mapping. Particularly interesting applications are those monitoring complex interactions, including wildlife habitats, disaster management, emergency response, asset tracking, healthcare and manufacturing process flow [12].

Sensor networks pose a number of unique technical challenges due to the following factors [11].

- *Ad hoc deployment.* Most sensor nodes are deployed in regions devoid of infrastructures. A typical way of deployment in a forest would be tossing the sensor nodes from an airplane. In such a situation, it is up to the nodes identifying connectivity and distribution.
- *Unattended operation.* In most cases, sensor networks have no human intervention. Hence, the nodes themselves are responsible for reconfiguration in case of any changes.
- *Untethering.* The sensor nodes are not connected to any external energy source. There is only a finite source of energy, which must be optimally used for processing and communication. Thus, as anticipated, communication should be minimized as much as possible to save energy.
- *Dynamic changes.* It is required that a WSN be adaptable to changing connectivity (due to addition of more nodes, failure of nodes, etc.) as well as changing environmental stimuli.

Solving the above problems means acting in the following areas [11].

- *Energy efficiency*. This is a dominant consideration, due to the limited source of energy allotted to the nodes. Many solutions, both hardware- and software-related, have been proposed to optimize energy usage.
- *Localization*. In most of the cases, sensor nodes are deployed in an ad hoc manner, so the nodes must identify themselves in a given space. This problem is referred to as localization.
- *Routing*. Communication costs play a great role in deciding the routing technique to be used. Traditional routing schemes are no longer useful because energy considerations demand that only essential minimal routing be done.
- *Energy optimization*. In a WSN can be accomplished by having energy awareness both during design and operation.

A sensor node usually consists of four subsystems, each of them allowing energy saving if properly managed:

- *Sensing device*. It consists of a group of sensors and actuators (in the simplest case, only one sensor). Energy consumption can be reduced by using low-power components and saving power at the cost of performance which is not required.
- *Power supply*. With a few exceptions (solar or piezoelectric energy), it consists of a battery which supplies power to the node (see Figure 4.12). The lifetime of a battery can be increased by proper management or switching the battery off when not in use.
- *Computing*. It is done by the microcontroller unit (MCU), which is responsible for the control of the sensors and execution of communication protocols. MCUs usually operate under various modes for power management purposes, but shuttling between these operating modes involves consumption of power, and this should be considered to increase the battery lifetime of each node.
- *Communication*. It is usually done by a short-range radio to get in touch with neighbouring nodes and the outside world. Radios can operate under the transmit, receive, idle and sleep modes. It is important to shut down the radio when it is not transmitting/receiving, because of the high power consumed in idle mode.

The nodes of a WSN are often randomly deployed – as in the case of deployment from an airplane – and assessing their coordinates becomes necessary. GPS can hardly be used in these cases, and most of the localization techniques are based on recursive tri- or multi-lateration [14]. The network can be organized as a hierarchy, with the nodes in the upper level being more complex and already knowing their location through some technique. These nodes then act as beacons by transmitting their position periodically; nodes that have not yet found their position listen to broadcasts from these beacons and use the information to calculate their own position. More information on localization techniques can be found in Ref. [11].

The energy constraints of WSNs pose severe limitations to routing protocols. In the conventional ones, a node stores the data it receives and then transmits it to all its neighbours. In this way, a given node may receive the same information from all its neighbours, this being only a waste of energy. Efficient routing protocols for WSNs can be proactive or reactive. The former try to maintain updated information among all nodes with one or more routing tables (files) containing the topology of the network immediately around a given node. In the latter protocols, routes are only created when they are needed.

A battery-aware routing protocol for streaming data transmission has recently been developed. It is based on a model allowing on-line battery capacity computation, this requiring a knowledge of the battery discharge behaviour [15]. This protocol avoids the disadvantage of previous protocols, that is the use of all the battery energy in some nodes. As a consequence of this, the routing path is to be switched to another node whose battery is still charged, this leading to delayed transmissions. Using the protocol of Ref. [15], the power consumption is more uniformly distributed. Simulations have shown that average battery energy on a node can be increased by up to 43%.

As is obvious, having an efficient battery and drawing most of its theoretical capacity is the basis for a long runtime of sensors and nets. The use of alkaline and Li-ion batteries is preferred. Other batteries include primary Li [16] and rechargeable Ni–Zn [17, 18]. Preferred primary Li batteries are based on the Li/$SOCl_2$ couple or on the hybrid obtained by adding a capacitor (HLC) to the above couple [3]. They can deliver high current pulses for a few seconds in a broad temperature range, that is 40–85°C.

Ni–Zn batteries seem especially suitable for wireless microsensors operating in remote areas and requiring high power. The need of building miniature systems involves using very small batteries that can be recharged by solar cells. Such batteries must have the dimensions of sensors and associated electronics (area: few mm^2, thickness 0.2–0.3 mm), and are constructed with the same technique used for MEMS and integrated circuits [18].

A recent investigation has evaluated the impact of key network design and environmental parameters on battery performance [19]. A sensor node containing the MICA2DOT mote and a coin-size Li battery was chosen to carry out this study. Battery life was found to depend on four parameters: (1) transmission power level; (2) sampling interval; (3) transmission time and (4) ambient temperature. Combining points 2 and 3 so to have a tougher duty cycle may reduce the energy delivered by a factor of up to 4; increasing the

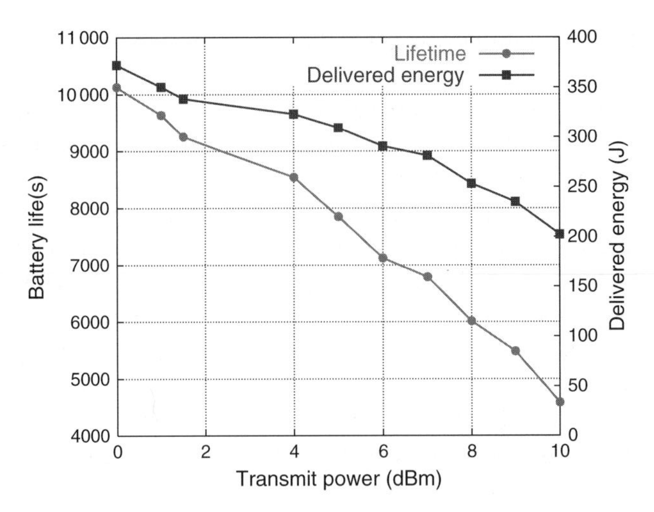

Figure 4.14. Effect of the transmission power level of a node on lifetime and energy of the battery powering the node. 0 dBm = 1 mW, 10 dBm = 10 mW.
Source: From Ref. [19].

temperature in the range −10 to 60°C significantly increases the energy delivered but also the self-discharge rate; finally, increasing the transmission power level may greatly reduce the amount of data sensed and transmitted before the battery is fully discharged. This last effect is shown in Figure 4.14: both the battery lifetime and the energy it can deliver rapidly decrease as the transmission power grows. For a power increase from 1 to 10 mW, the lifetime is reduced by 52%.

4.5. Alarms and Security Systems

Systems of this type encompass a wide application range [1]:

- Video surveillance
- Alarms
- Access control
- Remote level control
- Power line surveillance
- Pipeline inspection gauges
- Automatic barriers.

These systems are briefly described in the following, with indication of the type of transmission used to convey the information to a central receiving unit.

4.5.1. Portable Video Surveillance

Wireless video surveillance systems are used to transmit signals from video camera(s) to a receiving unit. The signals may be of the analog type (at radio-, video- or micro-frequency) or the digital IP network may be used. The simplest system is the one using a wireless video camera outside a building, sending radio signals to an internal receiver connected to a closed-circuit TV (CCTV). In this system, the distance between camera and receiver is up to 100 m at the common frequency of 2.4 GHz; however, the presence of walls and ceilings may reduce the distance to 50 m. Cameras with built-in video transmitters may reach a distance of 15 km.

Wireless systems of this type may have up to four cameras (four channels) per site; however, to limit interferences, typical of wireless systems, a maximum of three cameras per site is recommended. It is to be noted that wired systems can have up to 100 cameras per site and allow a resolution of up to 600 TV lines, whereas the wireless ones have a resolution limited to 420 TV lines. Their power sources greatly vary with duty: primary AA, D or 9 V batteries (including lithium) and rechargeable Ni-MH, Pb-acid or Li-ion batteries may be used. Cameras may have dimensions as small as pinholes. An example is shown in Figure 4.15, whose caption reports some specifications.

The more complex video security over IP (VSIP) is shown in Figure 4.16 with its basic components [20]. In this technique, digital technology allows audio/video data compression to reduce transmission bandwidth and memory space. Furthermore, coaxial cables are eliminated, as the cameras operate on digital networks by leveraging programmable DSP solutions. Thanks to the flexibility of DSP, compression standards may be changed, specific processing capabilities may be added, and different products may be developed on the same

Figure 4.15. Colour pinhole video camera. Dimensions: 8 × 8 mm (W × H); image sensor: ¼″ CMOS camera; resolution: 380 TV lines; 6–10 V battery; max. current: 35 mA.

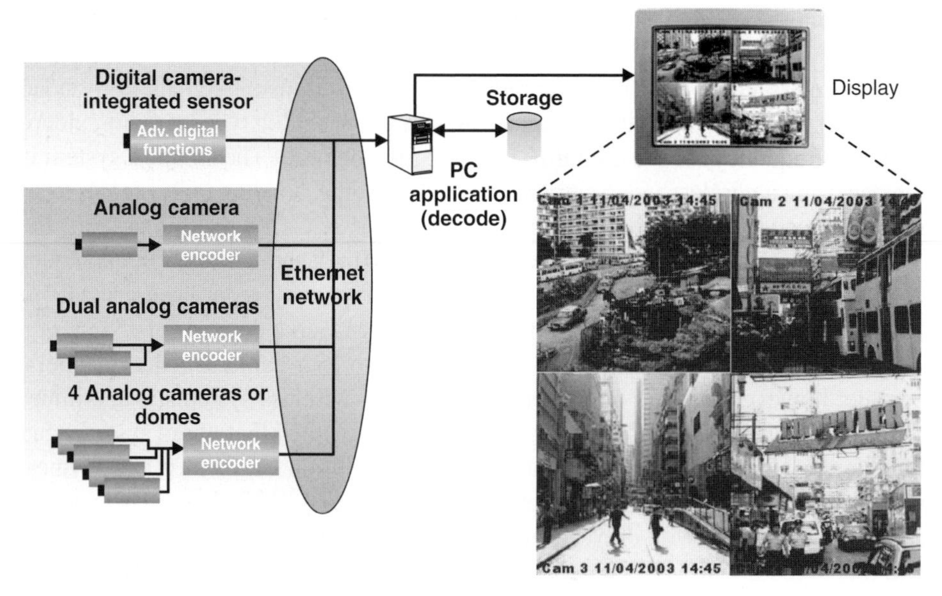

Figure 4.16. VSIP with main components.
Source: From Ref. [20].

hardware platform. The advanced digital functionality afforded by this technology may be incorporated directly into a standalone IP camera or into a network encoder that supplies digital intelligence for multiple standard analog surveillance cameras [20]. VSIP systems may be of the plug-in type with a backup battery, or may be made portable with a battery as the primary power source.

4.5.2. Wireless Alarms

Different events can be detected by alarms containing specific sensors connected via radio communication to central units: motion (of burglars), movement (of windows or doors), smoke or heat emission, leakage (of water or gas). Given the variety of types, the technical specifications and the power sources of these systems largely vary.

The operating currents are in the range 20–150 mA, the temperature may be as low as −40°C or as high as +70°C for outdoor applications, the runtime may be from 1 to 20 years.

These systems rely upon either alkaline or primary Li cells. The 9 V prismatic size is predominant in the alkaline family, while small cylindrical cells dominate the Li systems, with Li/MnO_2 and $Li/SOCl_2$ prevailing. The low self-discharge and the wide-operating temperature range of Li cells explain their success in an application where a 20-year life is now often required [1].

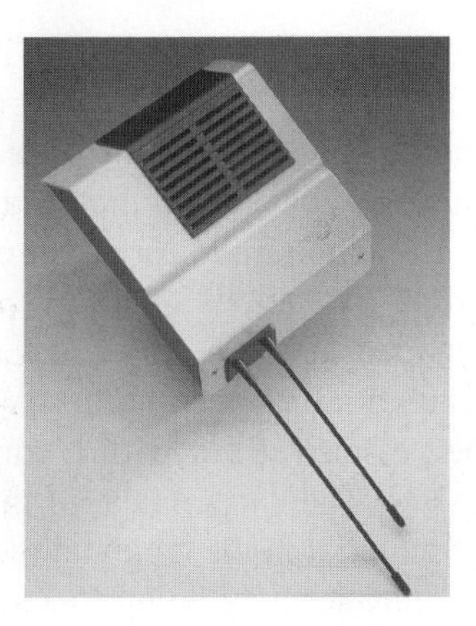

Figure 4.17. Central unit of a wireless alarm system.
Source: From Ref. [1].

The central unit is the brain of the whole alarm system, receiving information and, then, initiating actions (e.g. phone calls and siren actuation). An example of central unit is shown in Figure 4.17. High-end units have separate zones for every sensor, and trouble indicators (e.g. low battery). Their technical specifications include pulses from a few tens of mA to 2 A, operating temperature in the range 0/+50°C, runtime from 1 to 10 years. Alkaline or primary Li cells are used, the former being preferred for consumer alarm systems, the latter for industrial customers.

Alarm sirens may be installed inside or outside (bank outlets, factories, shops, museums, etc.). When in operation, they have to sustain current pulses of some amperes for several seconds or even minutes. Furthermore, a runtime of 5–10 years is needed. These requirements, in possible severe outdoor temperature conditions, have imposed the selection of high-rate primary lithium (e.g. two $Li/SOCl_2$ cells in series) in most alarm sirens used to protect commercial and industrial buildings. However, some manufacturers claim a 5-year operating time for their alkaline battery packs [1].

4.5.3. Remote Level Control

These systems can monitor the level of fluids or solid materials in reservoirs or tanks, for example gasoline in tanks and garbage in city bins [1].

Different types of sensors can be used to monitor the amount of fluid remaining in a tank. Pressure sensors are installed at the bottom of the fuel tank and transmit data, for example to a GSM box connected with a cable. This box transforms the data into SMS messages, which are received by a base station, where a computer estimates the quantity of fluid left. When the alert limit is reached, the tank is refilled.

In another system, the sensor (or tank unit – see Figure 4.18) transmits an ultrasonic sound wave that echoes back from the fluid surface. This sound wave echo is then converted into a height and sent to the indoor bench unit (Figure 4.18), which displays a height indication.

The capacitance measured at radio frequency may also be exploited to measure levels remotely and with high resolution. A probe mounted in a vessel forms a capacitor with the vessel wall, or with a concentric shield around the probe for non-metallic vessels. The capacitance of this arrangement is directly proportional to the level of material in the tank and may be measured to provide signals and controls. The controller may be up to 1 km away from the probe.

Technical specifications include pulses from a few tens of mA up to a few amperes, operating temperature in the range −40/+65°C, runtime of 2–5 years (sometimes 10).

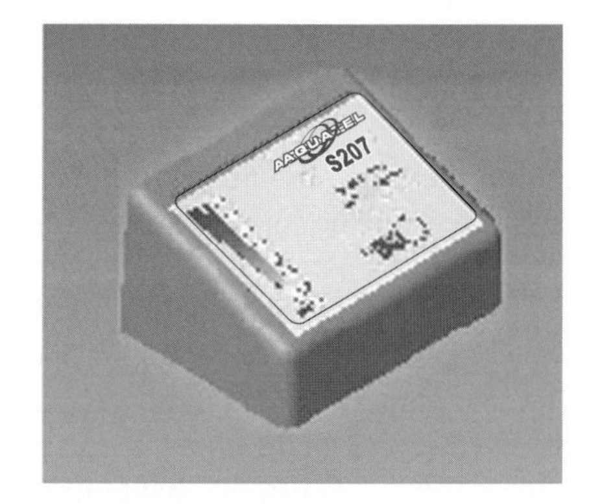

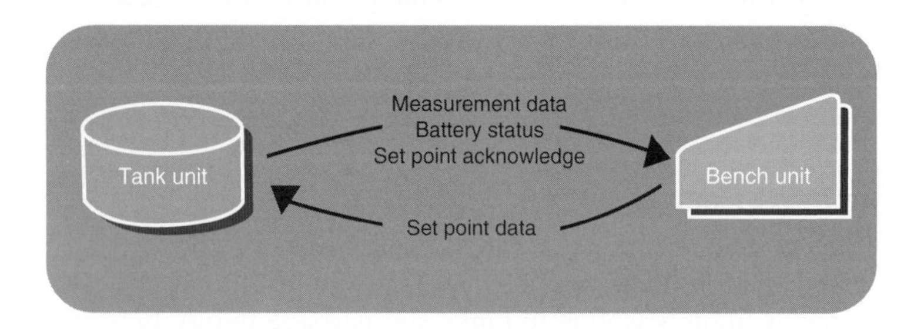

Figure 4.18. Fluid level control in a tank using the ultrasound method. Tank unit (top left); interior bench unit (top right); two-way communication (bottom). Maximum distance: 300 m; the system is powered by eight AA-alkaline batteries.
Source: Courtesy of Aquatel.

Older systems did not require high pulses and could use packs made with small, low rate, primary Li cells (Li/MnO_2 and Li/$SOCl_2$). Recent systems require the use of powerful packs made with the assembly of high-rate Li/$SOCl_2$ cells or with their low-rate counterparts plus a capacitor in parallel. For cost-saving reasons, packs made with alkaline cells are currently considered as well. A more complex system, as the one based on capacitance measurements, requires a 24 V lead-acid battery.

4.5.4. Power Line Surveillance

As outages can happen anywhere along power lines, their location must be detected rapidly. To do so, detectors fitted with flashing lights have been installed on power lines in several countries.

Technical specifications include: pulses of <500 mA, operating temperature in the range −40/+70°C, runtime of less than 1000 h, lifetime of 5–15 years.

Long operating lives and severe outdoor conditions recommend the use of 3.0 V or, preferably, 3.6 V primary Li cells of small or large size. When large capacities are needed, low-rate R20 Li/$SOCl_2$ cells, assembled in series and/or parallel, are preferred because of their high capacity (up to 19 Ah). A capacitor in parallel with the battery pack is often used to deliver current peaks [1].

The use of the so-called radio backscatter technology is currently under investigation [21]. It is based on battery-powered sensors, deployed along power lines, which do not transmit or receive radio signals, but reflect amplitude modulated (AM) sidebands. As these sensors do not require radio electronics circuitry, such as oscillators, mixers and amplifiers, their power consumption is limited, so that Li-coin cells can be used.

4.5.5. Pipeline Inspection Gauges (PIGs)

Although the most common tasks required to pipeline inspection gauges (PIGs) are geometry/diameter measurement and metal loss/corrosion detection, the information that can be provided by these devices covers a much wider range of inspection and troubleshooting needs, which include:

- Diameter/geometry measurements
- Curvature monitoring
- Pipeline profile
- Temperature/pressure recording
- Metal-loss/corrosion detection
- Photographic inspection

- Crack detection
- Wax deposition measurement
- Leak detection
- Bend measurement
- Product sampling
- Mapping

Intelligent (or smart) PIGs make use of different sensors and record the data, which is later analysed. These PIGs use such technologies as magnetic flux leakage (MFL) and ultrasound waves.

Depending on the size of the pipeline, the gauges can be very small or rather huge, like the one shown in Figure 4.19, which uses ultrasounds.

Technical specifications usually include:

- pulses: <10 A
- operating temperature: $-30/+90$°C
- runtime: <200 h.

Apart from some remote pipeline monitoring systems, which use solar energy in hazardous areas, large battery packs made with high-rate Li/$SOCl_2$ or Li/MnO_2 cells are preferred in this application. Superior drain capability, high operating voltage and wide temperature range explain this choice. In Figure 4.19, packs based on Li/MnO_2 are shown. For very demanding applications, they may contain up to 140 D-size cells for a total energy of 4400 Wh. To make this pack, 16 strings are connected in parallel: the pack uses one string until it can deliver energy, then it goes on using the remaining ones. Each string is endowed with a protection circuit that allows pack's operation in case of a string failure.

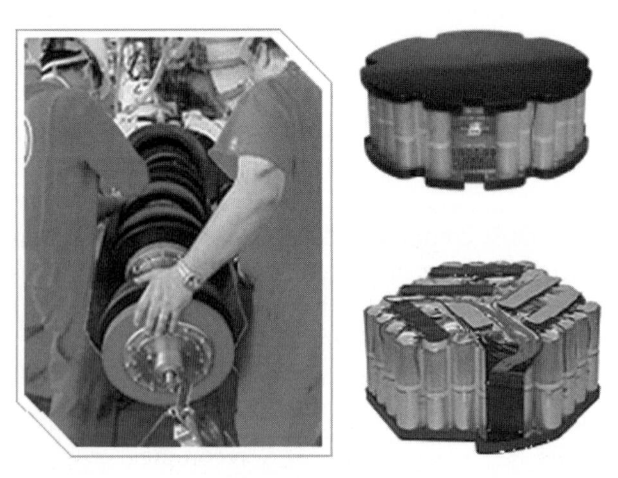

Figure 4.19. Left: an ultrasonic pipeline inspection gauge (PIG) (*source*: courtesy of Dacon); right: types of battery packs used in PIGs (*source*: courtesy of Ultralife).

4.5.6. Access Control Systems

Door locks, hotel safes, electronic keys, automatic blocking systems, ticketing equipment, etc., are examples of access control systems using batteries [1].

Being the applications so diverse, different power sources may be required. Alkaline cells are used in cheap applications and when easy replacement is possible. Primary Li-coin cells are used for backup memories – for example in safes – where μA currents are delivered for a long time. Cylindrical Li/MnO_2 or a low-rate $Li/SOCl_2$ cell, with a capacitor in parallel to deliver pulses (up to 3 A), actuate some locks. Prismatic 9 V Li/MnO_2 cells are preferentially used in digital locks. Runtimes of these cells are in the range 1–10 years.

Lead-acid batteries (6 or 12 V) are preferred for automatic barriers, for example for parking lots, which have to rely on backup batteries in case of main power outages. In a private parking, space can be reserved with the use of the so-called parking guard. When a car is not occupying the parking place, a 12-V battery-driven motor raises the vertical side of the barrier shown in Figure 4.20 The lead-acid battery ensures a power of 60 W at 1.6 A.

4.6. Automatic Assistance Systems

4.6.1. Emergency Lights

Self-contained emergency lights, that is those coming on in case of power outage, rely on batteries as power sources to comply with the safety requirements. The lights consist of one or more incandescent bulbs or one or more clusters of high-intensity LED.

Figure 4.20. Parking guard to reserve a private space.

Technical specifications include the following:

- lifetime: 4 years minimum in operating conditions
- operating temperature: from −20 to +60°C
- runtime: the light must provide lighting during 1–3 h, that is long enough for people to leave a building and for rescue teams to intervene
- trickle chargeable power source (that is batteries able to stand a permanent overcharge), which must be ready to operate at any time; permanent charge rate from C/20 to C/15; constant rapid recharge rate: C/10 maximum.

More than 30 years of field experience have demonstrated the reliability and reproducibility of Ni–Cd batteries. This explains why, today, these batteries are chosen for this application. Sintered or plastic-bonded positive electrode, and plastic bonded negative electrode technologies are used in the design of these cells (see Section 2.3.4.2). Their voltages are in the range 2.4–12 V, and their capacities in the range 4.0–4.5 Ah.

Ni-MH batteries have been tested for use in compact self-contained lights, in order to take advantage of their superior nominal capacity. This permits the use of slimmer battery packs than those made with Ni–Cd cells. However, some current technical drawbacks have limited their use:

- high-temperature conditions (above +40°C) shorten the life of these batteries, due to degradation of the metal hydride electrode
- the permanent overcharge situation, generally met in emergency lights, results in a temperature increase that reduces battery life
- the cost of the alloy used for the negative electrode is higher than that of Cd in Ni–Cd cells, this having a negative impact on the overall cell cost.

The higher cost of Li-ion batteries, in comparison with both Ni–Cd and Ni-MH, limits their use in this application, although their intrinsic performance could satisfy the emergency light requirements.

4.6.2. Beacons

Many automatic assistance systems are now available, which exploit satellite communications. Their signals can be sent by [1]:

- Emergency position indicating radio beacons (EPIRB); they are marine safety devices which are actuated in case of problems at sea

- Submarine personal locator beacons (SPLB)
- Emergency locator transmitters (ELT): for aircraft
- Personal locator beacons (PLB): for terrestrial use.

Telematics systems are described in the next Section.

Among the many types of beacons (this term includes any signalling device), the most famous are Argos and SARSAT/COSPAS (see List of Acronyms).

Argos beacons are not occasionally activated but emit on a regular time basis for continuous tracking. They are mainly utilized for environmental measurements (sea levels, stream speed and direction, surface and underwater equipment tracking, volumes of fish, pollution, etc.), but thousands of beacons and a network of satellites can also prove useful to locate sailors in distress. See more on the Argos system in Section 4.8.5.

SARSAT/COSPAS beacons, which can be transported anywhere thanks to their limited dimensions, are kept in standby mode. As soon as a beacon is activated, it transmits signals that, through a network of satellites, are received by a terrestrial base station. The position of the beacon is calculated and automatically sent to one of the control and rescue organizing sites located all over the world. More than 600 000 SARSAT-COSPAS beacons are used worldwide, most of them on-board boats.

Cellular phone networks are also used sometimes in case of emergency (for instance to locate stranded mountain trekkers) thanks to their location capability.

The SARSAT/COSPAS system includes ELTs (121.5 and 406 MHz), EPIRBs (121.5 and 406 MHz) and PLBs (406 MHz). The SARSAT/COSPAS beacons are often referred to as the "406 MHz distress beacons" (Figure 4.21) [1]. In the same figure, a complete search and rescue system based on these beacons is shown. The system utilizes two types of satellites: in low earth orbit (LEO) (LEOSAR system) and in GEO (GEOSAR system) (see more detail on LEO and GEO satellites in Section 4.9.2). The GEOSAR and LEOSAR capabilities are complementary: the former can provide almost immediate alerting if the signal is within its coverage; the latter provides coverage of the polar regions (not covered by GEOSAR), calculates the location using Doppler techniques, and is less susceptible to obstructions blocking the signal because the satellite is continuously moving vs the beacon. The alert from the satellites is sent to a local user terminal, which passes the information to a mission control centre that activates the rescue centre (see Figure 4.21).

Beacons equipped with GPS allow a more precise location of the signals and shorten the response time for rescuing.

Technical specifications of SARSAT/COSPAS beacons include:

- base current in active mode: <100 mA
- pulses: 0.5–2 A

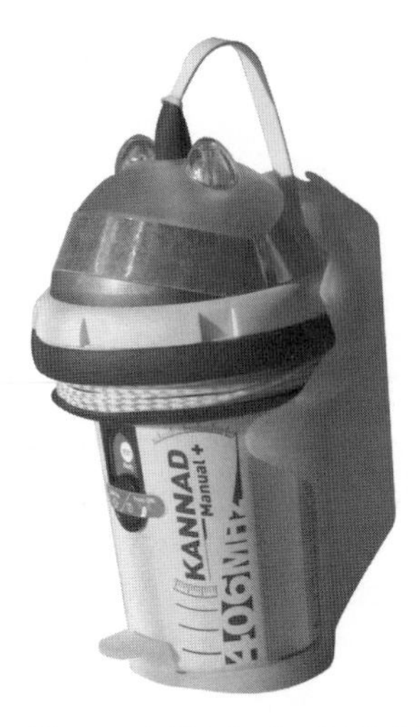

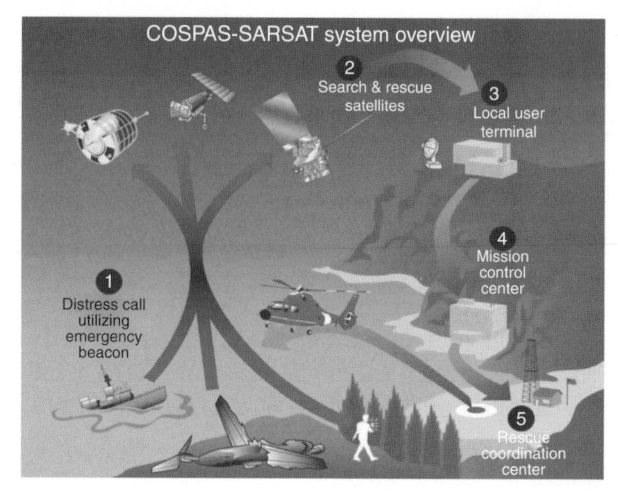

Figure 4.21. Left: SARSAT/COSPAS beacon. This model has a GPS receiver, in addition to the 406 MHz transmitter, for higher precision and faster detection. Weight: 1 kg; dimensions (without fixing bracket): diameter 14 cm, height 38 cm (with antenna deployed). Li/MnO_2 battery, 6 years of operation. Right: pictorial view of a search and rescue system using these beacons.
Source: Courtesy of NOAA (US National Oceanic and Atmospheric Organization).

- operating temperature: −20/+55°C (class 1) and −40/+55°C (class 2)
- runtime: 4 years of standby plus 48 h of transmission.

The success of primary batteries in these devices is due to their superior ability to stand long storage times, while remaining able to operate in a wide temperature range.

All the main Li systems are used (Li/MnO_2, Li/SO_2, $Li/SOCl_2$), with R14 and R20 high-rate cells assembled in battery packs of different sizes (from 2 cells up to 10).

Li/MnO_2 offers the advantage of a better resistance to Li passivation, while Li/SO_2 cells offer higher drain capabilities at very low temperatures. The higher operating voltage of $Li/SOCl_2$ cells allows using a lower number of cells to build the pack.

4.6.3. Automatic Crash Notification

Automatic crash notification (ACN) systems use wireless communications and GPS technologies to automatically notify an emergency medical service after an accident and give the rescue team its exact location. The system is activated when an airbag is deployed or an in-car button located on the dashboard is pressed. In its more advanced form, utilizing a group of sensors, advanced automatic crash notification (AACN) automatically calls a telematics service centre if the vehicle is involved in a crash, regardless of airbag deployment [22]. In Figure 4.22, a scheme of the basic components of this system is shown.

ACN is an example of vehicle telematics, which includes a number of other services, for example toll collection, fleet vehicle locations, stolen vehicle recovery, location-driven information, car accident prevention and vehicle's remote diagnostics. Future applications include on-demand navigation and on-board audio-visual entertainment. The heart of the telematics system in a vehicle is the telematics communication unit (TCU). It communicates location-specific information to a service centre, which provides assistance to a driver via cellular phone.

Most of the current assistance systems use the vehicle battery as the power source. However, in case of accident, this battery can be damaged and the communication interrupted. A backup battery may prove essential to maintain the communication; it has to provide an immediate high power of about 20 W for 5–15 min, operate in a wide temperature range (−40 to 75°C), be safe, be very reliable in harsh environmental conditions, and have a 10-year shelf life. Currently, only a few telematics systems offer a battery backup in the event that the main battery is disconnected or damaged during a crash. Quite possibly, backup batteries will be available in the next few years, once ACN systems become mandatory. Two different chemistries are designed for this application: Li/MnO_2 and $Li/SOCl_2$-HLC. Both can manage the high current pulses required soon after the crash.

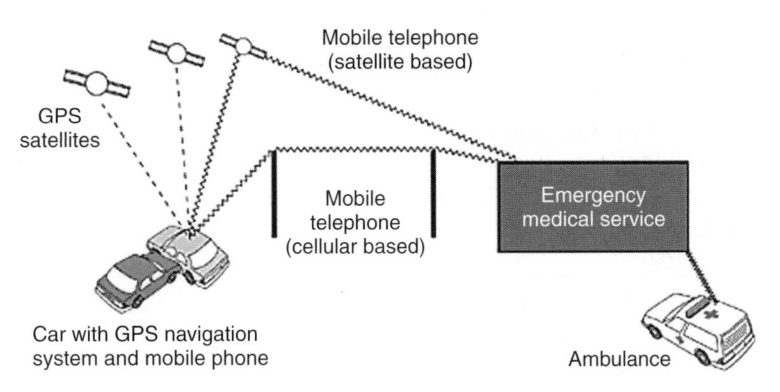

Figure 4.22. Car(s) with mobile phone and GPS navigation system may get in touch with a medical service after a crash.
Source: Courtesy of Volvo-Autoliv.

4.7. Oil Drilling

Down-hole drilling is a common practice in the oil and gas industry. It requires several types of measurements while drilling (MWD) and logging while drilling (LWD) without operation interruptions. Owing to the remote location of MWD tools, they must be powered by a battery pack placed as close as possible to the tool bit (Figure 4.23) [22].

The battery pack should meet several requirements depending on the specific application. The battery must be able to operate in very harsh environmental conditions: temperature range from −40 to 150°C (but in some cases up to 200°C), severe mechanical shocks and vibrations, and high pressure. The battery pack has to power electronics (memory backup and transmission), sensors and power tools. There is usually a single pack, but it can be divided into a few sections, each being used for different applications (Figure 4.23).

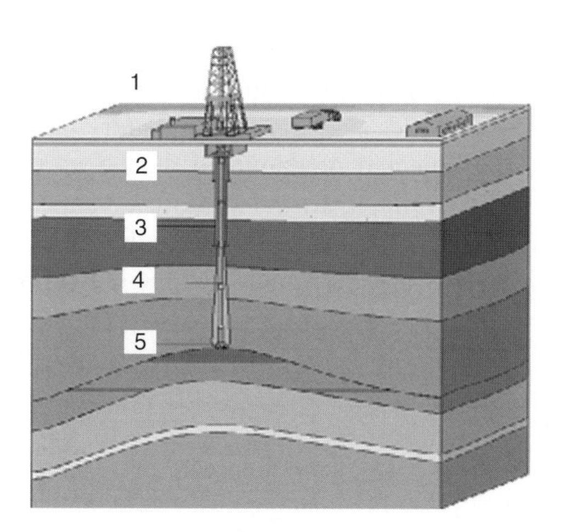

1. Derrick and drawworks; 2. & 3. Casing strings; 4. Drill pipe; 5. Drill bit.

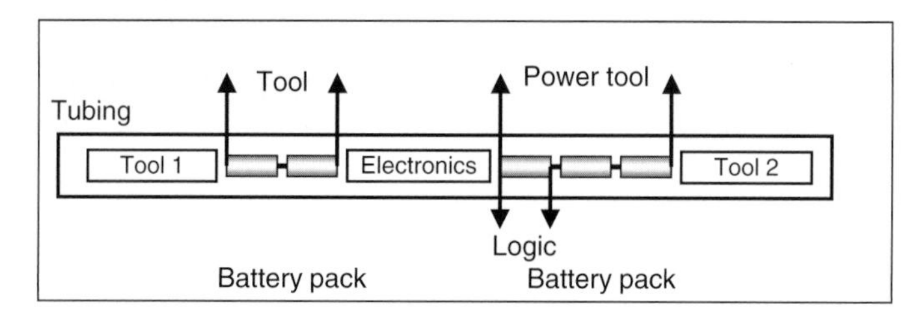

Figure 4.23. Up: components of a typical drilling rig; down: schematic battery pack location in the drilling bit (*source*: From Ref. [22]).

The current drained from the battery depends on the task: low rates are typical of memory backup, when the acquired data are stored in memory while drilling between transmission sessions; high drain rates are necessary, for instance, in the case of cutting tools or motors activation (see Table 4.2).

The energy-consuming transmission is done by means of mud pulse telemetry. Information about the tool orientation and inclination, measured by the accelerometer and magnetometer, is transmitted to the digital processor via the drilling mud circulation system. The same information can be transmitted, without mud pulse telemetry, by sending data to the surface from an electromagnetic (EM) MWD tool. The control information can be sent back to the tool electronics from the surface operator [22].

MWD sensors gather, store and transmit information including drilling temperature and pressure, soil formation resistance, drilling orientations and pipeline inspection data. The pressure and temperature sensors typically require currents of tens to hundreds of microamperes in memory backup mode and up to tens of microamperes while sampling and transmitting or switching between the sensors. The sampling rate is adjustable from seconds to hours.

Power-demanding operations comprise activation of sensors, shut-in tools and perforators for the formation penetration, shear tools for down-hole cutting procedures, spin-less flow meters, down-hole shut-off valves and different kinds of motors. The currents can be as high as several amperes for hundreds of milliseconds to several seconds. The high power required by these operations is a matter of concern, as the current drained from the battery causes a voltage drop and the output power decreases. In case the battery cannot maintain the minimum functional power, the operation will fail; in a shearing tool, for example, the tool motor will stall. Sleep mode with a few mA drain, incorporated between operating cycles, decreases the overall energy consumption and increases battery life.

In contrast to the data survey consisting of frequent functions, the power tool applications usually require only a few working cycles and could be executed during separate drilling operations. Sometimes such applications are intermittent, with long storage periods between working sessions.

The geometry of the drilling tool strictly limits the battery pack size and shape. Usually, the battery pack has a cylindrical shape and comprises, besides the batteries, appropriate safety circuitry, connectors and battery-to-battery wiring, and insulating sleeve/housing. In Figure 4.24, examples of battery packs are shown [23]. Packs can be formed by sections, sticks or may have an annular shape surrounding the drill pipe. It is to be noted that the last shape maximizes battery life, has fewer electrical connections and is more robust [24]. Tailoring the battery pack as a function of the tool does not always give the most efficient battery performance, but is practically unavoidable. This factor may shorten the application operating life and cause frequent and costly battery replacements.

Table 4.2. Some examples of down-hole drilling applications.

Application	Voltage (V)	Background Current (mA)	Low Pulse (mA)	High Pulse (mA)	Period	Temperature Range (°C)	Batteries in Use
Memory gauge	3	0.05–0.3	4–5 (<1 s)	10–50 (<1 s)	few seconds	−40 + 130	AA, C lithium, alkaline
Pressure-temperature gauge	3	0.1–0.2	4–5 (<1 s)		few seconds to hours	−40 + 150	C lithium, alkaline
Master controller (memory)	15	0.2				−40 + 150	C lithium
Sensing device	3	0.05	50 (few ms)		few seconds	−40 + 150	C lithium
MWD tool	30	70–80	250 (few seconds)	1200 (few seconds)	0.5–1 h	−40 + 120	DD Lithium
Power MWD timer	50/7	10		1500 (few seconds)	min to hours	−40 + 150	DD lithium
Shear tool	15	5	250 (few min)	5000 (few seconds)	4–6 cycles	−40 + 165	C lithium

Source: From Ref. [22].

Figure 4.24. Battery pack designs for downhole tools. Annular pack wrapping the drill pipe (top left); stick packs of D (or DD) cells (top right); separate sections of 3/2C cells (bottom). All packs based on $Li/SOCl_2$ cells. Temperature range: −40 to 180°C (200°C for stick packs).
Source: From Ref. [23].

As shown in Table 4.2, primary Li or alkaline cells (C- or AA-size) are mostly used; for the latter, the number of cells in series should be doubled due to the lower nominal voltage (1.5 V). The use of packs based on alkaline cells is limited by the high temperatures that can be reached (150 or even 200°C) and by the lower energy available (especially at low temperatures). Although alkaline batteries are cheaper than lithium batteries and possess a high-power capability, the drawbacks mentioned above make them not frequently used in down-hole drilling applications.

Preferred Li batteries are lithium/thionyl chloride ($Li/SOCl_2$) and lithium/sulphuryl chloride (Li/SO_2Cl_2, OCV: 3.91 V), capable of standing rather high temperatures. The maximum temperature is usually the most critical issue and significantly narrows the battery choice. Li batteries are available with different maximum temperatures: 125, 150, 165, 180 and 200°C. Some batteries with thionyl chloride can work in the broad range −40 to 180–200°C [23]. The issue of measuring and logging while drilling at high temperatures is more extensively treated at the end of this section.

The battery must also withstand drilling shocks (up to 3000 g) and vibrations (up to 40 g), and high-pressure conditions. This is especially important for high-power spirally wound and double anode Li batteries, because short circuit conditions may occur. The dual anode battery has a better performance than the bobbin one at moderate rates and is suited for very high vibration conditions.

Primary Li batteries with oxyhalide cathodes possess the highest energy density for these applications and exist in three designs: bobbin, spirally wound and dual anode. The cover laser welding, with the positive terminal enclosed in a glass-to-metal seal construction, hermetically seals these batteries and prevents electrolyte leaking or drying-out, particularly noticeable at high temperatures.

The volumetric energy density of the thionyl chloride system is very high, especially for the bobbin configuration (see also Table 2.3). Packs for downhole drilling reach 915 Wh/L, nearly three times higher than alkaline packs. If the spirally wound design is considered, the sulphuryl chloride system has a somewhat higher energy density.

Oxyhalide batteries manifest very low self-discharge rates in bobbin-type design: $\sim$1% capacity loss per year, thanks to the passivation layer formed on Li. High-rate spirally wound lithium thionyl chloride or sulphuryl chloride batteries are suitable for power tools applications, which require high current drain and power bursts. Drawbacks of these batteries, in comparison with the bobbin types, are: higher self discharge (5% per year), more evident passivation effect (voltage delay) because of the higher currents required, and less safety.

In recent years, the introduction of Li-ion rechargeable batteries has become significant. Rechargeable batteries could be a remedy for the primary battery packs substitution, and could be recharged by a mudflow-activated turbine. The turbine might also act a primary power source, while the battery could have a backup function. However, there are some points to consider:

- Li rechargeable batteries cannot withstand temperatures above 125°C even if specifically engineered
- Their energy density is lower than that of Li primary batteries: this means that the pack should be larger than a primary battery pack
- Safety may be an issue: all rechargeable Li-ion batteries are jelly rolled and heat dissipation may be less effective.

As anticipated, problems arise when performing downhole drilling at high temperatures, this regarding the battery pack as well as the electronics. Usually, MWD can be carried out up to a temperature of 175°C, while LWD has to be limited to temperatures below 150°C. However, in the last few years some projects have aimed at extending the temperature of MWD tools to 195°C, and that of LWD tools to 175°C with survivability to 200°C [25]. As expected, several hurdles have been found along this way, from a technical and an economic standpoint. High temperature tools tend to manifest shorter life, reduced performance and higher cost [25].

A new platform is needed for the electronics of these tools, and batteries different from those described above are also necessary. These issues have been particularly evaluated at Sandia National Laboratories (USA). Promising results have been obtained with tools based on silicon-on-insulator (SOI) technology instead of traditional CMOS technology. SOI electronic components have been demonstrated at temperatures of up to 250°C for hundreds to thousands of hours, and could allow raising the current temperature limit to 300°C [25, 26]. For high-temperature applications, ceramic packaging and gold wire bonding also show promise.

At these very high temperatures, liquid battery electrolytes have to be replaced by molten salts or solid electrolytes [26, 27]. For example, cells based on the Li(Si or Al)/FeS_2 couple, with an eutectic of the type CsBr–LiBr–KBr as an electrolyte, can work at temperatures above 250°C; cells working in the temperature range 150–250°C are under development. Cells based on solid electrolytes can work in a wide temperature range: from room temperature to 250°C or above; these electrolytes have a high resistivity and must be used as thin films in cells for memory backup.

4.8. Oceanography

Important oceanographic parameters, such as sea level, salinity, current, temperature, floor profile and tectonic conditions, may be measured with battery-powered equipment. The information, often collected in remote locations, is then sent to a central service through a wireless net.

Oceanographic applications are rather diverse, but they can be assigned to four groups [22]:

- Buoys of different kinds bearing oceanographic, geophysics and GPS equipment
- Underwater vehicles with scientific exploration payload
- Ocean floor equipment with different sensors for studying seismicity and bio-chemo-geo-hydrology properties
- Safety systems, requiring long standby and then a short-term high-power burst.

These devices support acquisition data survey, processing and storage, requiring low to moderate power for a long time; however, in the transmission stage or when changing a vehicle's location, the power demand can increase rapidly for a short time. The battery pack performance and life is critical: it could be requested to sustain high-power drains in a wide temperature range, −55 to 85°C, and to last several years.

In Table 4.3, examples of oceanographic applications are presented. As can be seen from the table, primary Li batteries are mainly used. Alkaline

Table 4.3. Examples of oceanographic applications powered by battery packs.

Application	Voltage (V)	Pulse Load	Pack Capacity	Temperature Range (°C)	Life Time	Typical Battery
1. Sea-floor seismometer	~15	3 A 1 s	300 Ah 3.9 kg	About 0	62 days	PulsesPlus[a]
2. Sea-floor seismometer	27	5 A 1 m	50 Ah	About 0	7 days	Li (spiral wound)
			48 Ah 2.5 kg		10 days	PulsesPlus
Safety inflation system	6	4.5 A 20 ms	2.3 Ah 0.12 kg	−34 to +54	7 years	PulsesPlus
GPS seismic sensor	~7	Few watts	50 Ah	About 0	1 year	Li (spiral wound)
		Few watts	80 Ah 3.2 kg	About 0	2 years	PulsesPlus
GPS ice buoy	~15	Few watts	54 kg	−40 to +85	1 year	Alkaline
			152 Ah 3.2 kg	−40 to +85	1 year	PulsesPlus
GPS buoy	~15	2 A 10 s	960 Ah 19.4 kg	−40 to +85	6 months	PulsesPlus
Underwater glider	~15	Few watts	30 Ah 1 kg	−40 to +85	45 days	PulsesPlus, Li/CSC[b]
Current meter	9	1.27 A 1 s	7.8 Ah[c] 70 Wh	0 to +35	~1 month	Alkaline
	~12		17 Ah[c] 199 Wh		3 months	PulsesPlus
Sensor system	6.2	10 mA	137 Ah	20	13 months	Li (bobbin)

[a] Tadiran's brand name for the $Li/SOCl_2$-hybrid layer capacitor cell.

[b] Li/SO_2Cl_2 cell with Cl_2 as an additive (CSC: chlorinated sulphuryl chloride). OCV: 3.95 V, temperature range: −30 to 90°C.

[c] Same battery pack volume.

Source: Adapted from Ref. [22].

batteries possess high-power capabilities but low-energy densities and low working voltage in comparison to Li batteries. The $Li/SOCl_2$ system is the most widely used. In the modification comprising a bobbin-type cell and a hybrid layer capacitor for sustaining high pulses, it can be built as high-capacity and high-energy cells (PulsesPlus by Tadiran). In order to maintain the same power and energy, the alkaline pack should be significantly larger and heavier. This is especially important in under-water vehicle applications, where weights and volumes are primary issues for the designer.

4.8.1. Current Meters

Measuring the current velocity in oceans, rivers or channels is a challenging task. Much has been learned about water currents using traditional electromechanical meters. However, flow distortion, large sample volumes and impeller stalling have limited the use of these instruments for surface and deep water measurements. New acoustic measurement techniques offer the potential to overcome the inertial limitations of electromechanical sensors, allow accurate measurement of slow and fast currents, and only require small sample volumes [28].

Acoustic meters can measure velocity and direction of current and waves. They transmit an acoustic signal along four paths, at the typical frequency of 1 MHz, either in pulses or continuous wave bursts. For an accurate representation of speed and direction, the phase-shift technology is used. The total phase shift, including that of the receiver, is first measured in one direction, and then in the opposite direction where the total phase shift is measured again, using the same receiver. The current velocity is proportional to the difference in phase for the two directions. A very low-power circuit for measuring phase shift at the carrier frequency is used [28]. A microprocessor allows determining which axis is contaminated by flow interaction with the centre support strut and rejects data from this axis [28]. The internal three-dimensional (3D) magnetometer and 2D tilt sensor allow measuring the earth's magnetic field and the meter's angle; in this way, the current direction and tilt can be determined.

3D acoustic current meters (ACM) are shown in Figure 4.25. The four paths ("fingers") are visible; each houses two transceivers to create acoustic paths. Speeds in the range 0–300 cm/s can be measured with 2% accuracy.

These meters can log data and simultaneously transmit it. Temperature and pressure measurements are also possible. They can be used at a depth down to 1000 m (standard) or to 7000 m with titanium housing. To operate this instrument for some months, the power consumption needs to be very low (few tens of mA). Two features of the measurement technique described above help reducing power consumption. First, the technique uses direct acoustic paths and not reflected sound (as in the Doppler effect – see below). This reduces the power dissipated

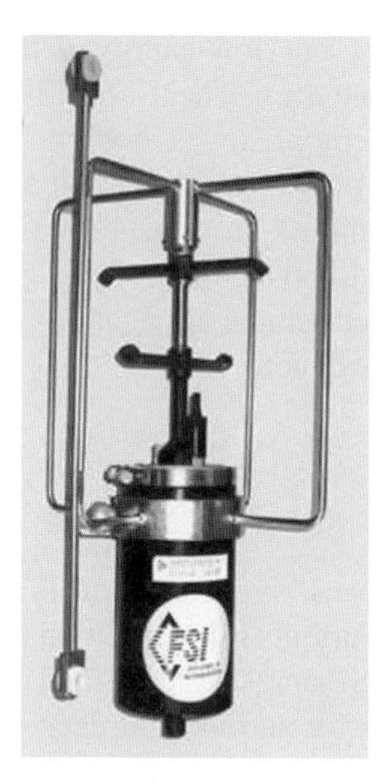

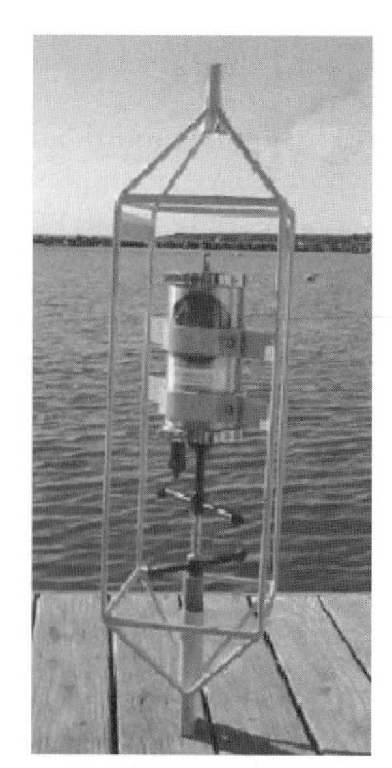

Figure 4.25. Left: 3D acoustic current meter with four acoustic paths ("fingers") protected by a frame; the fingers are the horizontal rods attached to the central axis (strut); right: 3D meter mounted on a larger frame, ready for deployment.
Source: From Ref. [28].

into the water and thus the power required. Second, the instrument measures phase shifts of the acoustic signals along the paths, not the time of travel. Measuring phase shift can be accomplished with slower circuits than measuring time of travel; therefore, these circuits require less power to operate [29].

Deployment times of 1–3 months are ensured by primary alkaline or C-size Li batteries (see Table 4.3). The equipment can be placed on the sea floor, attached to a buoy or mounted at the ships bottom.

Another type of acoustic current meter is the acoustic Doppler current profiler (ADCP). It sends sound pulses of 40–3000 kHz which are reflected by small particles in water according to the Doppler effect (see also Section 4.2.5). The Doppler shift in frequency of the reflected sound is used to calculate the current speed.

4.8.2. GPS Buoys

Buoys deployed in remote oceanographic sites allow collection of important environmental data with minimal or no human intervention.

Examples are provided in Figures 4.26 and 4.27. The off-shore buoy of Figure 4.26 serves primarily as a calibrating tool for satellite radar

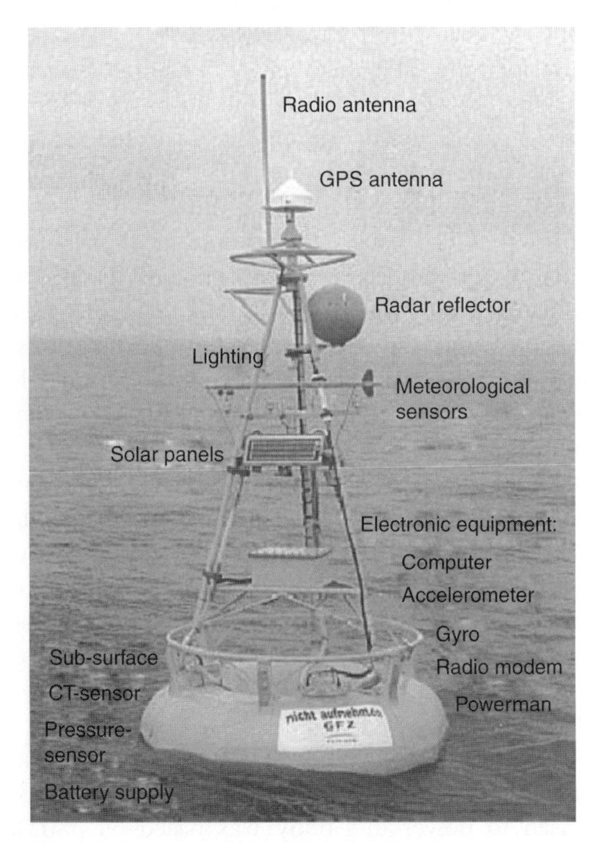

Figure 4.26. Off-shore moored buoy (See colour plate 2).
Source: From Ref. [30].

Figure 4.27. GPS/ice buoy being retrieved by a helicopter in the Arctic Sea. This buoy can measure wind, temperature, sunlight and ice thickness.
Source: From Ref. [30].

altimetry [30]. Besides the GPS receiver, it is equipped with a multi-sensor system for observing the sea level with high accuracy. Each time a satellite passes the buoy, a multi-sensor session is recorded, at the 10-Hz rate, to supplement the GPS measurements. Tilt angles as well as three-axis accelerations are monitored by a dynamic motion unit. An under-water pressure gauge records pressure values at the bottom of the buoy. Information about air pressure, water temperature and conductivity are provided by the meteorological sensors (10 min rate). In addition, analysis of both the buoy's accelerometer and the GPS data allow estimations of the wave height. The buoy is moored not far from land, so that a bidirectional HF radio link between the buoy and a land station allows communication and data transmission [30].

Buoys of this type can be used as part of a tsunami early warning system, as they can help determining the tsunami wave front and the wave height. Sensors for measuring the ocean bottom pressure can send useful data to the buoy if seismic activity is occurring.

As can be seen in Table 4.3, GPS buoys need battery packs of relevant capacity, for example 960 Ah, this granting some 6 months of operation. High current pulses, up to 2 A, may be needed. A battery of this type is shown in Figure 4.28 (left).

The GPS/ice buoy of Figure 4.27 has been used to measure the effect of global climate change on ice floating on the Arctic Ocean [31]. The battery pack formerly used to power this buoy was based on 380 D alkaline cells and weighed 54 kg. Later, a pack based on D Li/$SOCl_2$ cells and hybrid layer capacitors (3.2 kg) was adopted (see Figure 4.28 (right)). This impressive difference in volume and weight gives a great advantage for transportation and handling.

More details on the GPS technology can be found in Section 4.9.2.

4.8.3. Seismometry

The heart of any seismometer for oceanographic applications is the ocean bottom seismometer (OBS). This device is sensitive to the ground motion and is similar to a land seismometer. It can be small enough to fit in a pressure container (a glass, or aluminium, sphere resistant to high pressure (see Figure 4.29)), affording deployment at depths up to 7000 m on the ocean floor.

The OBS can be used either as a temporary network – for local seismicity studies (location of earthquakes in an active region) – or for continuous seismic work to find characteristics of the earth's deep layers (velocity of seismic waves,

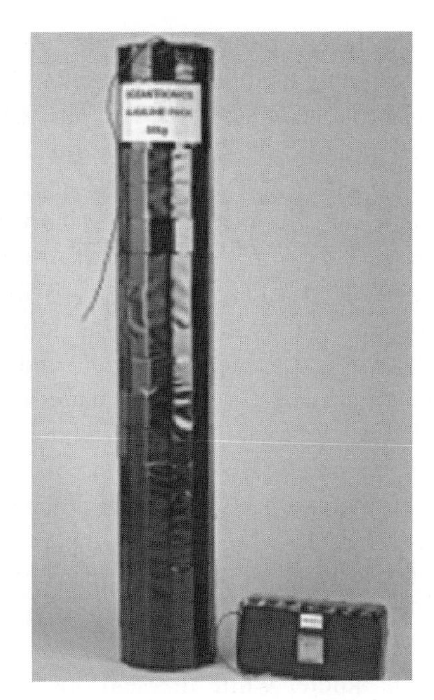

Figure 4.28. Left: GPS-buoy battery pack formed by 96 DD-size $Li/SOCl_2$ cells and 12 hybrid layer capacitors (960 Ah, 14.4 V, 19.5 kg); right: comparison of batteries used in GP/ice buoys: on the left, an alkaline pack with 380 D cells (weight, 54 kg); on the right, a pack formed by 32 $Li/SOCl_2$ D cells and 4 hybrid layer capacitors (weight, 3.2 kg).
Source: From Ref. [22].

Figure 4.29. Left: an OBS (ocean bottom seismometer) deployed on the ocean floor; right: OBS surfaced after receiving an acoustic signal.
Source: Courtesy of IRD, Paris, France.

geometry of layers, etc.). Wide-angle seismic work (also called seismic refraction) consists in generating seismic waves with an artificial source at different distances from one or several OBS. The velocity of propagation of the waves in the subsoil and the amplitude of the seismic signals provide information on the physical properties of the geological environment [32].

OBSs can be divided into short- (>1 Hz) or long-period (<1 Hz) instruments. The former are used in few-month deployments, while with the latter deployments up to 3 years are possible. The sensors consist of geophones (seismometers) and hydrophones; the geophones can have a maximum of three components, for example one vertical and two horizontal. Alkaline, primary Li or combined primary Li-capacitor batteries can be used.

Pressure, temperature, pore water chemistry and biology of the sea floor can be measured, in addition to seismic data, with a more sophisticated instrumentation [33]. The hypothesis is that seismic events change the stress in the rock, which affects the pressure on fluids in the pores of the rock. Therefore, borehole fluid pressure (and chemistry and biology) may provide precursors to the seismic activity. This technique also allows seeing small events, such as nano- and micro-earthquakes: this is possible as the borehole is quieter while the seafloor is a noisy seismic environment.

A scheme of this technology is shown in Figure 4.30. The following points may be stressed:

- The borehole array is hardwired to a junction box at the seafloor. The box, which is mounted on the wellhead, connects the various pressure cases and provides an access panel for the underwater mateable connectors. In addition to the downhole cable, the logging cable uplink and an acoustic communication unit are hard-wired into the junction box.
- The data acquisition case contains telemetry unit, computer, hard disk for data storage, clock and power control unit (with an 8.7-kWh battery pack on board). In autonomous recording mode, this unit would be replaced each time data are recovered by a remotely operated device (ROV).
- The borehole contains a string of three-component geophones. The number of geophones is typically four. The geophone sondes are connected by armoured coaxial cable. Data is transmitted by a down-hole telemetry unit.
- Additional batteries can be plugged-in and replaced through the junction box.

For installations located far from an observatory network, sensors must run in autonomous mode. In this case, power is mainly derived from batteries. Running a 10-W seismometer/data logger for a year on the seafloor requires roughly 1000 DD-size Li primary cells. Power cycling of high current drain loads such as computers and disk drives can significantly increase the battery count.

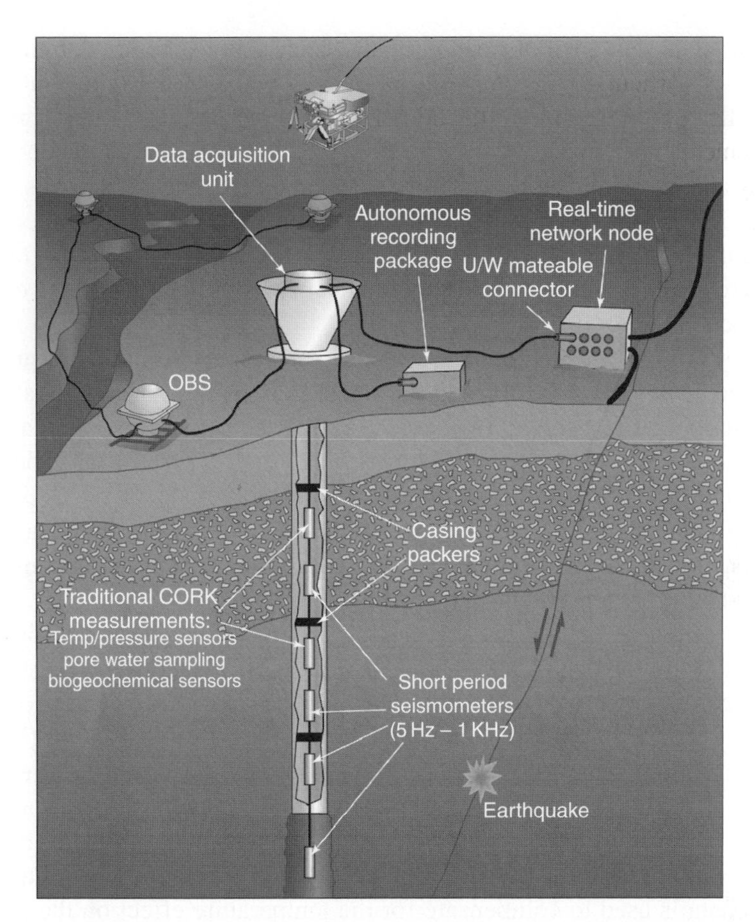

Figure 4.30. Equipment for the simultaneous monitoring of earthquakes and other physical parameters on the seafloor. CORK stands for Circulation Obviation Retrofit Kit.
Source: Courtesy of Woods Hole Oceanographic Institution.

4.8.3.1. Tsunami Detection

The Indian Ocean tsunami of December 2004 killed more than 300 000 people – more than other tsunamis ever recorded. This tragedy has spurred initiatives to install a global tsunami warning system. The strategy is to develop a forecast system to interpret earthquake and sea-level data and to preview the tsunami impact on coastal populations. This system has to provide decisive elements to people in tsunami warning centres who have to make rapid decisions: either ordering evacuations to prevent devastating fatalities or avoiding needless evacuations that could prove dangerous and expensive.

The US National Oceanographic and Atmospheric Administration (NOAA) is particularly active, with its Pacific Marine Environmental Laboratory (PMEL), in developing a technology based on the integration of real-time measurements and modelling approaches [34]. Sea-level data in deep ocean is made by a network of deep-ocean assessment and reporting of tsunamis (DART) systems. By the end of 2007, 39 network stations were built; they are composed of portable elements that will be illustrated in the following. Transportability is a prime requisite to study a phenomenon that hardly allows planning a choice of deployment sites.

It is critical for a system created to forecast a marine earthquake and subsequent tsunami to be located on the ocean floor, rather than at coastal port or harbour sites; this allows detecting the tsunami as it propagates from the point of origin to the coasts. Furthermore, such a system must be able to process and deliver relevant data in a few minutes [34].

Two physical components form the DART system on the ocean side: a tsunameter on the ocean floor and a surface buoy with satellite telecommunications capability (see Figure 4.31).

The principle behind the forecasting ability of this system is that earthquake waves travel much faster than tsunami waves. Therefore, the vertical shift of the ocean floor caused by an earthquake, and the associated change of bottom pressure, is the symptom allowing to anticipate the onset of tsunami tides.

The bottom pressure is read by the tsunameter according to the scheme of Figure 4.32: the pressure sensor sends data to a computer that runs a tsunami detection algorithm and sends/receives commands/data to/from the buoy via acoustic telemetry. The sensor indicates both pressure and temperature: thermal data is used to compensate for the temperature effect on the pressure-sensing element. The measurement sensitivity is very high: less than 1 mm in 6000 m (2×10^{-7}).

The tsunameter operates either in standard mode or event mode. The former is a low-power, scheduled transmission mode reporting data every 6 h; information includes average water column height, battery voltages, status indicator and time. The event mode reports earthquakes and tsunamis when a threshold is exceeded. Waveform data is transmitted in <3 min.

The tsunameter's pressure measurement/computer system uses alkaline D-size cells assembled in a battery pack of 1560 Wh. The tsunameter's acoustic modem is powered by similar battery packs that can deliver over 2000 Wh of energy. These batteries are designed to last for 4 years on the seafloor; of course, this depends on the number of events that may occur and the volume of data requested from the shore. A battery monitoring system is in place to maximize battery's life.

The surface buoy, shown in Figure 4.33, ensures a bidirectional link with the tsunameter, through underwater transducers, and with the Iridium satellite

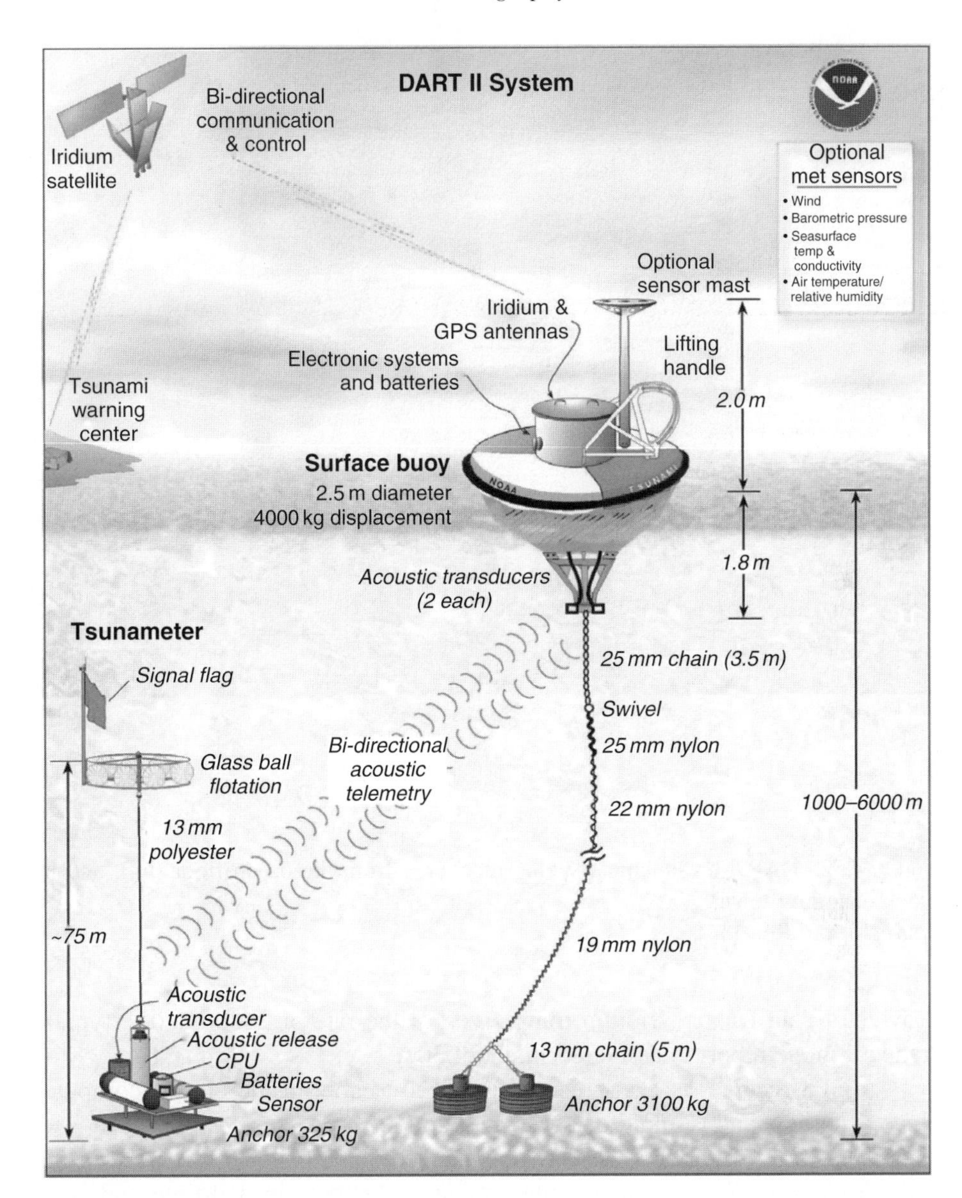

Figure 4.31. System for tsunami detection (DART). Its main components are: a tsunameter moored on the ocean floor, a surface buoy able to communicate with satellites (optionally endowed with meteorological sensors), satellite network, and tsunami warning centre.
Source: From Ref. [34].

network. The Iridium constellation is made up of 66 satellites in big LEO orbits (780 km from earth). Communications with satellites are either in standard mode or event mode, whichever is indicated by the tsunameter. Connectivity is

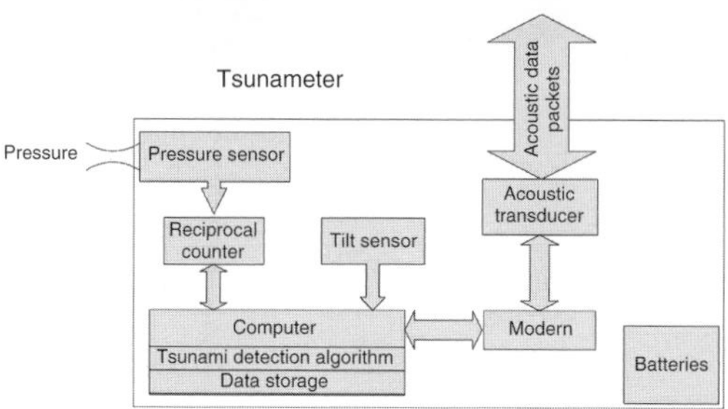

Figure 4.32. DART II tsunameter being recovered from the ocean floor (up), and its block diagram (down).
Source: From Ref. [34].

provided by an L-band Iridium transceiver; at the rate of 2400 baud, a typical standard mode report is delivered in about 30 s.

A GPS receiver is also present: it is used to control the buoy's computer clock accuracy, and to monitor the exact buoy location.

Data from a DART system is downloaded and stored on a server via the Iridium gateway. The warning centres monitor this data in real-time and issue alarms if needed; in addition, the data is posted to a web server and can be viewed on the Internet.

The buoy's batteries are made up of packs of D-size alkaline cells: those for the computer and iridium transceiver have energy of 2560 Wh, while those of the acoustic modem are of 1800 Wh. These batteries can power the buoy for at least 2 years. The potentially dangerous build up of hydrogen gas naturally vented from alkaline cells is mitigated with hydrogen getters and pressure-relief valves [34].

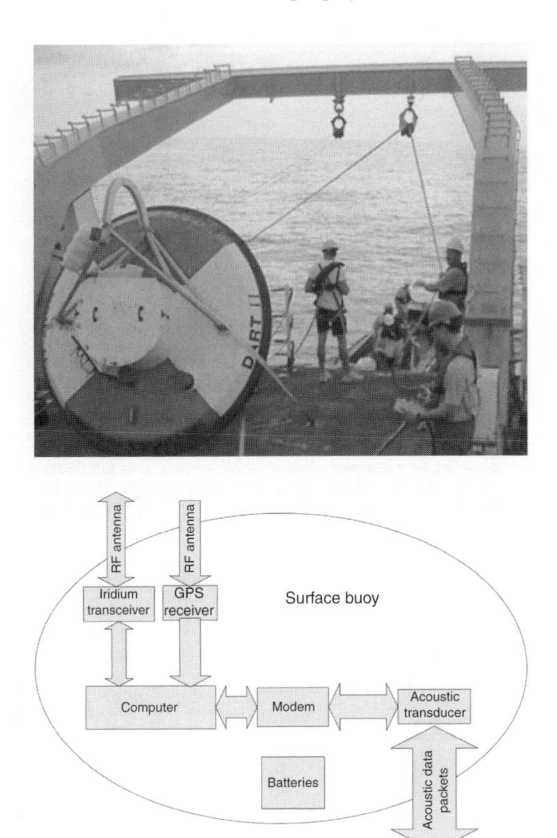

Figure 4.33. DART II buoy ready for deployment (up) and its block diagram (down). *Source*: From Ref. [34].

4.8.4. Underwater Gliders

These vehicles, also called autonomous underwater vehicles (AUV), gather data including time- and location-specific readings of temperature, pressure and salinity. Motion of these vehicles is determined by changes in buoyancy. In Figure 4.34, the glider developed by Scripp & Woods Hole scientists, "Spray", is shown together with its typical upwards/downwards gliding [35].

In this technology, primary Li battery power and a hydraulic pump are used to periodically change the glider's volume by a few hundred cubic centimeters in order to generate the buoyancy changes causing the forward up or down gliding. Indeed, the glider alternately floats upward and sinks downward through the ocean, all the time gliding forward. Heading and ascent/descent rate are controlled by moving weight (battery packs) inside the hull to change roll and pitch.

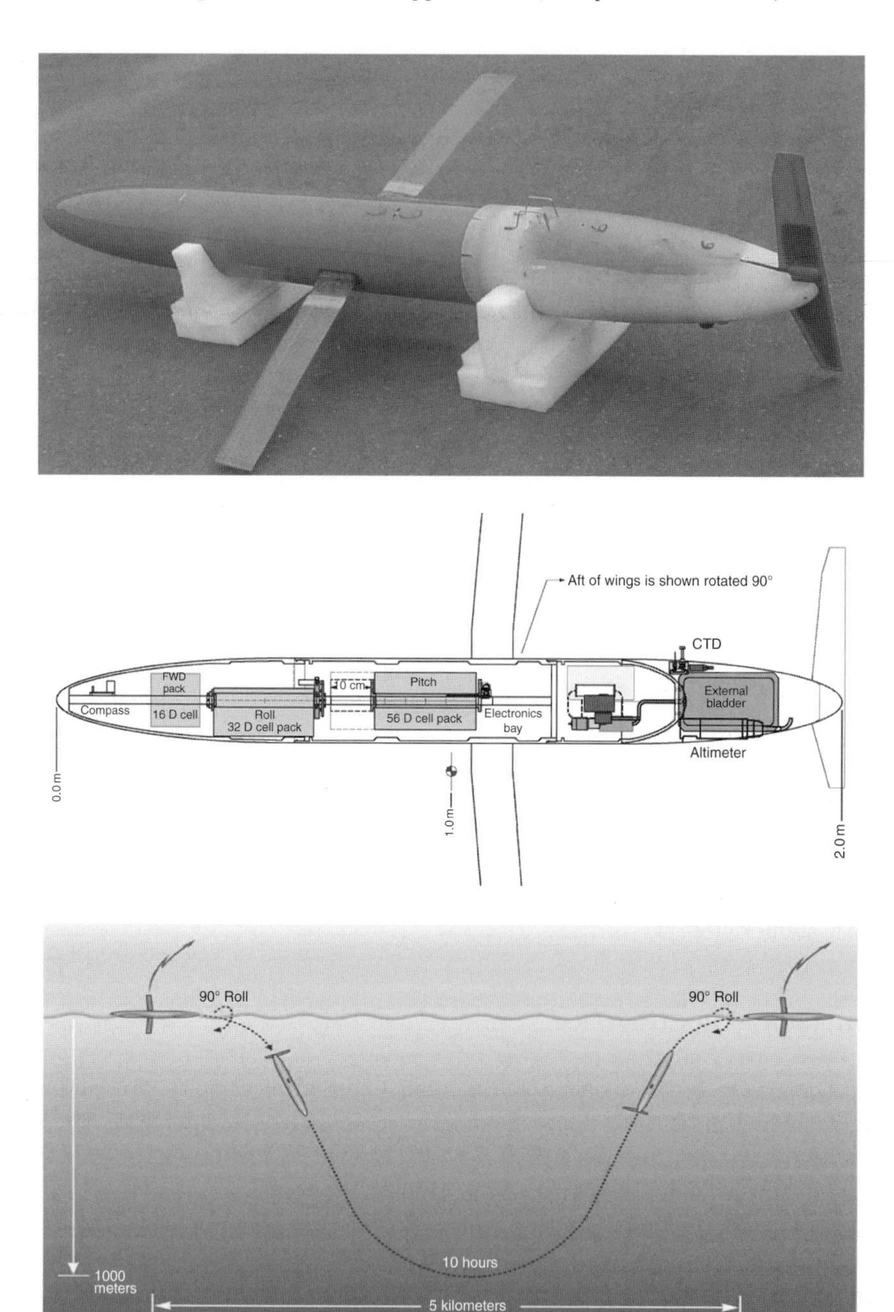

Figure 4.34. An underwater glider (top); section showing instruments and batteries (middle); representation of the glider's movement (bottom). Vital statistics include: length, 200 cm; diameter, 20 cm; wing span, 110 cm; mass, 51.8 kg; payload, 3.5 kg. *Source*: From Ref. [35].

Depth capabilities range from 200 to 1500 m while operating lifetimes range from a few weeks to several months. These long durations are possible because the gliders move very slowly, typically 25 cm/s or 0.5 knots. In Figure 4.34, the glider developed by Scripp & Woods Hole scientists, “Spray”, is shown together with its typical upwards/downwards gliding [35].

The Spray’s aluminium pressure case is made in three pieces and houses batteries, hydraulic system, compass, GPS receiver, Iridium transceiver and the microprocessor controller. An aft-flooded bay houses the external bladders that expand to increase buoyancy. This bay also houses the sensor suite selected for each mission. Examples of sensors are conductivity-temperature-depth (CTD) sensor, optical backscatter sensor, chlorophyll fluorometer, acoustic altimeter for bottom avoidance and acoustic Doppler current profiler.

This glider is powered by DD-size Li/CSC cells, that is primary cells with chlorinated sulphuryl chloride (SO_2Cl_2) as a cathode (see footnote of Table 4.3). In these cells, Cl_2 is added to reduce the voltage delay caused by Li passivation.

Li-ion batteries may also be used. For instance, SAFT powers with these batteries an AUV by BAE Systems (Talisman M), which can carry payloads of 500 kg or more and can operate at depths of up to 300 m. This AUV is equipped with two or four battery packs, depending on its application, which allow a 24-h continuous operation. Each battery module, built with VL45E cells (see Table 5.16), has a voltage of 320 V and a capacity of 45 Ah.

AUVs thrusters can also use lithium polymer batteries (e.g. packs of 14.8 V, 8 Ah).

4.8.5. Location by Argos System

The Argos system can be used to track any object, person or animal. However, it is of great help especially in oceanographic applications, which represent more than 50% of its total applications. For this reason, the Argos system is introduced here in advance of the more general treatment of tracking (Section 4.9).

This system consists of six polar orbiting satellites equipped with receivers centred at 401.6 MHz. From altitudes of 740 to 850 km, these satellites have 5000-km visibility circles, as shown in Figure 4.35. During its orbit, a satellite sweeps a wide area around the earth, receiving signals from several transmitters, known as platform transmitter terminals (PTTs). The number of daily passes over a PTT increases with latitude: at the earth’s poles, a PTT would be within the visibility circle of the six satellites ~80 times per day, whereas at the equator, a PTT would only be visible ~20 times per day. Standard Argos processing uses data from three satellites [36].

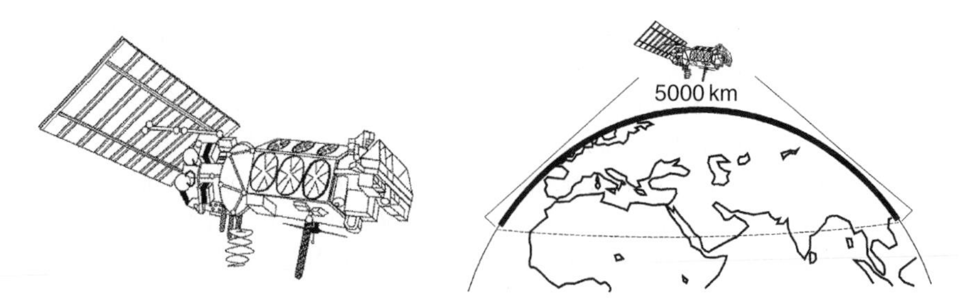

Figure 4.35. A polar orbiting environmental satellite carrying Argos instruments (left); satellite's visibility area (right). At a given time, a satellite sees all transmitters located within a 5000-km arc.

Argos locations are calculated by measuring the Doppler shift (frequency change) between PTT transmissions. The average duration of transmitter visibility by the satellite (pass duration) is 10 min. During this time, a minimum of two PTT transmissions must be received by the satellite to calculate a location. The satellite-beacon (PTT) interaction is shown in Figure 4.36 (left).

PTTs are highly stable UHF transmitters operating on 401.6 MHz. If location only is needed, a transmission length of 360 ms is sufficient, whereas transmission of sensor data may require up to 920 ms. Transmissions are

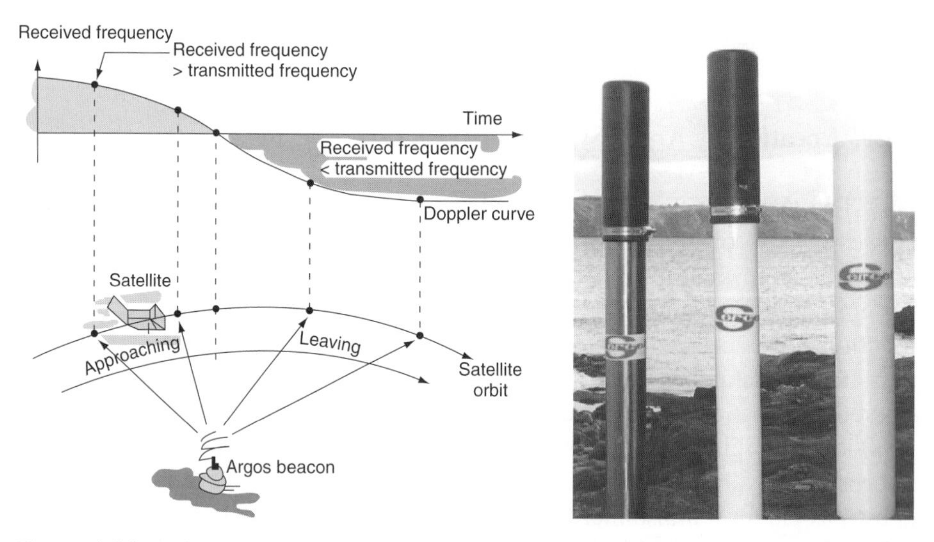

Figure 4.36. Left: scheme of the Argos system (Argos stands for Advanced Research and Global Observation Satellite) (*source*: courtesy of Service Argos); right: Argos beacons to monitor underwater equipment; dimensions: diameter 7–9 cm, length 53–77 cm; power source: 3 D-size $Li/SOCl_2$ batteries (*source*: courtesy of Sercel).

repeated at set intervals, typically between 20 and 200 s. PTTs can be "put to sleep" to save battery power and extend transmission life. In this sleep mode, PTTs draw very little current, therefore a typical 1:1 duty cycle, for example 6 h *on* – 6 h *off*, almost doubles the transmission life. Packaging involves matching PTTs with suitable antennas and batteries and encasing them in waterproof polymers to ensure they can withstand adverse environments. Examples of Argos beacons for underwater equipment tracking are shown in Figure 4.36 (right).

Two main ground-receiving stations provide global coverage by interrogating satellites once per orbit. The location accuracy depends upon these factors [36]:

- Satellite/PTT geometry during satellite pass
- Number of PTT messages received by the satellite during the pass
- PTT frequency stability
- PTT power output
- Electromagnetic interference at the PTT location.

With the best accuracy, the PTT is located within 150 m. A better resolution is possible if GPS receivers are also used, because they continuously recalculate position fixes. Of course, use of a GPS receiver impacts on power requirements and costs. Primary Li batteries are the power sources of choice (see also Section 4.6.2).

While the Argos system is to date the most used by the oceanographic community, other communication satellites may have similar applications. Argos satellites are placed in low earth orbits (LEO); other well-known LEO satellites are Iridium and Globalstar. Satellites may also be on a mid-altitude earth orbit (MEO, 4000–36 000 km) or geostationary earth orbit (GEO, ~35 000 km). Examples of MEO satellites are those belonging to the GPS' constellation, while Meteosat is an example of GEO satellites [37]. More details on the characteristics of these satellites can be found in Section 4.9.2.

4.9. Tracking and Monitoring Systems

In this section, systems for tracking/monitoring fixed or mobile goods, persons or animals will be described.

Surveillance of transported parcels or luggage, study of animal migration, detection of machine failures, data collection on production lines, location of trucks, issues and returns system in public libraries: all these examples illustrate the fast growing interest in automated tracking/monitoring, due to the noticeable productivity gains and service improvements that this process brings.

This tracking/monitoring action could not work, especially when mobile assets are concerned, if dedicated communication terminals, powered by efficient power sources, were not available.

Functions to be carried out for an efficient monitoring include [1]:

- identification, thanks to, for example, transponders (radio-frequency identification (RFID) tags), or bar codes
- precise remote location, mostly using a satellite network (GPS system) or, sometimes, terrestrial base stations
- communication using either satellites, terrestrial cellular networks (GSM, GPRS) or, rarely, radio systems
- messaging with SMS or e-mails.

The two main systems, that is the one based on RFID and the one based on satellite technology, are described in the following.

4.9.1. Radio-Frequency Identification

ID tags can be found not only in mobile monitoring systems but also in alarm, tracking and access control systems.

The RFID technology uses radio frequencies: a smart tag located on the object contains all the necessary information while an emitting/receiving reader allows the collection of data to be analysed and treated by the information system. The scheme of an RFID system is shown in Figure 4.37.

The higher cost of RFID vs other automatic identification technologies (e.g. bar coding), is offset by many advantages, for example re-use and high-speed read/write possibilities, no need for the tags to be visible, larger memory. In fact, these smart tags are already widely present today, for example in building access badges, car locking systems, highway toll systems, goods delivery trucks. In warehouses, tag readers can be mounted on forklifts to read RFID tags anywhere within reach of the forklift. In an extreme and still controversial application, RFID chips can be implanted in human bodies, for example for controlling access in highly reserved areas or for medical purposes.

RFID tags may not contain a battery (passive tags) and present the advantage of a smaller size, but battery-powered larger tags (active tags) are more versatile. For instance, they can incorporate sensors in order to relay interesting information, for example temperature changes when used to track food items, options when attached to a car frame during the assembly process, or security-related information when attached to sea containers [1].

Passive tags, which can be as tiny as a grain of size, only contain an IC and an antenna (plus capacitors and diodes); they modulate and backscatter RF

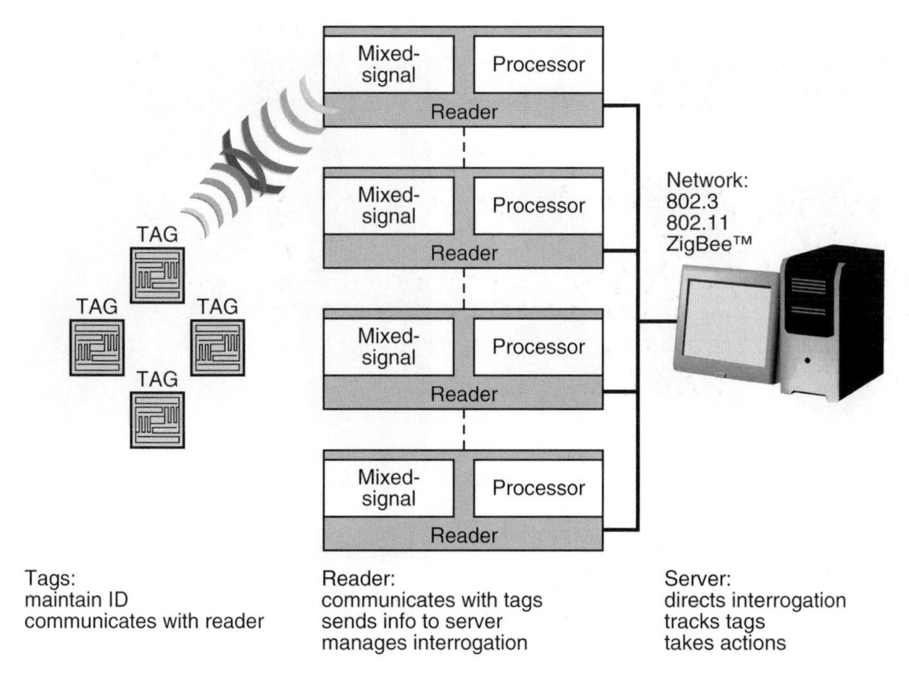

Figure 4.37. Scheme of an RFID system showing tags, readers and the server unit. *Source*: From Ref. [38].

energy transmitted by the reader's antenna, and transmit/receive data within a limited communication distance. Active tags, in addition to a battery, may have a transmitter allowing RF signal transmissions up to several meters (typically 70–90 m in open air). Examples of tags are shown in Figure 4.38.

Portable ear tag readers for cattle's identification are shown in Figure 4.39. They are powered by Li-ion batteries and have memory and wireless connection capabilities.

Active tags have base currents $<100\,\mu A$, pulse currents from a few mA to 250 mA, operating temperature in the range -40 (or lower)/$+85$°C (or higher), and runtimes from a few months to 10 years.

Primary Li cells are preferred to the alkaline ones. In less demanding applications, for example anti-theft tags in stores, coin cells are sufficient, whereas cylindrical $Li/SOCl_2$ cells are needed for tags to be used in long-lasting, outdoor, multi-sensor applications. In harsh conditions, for example extended exposure to very high temperature, these $Li/SOCl_2$ cells may experience huge Li passivation and severe voltage drop during peak current periods. Consequently, the voltage may drop below the cutoff value and the tag may stop working. In these cases, either a high-rate, spirally wound cell or a low-rate cell fitted with a capacitor in parallel may be used.

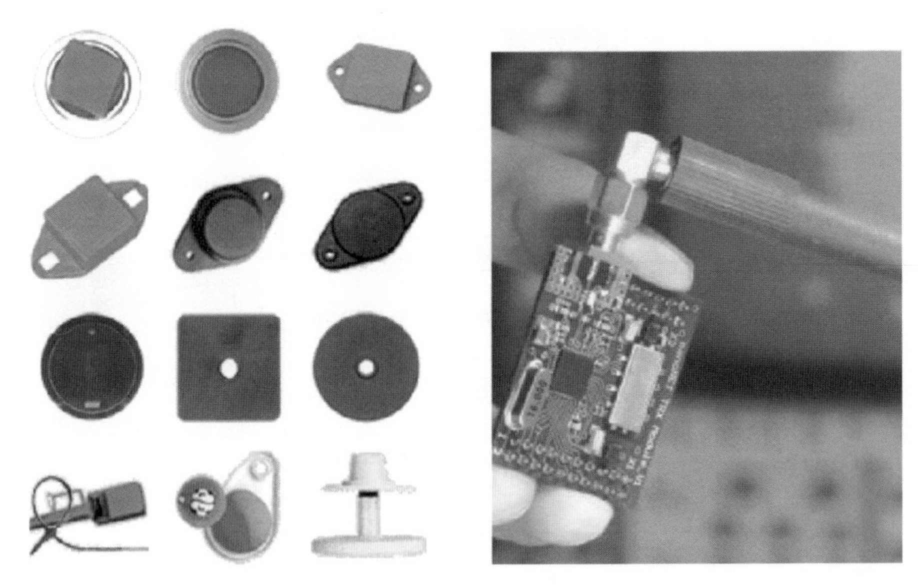

Figure 4.38. Left: examples of passive tags: they often have the size of a key fob; right: electronics and antenna of an active tag.

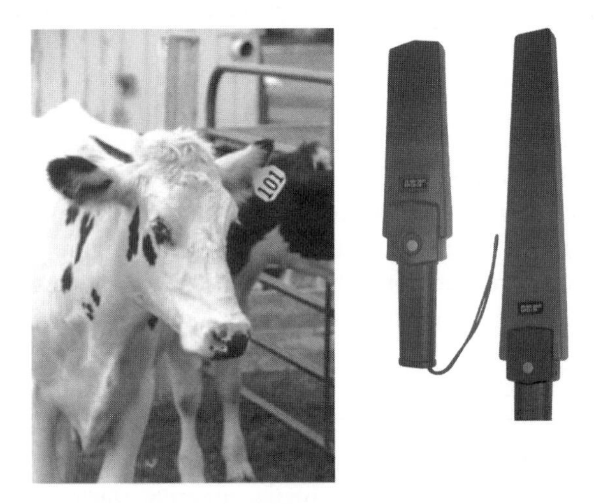

Figure 4.39. Left: an RFID ear tag (of the passive type) containing information for identifying a beef in a herd (*source*: courtesy of RFIDtagsTCC); right: portable readers, powered by Li-ion batteries, with integrated Bluetooth and memory, designed for livestock identification and monitoring (*source*: courtesy of iD TRONIC).

Typical characteristics of large, active RFID tags are (a) $\sim$3 kg weight; (b) transmission power of $\sim$1 W and (c) power source's life of $\sim$6 months. For these tags, a 2/3A Li/MnO_2 cell is convenient. The capacity of this cell is about 1.3 Ah, and it can deliver at least 13 000 transmissions of 1-s duration in 6 months of operation. High- or low-rate $Li/SOCl_2$ cells are also suitable.

4.9.1.1. Electronic Toll Collection

A major component of electronic toll collection (ETC) is the automated vehicle identification (AVI), that is the process of determining the identity of a vehicle subject to toll [22]. Most of the AVI systems rely on RFID: a reading device located at the tollgate communicates with a RFID transponder (tag) on the vehicle (Figure 4.40). This tag (mostly of the active type, but passive tags are also used) can receive, store and transmit data; it has excellent accuracy and can be read without the need of stopping or slowing down the vehicle.

An active tag with a transmitter is called RF-active tag; an RF-passive tag has no transmitter. This latter may also be built in passive (no battery) configuration [38]. RF-active tags have these advantages: high-speed tag read/write cycles can be achieved with relatively short and low-level emission of RF energy from the interrogator; long-range operations are possible. Advantages of RF-passive tags are choice between active and passive tags; high flexibility in terms of RF: the interrogator determines the RF, power and stability. This latter characteristic allows using various frequencies to avoid interferences or to comply with the RF requirements in different locations.

Traditionally, interrogator systems designed for one-tag technology were incompatible with the other. However, multi-protocol interrogators have been recently introduced, which allow innovative applications to use multiple tag technologies simultaneously [39].

Another technology of road toll collection is the one based on optical or electronic reading of a barcode carried by the vehicle. The optical systems use

Figure 4.40. Transponder on a vehicle for electronic toll collection.
Source: From Ref. [40].

the reflection of light, that is a laser, from barcoded tags. These systems are sensitive to the tag's position relative to the reader and to atmospheric conditions, that is fog, rain, dust and snow. The optical systems are more useful in parking lots, where environmental and speed constrains are of less concern.

To avoid the need for transponders, some systems use automatic number plate recognition by means of cameras capturing images of the vehicle passing through the tollgate. This allows drivers to use the facility without any advanced interaction with the toll authority. The disadvantage is that fully automatic recognition requires a manual review stage to avoid billing errors; furthermore, the transaction processing can have a significant cost [40].

The active-tag ETC systems use an independent power source. A typical load profile consists of small standby currents of 5–30 μA coupled with very short pulses of 10–50 mA. Out of the tollgate area, the current drawn is mainly the standby current; as the vehicle approaches the toll area, the transponder is activated and communicates with the reader through current pulses. Typical operating voltage of a RFID tag is in the range 3.0–3.6 V with a minimum cutoff voltage of 2.7–2.8 V. The required battery-operating life is highly depending on the ETC equipment and can range from 2 to 10 years. Operating temperatures in the range −40 to 85°C must be faced.

Owing to the relatively high-voltage requirements under the pulse load, the battery most widely used is $Li/SOCl_2$ (bobbin or bobbin+capacitor cell type). This system can also provide the widest operating temperature range. Li/MnO_2 or alkaline batteries may also be used.

4.9.2. Satellite Tracking

Precise tracking of living beings, cars or mobile assets, such as containers carried by trucks, ships, trains and airplanes, is preferentially carried out with the use of GPS satellites and GPS tracking units. Once a location has been determined, it may be stored in the unit or may be transmitted by the unit to a central station through GPRS, radio or satellite modem. Transmission to a communication satellite is described hereafter.

The general tracking scheme may be seen in Figure 4.41. The transmitter of the tracking unit sends the position determined by GPS satellites to a communication satellite, for example of the Globalstar constellation (see later). The satellite relays this information to a station that delivers it to the user [22]. Different types of transmitters are used according to the satellite type.

A brief description of satellites grouped according to the their orbit height is given here, followed by details on the GPS constellation. As already pointed out when dealing with Argos and COSPAS-SARSAT rescue systems, a number of satellites can be used independently of GPS to localize, but their accuracy is limited.

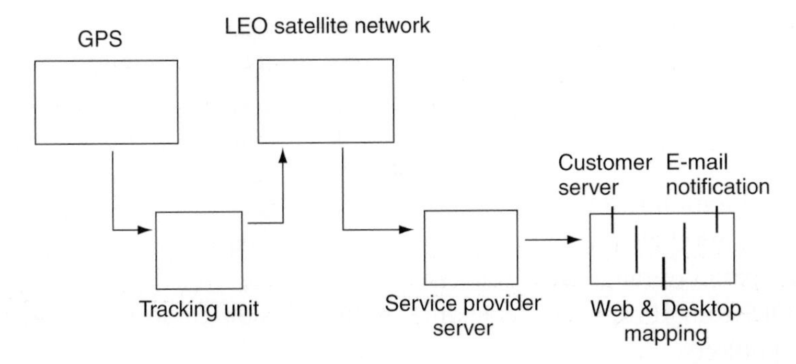

Figure 4.41. Scheme of the tracking procedure utilizing the GPS satellite constellation, a network of communication satellites (LEO in this case), a main station on earth, and the customer's station.

4.9.2.1. Low Earth Orbit Satellites

LEO satellites surround earth at a relatively low altitude and complete their orbits in about one and a half hour. These orbits are either elliptical or, more usually, circular at a height from 400 to 2000 km, with an inclination of 0 to 90° vs the equatorial plane. A single satellite in a polar orbit provides in principle coverage to the entire globe, although there are long periods during which the satellite is out of view of a particular ground station. Accessibility can of course be improved by deploying more than one satellite in different orbital planes.

These satellites can be further subdivided into big LEO and little LEO satellites. The former offer voice, fax, telex, paging and data capability, whereas the latter offer data capability only.

Their low altitude enables them to cover only small areas at a time and, due to their high velocity (>25 000 km/h), the periods of communications are short with many dead periods. A global communications system using this type of orbit requires a large number of satellites in different inclined orbits. Satellites in LEO are affected by atmospheric drag, which causes the orbit to deteriorate gradually, and their typical life is 5–8 years. However, launches of these satellites are much less costly than into GEO orbit, due to the shorter distance and their much lighter weight. Moreover, multiple LEO satellites can be launched at a time whereas only two GEO satellites can be simultaneously launched. In addition, this system has low-power consumption during transmitters' communications.

Most of the LEO satellites (except the constellations for direct telecommunication systems, for example Iridium and Globalstar) are used for earth or space observation. Examples of these satellites are the Hubble Space Telescope, the Spot family (earth imaging and survey), military observation satellites, the satellites of the Argos system (see Section 4.8.5), and those of the

COSPAS-SARSAT search and rescue system (see Section 4.6.2) [41]. The satellites of these last two systems have polar orbits and belong to the class of little LEOs.

The main parameters for the batteries to be used in LEO missions are:

- Life duration from 2 to 10 years
- About 5000 charge and discharge cycles per year
- Temperature range from 0 to +40°C
- DOD from 10 to 40%, depending on the mission duration and the battery technology
- High vibration levels in relation with launcher characteristics.

The battery weight is not a key factor compared to GEO satellites, as the launch expenses benefit from the lower altitude.

4.9.2.2. Geo-stationary Earth Orbit Satellites

GEO satellites orbit the earth, in the equatorial plane, at an altitude of 35 786 km. Their orbital period exactly matches the rotation of the earth; therefore, the satellites appear to stay stationary over the same point.

Their high altitude enables them to cover a very large communication area and, therefore, three GEO satellites can cover almost all earth's area except for small parts at the poles.

These characteristics enable continuous communication and tracking ability without any dead periods. However, in a voice communication system, there is a round-trip delay of approximately 250 ms due to the huge earth-satellite distance [41]. Furthermore, contrary to the LEOs' communications, GEO transmitters require a high power level.

These satellites are used for telecommunications in fleets covering the earth (e.g. direct home TV broadcasting, voice and Internet: Intelsat, Eutelsat, Astra satellites). GEO satellites are also used for earth observation purposes; indeed, many meteorological satellites (e.g. Meteosat and NOAA) are also positioned in geostationary orbits [41].

Depending on the application, their lifetime is from 12 to 18 years. This corresponds for the battery to 1080–1620 charge/discharge cycles. The key parameters for the batteries to be used in GEO missions are:

- Life duration up to 18 years
- Up to 1620 charge and deep-discharge cycles
- Use of the battery for the electric thrusters used to reposition the satellite (once to twice a day)

- Capability to stand high vibration levels during launch
- High specific energy to reduce weight, as the battery represents a big part of satellite weight (from 10 to 20% depending on the technologies)
- High reliability.

4.9.2.3. Medium Earth Orbit Satellites

MEO satellites orbit the earth at a medium altitude (4000–36 000 km). Their orbit is elliptic, and not circular like that of GEO and most of LEO satellites. The maximum time during which a satellite in MEO orbit is above the local horizon, for an observer on the earth, is of the order of a few hours. A global communication system using this type of orbit requires a modest number of satellites in two to three orbital planes to achieve global coverage. MEO satellites are operated in a similar way to LEO systems.

The satellites forming the GPS constellation, widely used for precise tracking, move in a MEO orbit at ~20 000 km. Other examples are the Galileo system: 30 satellites in three orbits at 23 600 km, inclined 56° vs the equator's plane, and the global navigation satellite system (GLONASS) constellation: 21 satellites in three orbital planes at 19 100 km with an inclination of 64.8°.

To understand the satellite's battery duty, the satellite can be thought of as consisting of payload and platform. The payload comprises all equipment, that is transponders and antennae, delivering services to the user; the platform supplies all the services to the payload, including electric power. This power supply is based on a system including solar panels, batteries and a power control unit (PCU). When the satellite is in sunlight (orbit day), the solar panels provide power to the satellite through the PCU; in addition, the panels recharge the batteries during daylight. When the satellite enters in the earth shadow (eclipse), as the solar panels are no longer efficient, the batteries supply power to platform and payload. Therefore, the batteries are discharged during the eclipse and recharged during the next sunlight period.

Ni–Cd, Ni–H_2 and Li-ion batteries may be used on board satellites. The first system can now be considered obsolete (although still used in LEO satellites), the Ni–H_2 system has been used since mid-1980s and is the most used at present, while the Li-ion technology is rapidly emerging. This last battery is used, for instance in the Galileo constellation and in meteorological LEO and GEO satellites (see Section 4.10). The satellite's size and, consequently, the power demand during eclipses have been continuously growing since the 1980s. By the early 1990s, the power demand was in the range 2 to 3 kW. Today, the power demand is in the range 15–30 kW.

In Table 4.4, characteristics of these batteries and advantages/disadvantages of Li-ion vs Ni–Cd and Ni-MH are outlined.

Table 4.4. Characteristics of batteries used in communication satellites, and advantages/disadvantages of the Li-ion technology vs Ni–Cd and Ni–H_2.

	Ni–Cd	Ni–H_2	Li-ion	Advantages/Disadvantages of Li-ion Batteries
Specific energy (Wh/kg)	30	60	125	Weight saving
Energy efficiency (%)	72	70	96	Less solar panels
Thermal power (scale: 1–10)	8	10	3	Radiator reduction, heat pipes sizes
Self discharge (%/month)	10	60–80	1	No trickle charge and simple management at launch
Temp. range (°C)	0–40	−15–20	10–35	Management at ambient temperature
Memory effect	Yes	Yes	No	No reconditioning
Energy gauge/ monitor	No	Pressure	Voltage	Better SOC determination
Charge management	CC	CC	CC, CV + balancing	More complex charge management
Modularity	No	No	Yes	One cell design, cells can be set in parallel

Source: Adapted from Ref. [41].

The specific advantages of Li-ion batteries, in terms of weight saving, thermal dissipation and simplification of the launch procedure, are depicted in Figure 4.42. The use of Li-ion affords weight reduction (~50% vs Ni–H_2), and reduction of the solar panel and radiator size because of the much lower thermal dissipation. Furthermore, thanks to its very low self-discharge, Li-ion batteries simplify launch operations, as no further charge is required to keep the battery fully charged when the satellite is on the launch platform. As pointed out in the previous sections, the Li-ion technology requires a reliable electronic control for correct operation.

More information on these batteries can be found in Ref. [41].

Power requirements and other characteristics of the transmitters of commercial tracking units are reported in Table 4.5 Background currents are relatively low, and in some cases no background current is needed; if accessory systems, such as data loggers, are present, several mAs are required. During transmissions, currents of tens to hundreds of mAs are needed, with peak currents of up to 2.5 A for a few ms (GSM and CDMA transmission profiles). Transmission times vary between several seconds to several minutes every few hours, up to one transmission a week. The operating life is normally in the range 2–10 years.

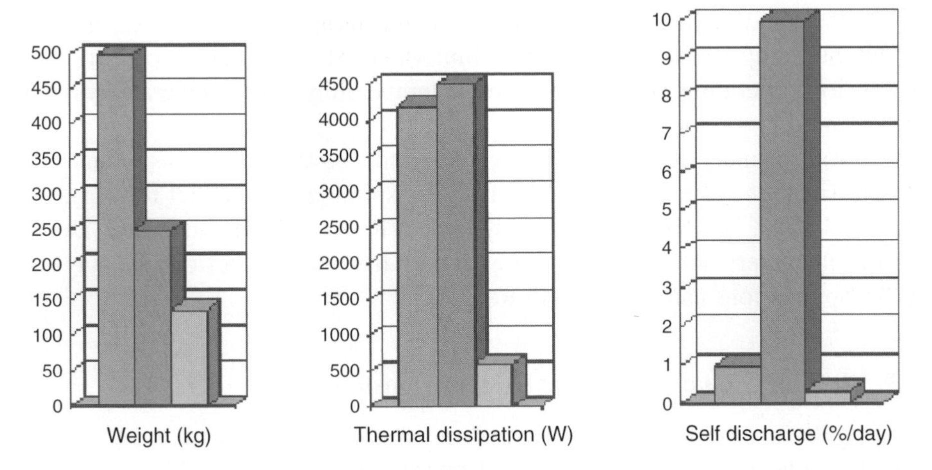

Figure 4.42. Comparison of the three satellite battery technologies (for 8 kW satellite power). Left bar: Ni–Cd; centre bar: Ni–H_2; right bar: Li-ion. Self discharge rates measured at 30°C: this explains the high value for Ni–H_2.
Source: From Ref. [41].

Table 4.5. Characteristics of different transmitters for satellite tracking applications.

Application (no.)	Operating Voltage (V)	Background Current (mA)	Peak Current (A)/ Transmission Type	Working Temperature Range (°C)	Operating Life (Years)
1	7.2	0.4	2.2 CDMA	−40 to 85	3
2	12	0	2 GSM	−20 to 70	2
3	12	15	1 GSM	−25 to 70	10
4	7.2	0	2 GSM	−40 to 85	2
5	8	0.05	2.5 GSM	−30 to 72	0.3
6	12.4	0	2 CDMA	−20 to 60	2
7	36	0.1	2.5 GSM	−40 to 85	2
8	6	0.04	1 CDMA	−40 to 60	3
9	7.2	0.1	2 GSM	−35 to 55	3
10	14.4	0.035	0.725	−35 to 85	3

CDMA, code division multiple access (with this technology, cellular phones using the UMTS standard can share the same frequency channel); GSM, global system for mobile communication.
Source: From Ref. [22].

The transmission currents, duration and frequency depend on the satellite type: communication to LEO satellites usually requires relatively low currents with short communication times and frequency, whereas communication to GEO satellites requires higher currents for frequent and long transmissions [22].

The maximum temperature range for transmitter's batteries is – 40°C to 85°C, but temperatures beyond these limits may also be experienced. In many cases, long operating times at the extreme temperatures (up to several days) may be required.

Both primary and secondary batteries can be used for transmitters. The former include alkaline, Li/MnO_2, Li/SO_2, spiral and bobbin Li/$SOCl_2$, and bobbin Li/$SOCl_2$ + HLC. The latter include VRLA, Ni-MH and Li-ion. Details on the characteristics of these batteries and on their advantages/disadvantages in these applications can be found in Ref. [22].

4.9.2.4. The Global Positioning System Constellation

GPS is a satellite-based navigation system made up of a network of satellites placed into orbit by the US Department of Defence. GPS was originally intended for military applications, but in the 1980s the system was made available for civilian use.

GPS satellites circle the earth twice a day in an MEO orbit, at ~14 000 km/h, while transmitting information signals. In Figure 4.43, a constellation of 24 satellites is shown. By March 2008 the number of active satellites was brought to 32. The additional satellites improve the precision of GPS receiver calculations by providing redundant measurements. With 32 satellites, the constellation

Figure 4.43. Representation of the GPS constellation: 24 satellites orbiting at about 20 200 km from earth in six MEO orbits inclined 55° vs the equator's plane. Each satellite lasts ~10 years, has a weight of ~900 kg, and is 5.3 m across with the solar panels extended. The number of satellites has been increased to 32 by March 2008.

was changed to a non-uniform arrangement, this improving reliability and availability of the system when multiple satellites fail.

GPS receivers take the GPS' information and use triangulation to calculate their exact location. Essentially, the GPS receiver compares the time a signal was transmitted by a satellite with the time it was received. The time difference tells the GPS receiver how far away the satellite is. Each satellite has an atomic clock, and continually transmits messages containing the exact time, the location of the satellite (the ephemeris) and the general system health (the almanac). The receiver, using its own clock, carefully measures the reception time of each message, and has to use the signal of at least three satellites to calculate a 2D position (latitude and longitude) and track movement. With four or more satellites in view, the receiver can determine its 3D position (latitude, longitude and altitude). Once the position has been determined, the GPS unit can calculate other parameters, such as speed, bearing, track, trip distance, etc.

Today's GPS receivers are extremely accurate (to 10–20 m), thanks to their parallel multi-channel design; a 12-channel receiver, for instance, can lock onto 12 satellites. Certain atmospheric factors and other sources of error can affect the receiver's accuracy.

To describe the full GPS system, three segments, that is space, control and user must be considered. The space segment, comprising the GPS satellites, or space vehicles, has been described above. The control segment is represented by five monitoring stations, three ground antennas, and a master control station located in Colorado Springs (USA). The master station contacts each satellite regularly with a navigational update. This allows synchronization of the atomic clocks on board the satellites to within a few nanoseconds, and adjustment of the satellite's internal orbital model. The user segment of the GPS system is the receiver. In general, GPS receivers are composed of an antenna capturing the frequencies transmitted by the satellites, processors, and a highly stable clock (often a crystal oscillator). They may also include a display for providing location and other information.

A more detailed description of the receiver's core subsystem may be helpful [42, 43].

GPS satellites broadcast a data stream at the primary frequency L1 of 1.57 GHz, which carries the coarse-acquisition (C/A) encoded signal to be captured by the receiver's antenna. The GPS receiver measures the time of arrival of the C/A code to a fraction of a millisecond. Apart from the antenna, a GPS receiver typically include the following components (see Figure 4.44):

- RF block for receiving the GPS data
- Digital signal processor (DSP), where the data are converted into digital form
- Microprocessor (MCU)

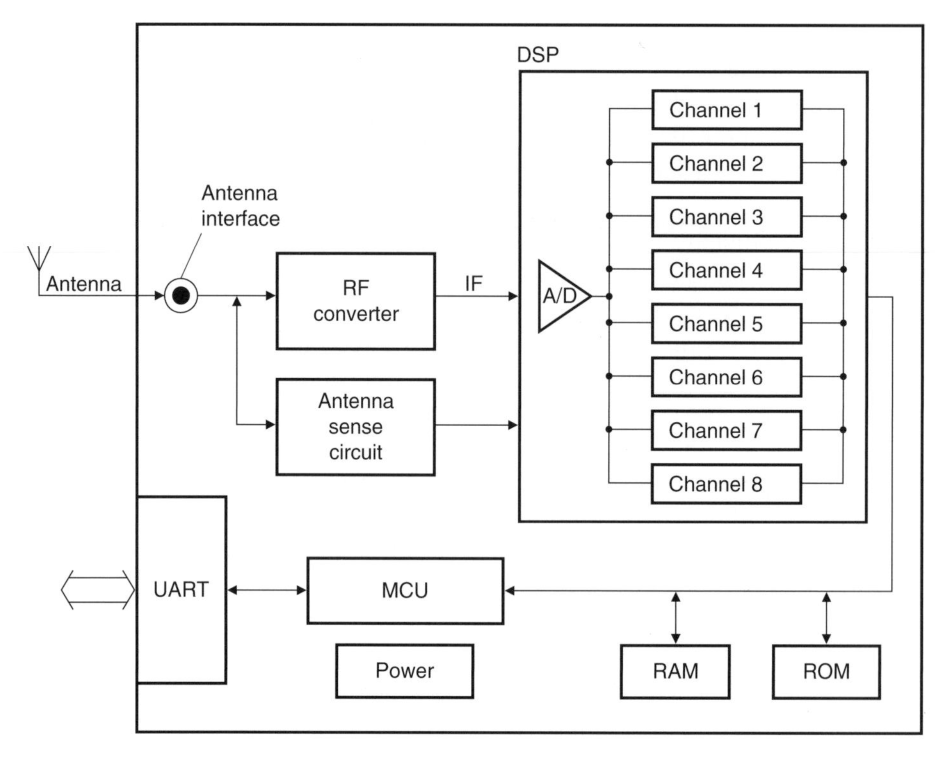

Figure 4.44. GPS receiver's block diagram.
Source: From Ref. [42].

- ROM to store the firmware for processing the data
- RAM
- Battery, usually Li-ion
- Universal asynchronous receiver/transmitter (UART) to send data to the GPS application.

The UART provides serial asynchronous receive data synchronization, parallel-to-serial and serial-to-parallel data conversion for both the transmitter and receiver sections. The data interface could also be an USB – see the more extended block diagram of Figure 3.44; however, an UART has a simpler transport protocol, this requiring much less software and allowing more effective hardware solutions [42].

GPS receivers may range from cheap handheld orientation devices to animal tracking collars and to expensive asset trackers. It is to be noted that some professional receivers can use signals from two or three satellite constellations, for example GPS + GLONASS or GPS + GLONASS + Galileo. Preferred batteries are Li-ion, but alkaline batteries can also be used.

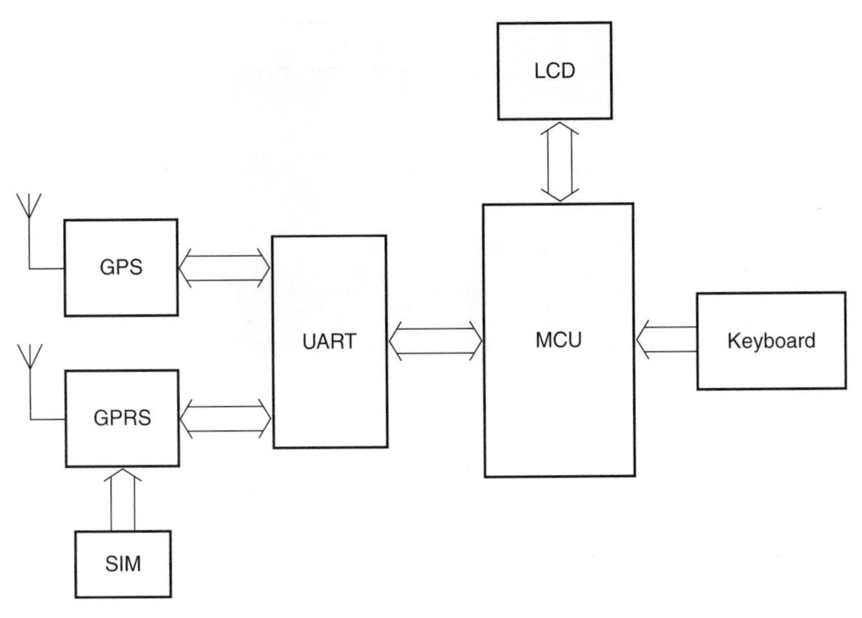

Figure 4.45. Block diagram of a car navigation system.
Source: From Ref. [42].

The block diagram of the popular GPS-based car navigation system can be simplified as shown in Figure 4.45. One may notice the keyboard for programming the navigator, the LCD for viewing data and route, and the GSM/GPRS for communications. With signals from at least three satellites, the vehicle position may be determined with good accuracy.

An example of the use of GPS in wildlife investigations is shown in Figure 4.46 (see caption for details).

Mobile asset tracking, that is goods on trucks or large containers on cargos, requires sophisticated receivers often providing not only a position, but also sensor-based information. They are based on a GPS receiver and a transmitter to communicate with a satellite system, for example one belonging to the LEO category. A tracker of this kind is shown in Figure 4.47 [44]. It works with the technology shown in Figure 4.41; more details are provided in the following.

The low-power, 12-channel GPS receiver allows determining the real-time location of the asset; this information, plus data collected by the sensor system, is transmitted via one-way communication to the Globalstar LEO satellite system, which relays messages to a central station. Finally, the data are transferred to the user via the Internet or telephone network. The tracking units are attached to trucks or container, as shown in Figure 4.47. The simplex transmitter

Figure 4.46. Up: lynx fitted with a GPS receiver – GSM transmitter. Down: how the device works: the position of the lynx is determined via GPS; the lynx transmits data over the GSM to a mobile phone and its position is visualized on a computer.
Source: Courtesy of the Bavarian Forest National Park.

allows a low-power, one-way connection with the Globalstar system, without spending power in listening or negotiating. Furthermore, being the satellites in LEO, the short transmission distance also allows saving power. The device is normally in sleep mode, where it consumes $<10\,\mu A$. It wakes according to a schedule or when requested by alarms or sensors; then, the GPS module is activated to ascertain location. The next step is GPS disabling and transmission via Globalstar satellites. These satellite are placed at an altitude of 1414 km in eight orbital planes of six satellites each; four more satellites are in orbit as spares.

The battery pack of the tracker of Figure 4.47 is formed by eight AA Li/$SOCl_2$ cells plus a HLC capacitor. With this pack, the tracker may deliver two location reports a day for 7 years of continuous service. This hybrid battery can stand the high current pulses required (up to 1.2 A for up to 1.3 s).

Tracking devices not endowed with GPS are also available. They send information in real time at preset intervals to a satellite network, for example Globalstar, which then transfer data to a ground station.

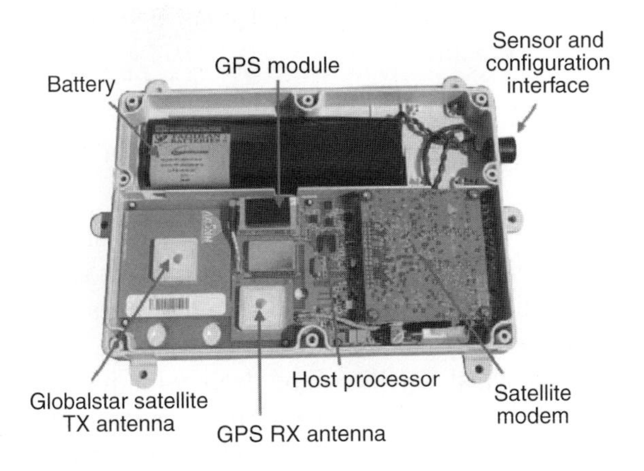

Figure 4.47. Up: asset tracking device based on a GPS receiver and a Globalstar simplex transmitter (courtesy of Axxon LLC); down: two examples of its location (white rectangles).
Source: From Ref. [44].

4.10. Meteorology and Atmospheric Science

Meteorological phenomena are events related to atmospheric variables, that is pressure, temperature, water vapour. Their changes and interactions allow weather forecasting, which has become increasingly accurate in the last decades thanks to the availability of sophisticated instrumentation.

Parallel to the weather science, a study of the atmosphere including such variables as ultraviolet and X-rays, particles of various nature, magnetic phenomena, etc., is important to improve our environmental knowledge and eventually the quality of our life. A major contribution in this field has come from meteorological satellites.

4.10.1. Meteorological Satellites

Satellites can be classified according to their mission: scientific research, weather, communications, navigation, earth observation and military [45]. As is obvious, some satellites may accomplish different missions at the same time, this depending on their payload and schedule.

Here, meteorological satellites are dealt with in some detail. Weather satellites observe the atmospheric conditions over large areas. Some of them travel in a polar orbit, from which they make close, detailed observations of weather over the entire earth. Their instruments measure cloud cover, temperature, air pressure, precipitation, wind, water-vapour content and the chemical composition of the atmosphere. Because these satellites always observe earth at the same local time of day (as they are sun-synchronous), weather data collected under constant sunlight conditions can be compared. The network of weather satellites in these orbits also function as a search and rescue system, as already pointed out in the previous sections.

Other weather satellites are placed in high altitude, geosynchronous orbits. From these orbits, they can always observe weather activity over nearly half the earth's surface at the same time. These satellites photograph changing cloud formations. They also produce infrared (IR) images, which show the amount of heat coming from earth and the clouds [45].

In Figure 4.48, the network of GEO and LEO satellites exploring earth and its atmosphere is depicted. Details on the orbital characteristics of these satellites can be found in Section 4.9.2. As of April 2007, five main GEO satellites were active and could provide images in near real time: their data can be merged into datasets covering all of earth. The satellites are Meteosat-9 (covering Europe and Africa), GOES-East (also known as GOES-12, North and South America), GOES-West (also known as GOES-11, Pacific Ocean), Himawari-6 (also known as MTSAT-1R, East Asia and Australia), Meteosat-7 (Indian Ocean and Asia). All of these satellites have imagers (see later) that view earth in both visible (VIS) and IR wavelengths. VIS-wavelength views see day and night, while IR-wavelength views can show cloud patterns even at night time. Only one of these satellites, Meteosat-9, can take colour pictures; Meteosat-9 supplanted Meteosat-8 in this capacity in April 2007. All other satellites' VIS-wavelength views are only black and white. A nice view of Africa by Meteosat-8 is shown in Figure 4.49.

Weather satellites in polar orbits circle the earth at an altitude of 850–1200 km in a north to south direction or vice versa. As anticipated, these

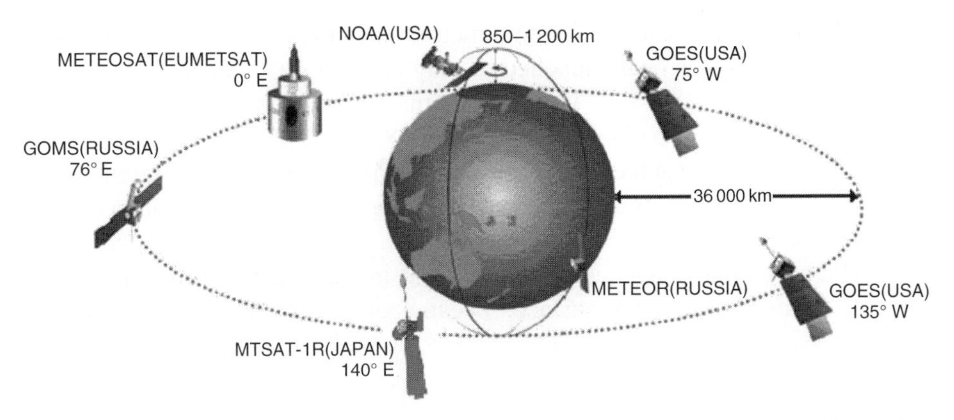

Figure 4.48. Representation of satellites providing meteorological information. Most of them are in a geostationary orbit at ~36 000 km from earth, while some use LEO orbits at ~1000 km.
Source: Courtesy of Japan Meteorological Agency.

Figure 4.49. Meteosat-8 image of earth taken on 1 January 2007. Meteosat-8, located over Africa, has been supplanted by Meteosat-9 in April 2007.
Source: Courtesy of EUMETSAT, NERC Satellite Receiving Station, University of Dundee, UK.

satellites are in sun-synchronous orbits, that is they are able to observe any place on earth and view every location twice a day. Polar orbiting weather satellites offer a much better resolution than GEO satellites as they are closer to the earth. The NOAA series (USA), the European Metop-A, and the Russian Meteor were the main satellites orbiting by October 2007.

A geostationary satellite for both meteorological and environmental observations, GOES-N, has been launched in May 2006 by NOAA and NASA in the USA (Figure 4.50) [46, 47]. GOES-N is the first of three satellites built by Boeing; GOES-O will be launched late in 2008, and GOES-P late in 2009. The three satellites are the same, only their position in space will be different. Their main payload components are indicated in the figure and briefly explained in the following.

The terrestrial Imager is a multispectral instrument that produces visible and IR images of earth's surface, oceans, cloud cover and severe storm developments.

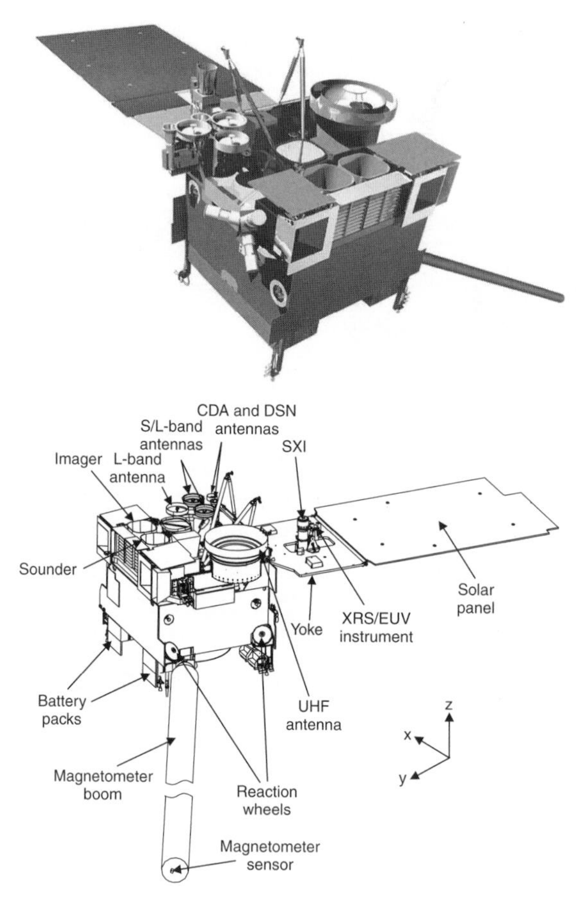

Figure 4.50. Geostationary Operational Environmental Satellite by NOAA-NASA (GOES-N). Up: pictorial view; down: main components. Some characteristics: solar panel (GaAs) providing 2.3 kW; batteries, 24-cell Ni–H_2 (123 Ah); weight, 1.8 tons in orbit; expected life, 10 years.
Source: From Refs. [46, 47].

The multispectral Sounder provides vertical temperature and moisture profiles of the atmosphere, augmenting data from the Imager. Sounder's data are also used in computer models, which produce mid- and long-range weather forecasts. In addition to the terrestrial Imager, the satellite incorporates a Solar X-ray Imager (SXI) to monitor the sun's X-rays for the early detection of solar flares and other phenomena. Solar flares affect not only the safety of astronauts, but also military and commercial satellite communications.

The satellite is equipped with space environment monitoring instruments, which monitor X-rays (XRS), extreme ultraviolet (EUV), and particle emissions – including solar protons, α-particles and electrons. These instruments include a magnetometer that samples the earth's magnetosphere.

The GOES-N, -O, -P satellites, representing the evolution of a series of satellites (GOES 1–12), work to scan small areas as well as all of North and South America, and global scenes (called full-disk images) [46].

They use Ni–H_2 batteries, as other GEO satellites also do. Li-ion technology is currently being validated for these satellites, and has been chosen for LEO satellites, for example the Franco-American Calipso and the Russian Meteor-3M. SAFT (France) is providing modules of these batteries based on cells having energies of 100 Wh and 140 Wh for Calipso and Meteor, respectively.

4.10.2. Launchers

Satellites are brought to space by launchers, an example of which can be seen in Figure 4.51. Launchers need power sources with high specific energy and power, whose functions are to supply power to the electronic equipment, to activate pyrotechnics devices, and to supply power for the control of actuators.

In Figure 4.51, batteries for the Atlas V launcher are shown. All of them are Ag–Zn batteries, produced by Yardney: two batteries are for the main booster power, another four for the flight termination system, and six more batteries supply power to fire various ordnance functions, such as stage separation [49].

These batteries are kept in dry storage and activated, by electrolyte injection, just before launch. Storage can last 4–5 years, but after activation the battery life is 6–12 months. These batteries feature specific energies of ~100 Wh/kg, very high specific powers (140–200 W/kg), and excellent resistance to mechanical stress, of great importance to stand launch vibrations [50].

Other batteries frequently used in launchers are Ni–Cd and Li-ion. SAFT produces Ni–Cd (and Ni–Zn) batteries for the European launcher Ariane, and Li-ion batteries, for example for the Korean launcher Kari [41].

Primary Li batteries (Li/SO_2, Li/$SOCl_2$ and Li/MnO_2) are also used [41].

Figure 4.51. Left: Atlas V launcher (United Launch Alliance) in the 401 format (*source:* From Ref. [48]); right: silver-zinc batteries for this launcher, by Yardney (*source*: From Ref. [49]).

4.10.3. Portable Weather and Ambient Monitoring Stations

In remote areas, more or less sophisticated portable weather stations can be deployed. Depending on the type of sensors available, they can provide information on different meteorological parameters. For instance, the station of Figure 4.52(left) can monitor wind speed/direction, air, soil and crop temperature, precipitation, solar radiation, relative humidity, barometric pressure and ground water level. Data are recorded and summarized hourly and daily by data loggers.

Stations of this type may work with alkaline AA batteries for a 24-h operation time, or for longer time with external batteries, for example a 12 V, 7 Ah Lead-acid battery.

Air quality may be monitored with portable air samplers; particulate matter and/or gases, for example CO and NO_x, may be measured. A 12-Ah lead-acid battery grants 1-day operation.

Fire weather stations monitor, record and transmit meteorological data relevant to fire danger prediction. They may be equipped with standard meteorological sensors (see above) and specific fuel sensors. Some equipment can

Figure 4.52. Left: portable weather station for measuring different meteorological parameters; right: fire station (See colour plate 3) (*source*: courtesy of Campbell Scientific).

monitor conditions near fire lines (see Figure 4.52 (right)) and can be used for fire research during prescribed burns, that is those for agricultural purposes.

Data can be transmitted from the stations over a variety of telemetry options including satellite transmitters, telephone, cellular phone and radio.

4.11. Aerospace Applications

4.11.1. Aircraft

Aircraft equipment powered by batteries includes emergency lighting units, avionics equipment, emergency escape deployment, communication system and inertial reference/navigation system. The on-board generator provides the necessary power but, prior to engine starting or during an emergency, batteries have to take over. These batteries are normally grouped under the name of emergency batteries. The main aircraft battery is used for engine or auxiliary power unit (APU) starting, requiring high peak power capability for a short duration of typically 15–30 s. It also provides the electrical energy necessary, in case of generator failure, to assure return to ground in 30–60 min. In a shorter timeframe, the battery can act as a capacitive filter helping to regulate the aircraft's DC network, reducing the effect of transients, spikes and superimposed AC that may arise from the generating source [41].

In candidate batteries for aircraft applications, several requirements should be considered: capacity, stability in various environments, depth of discharge (DOD), lifetime, ruggedness and weight. Selection of the optimum battery for these applications allows a safe, effective, efficient and economical power storage capability. The optimum battery also facilitates starting operations, minimizes impacts to resources, supports contingency operations and meets demanding loads.

Two chemistries are mostly in use on aircraft today, that is Ni–Cd and lead-acid [51, 52]. Vented lead-acid and Ni–Cd batteries have been used since the very earliest days of flying. The former continue to be used in light aircraft or general aviation but the introduction of valve regulated lead-acid (VRLA) batteries has provided strong competition. On the other hand, vented Ni–Cd batteries dominate the larger aircraft and helicopter applications, while some overlap with lead-acid technology is found in smaller aircraft such as business jets. In Figure 4.53, a Ni–Cd main battery and a lead-acid emergency battery are shown. Li-ion offers large weight savings and has already been adopted in advanced military programs.

Main 24 V aircraft batteries are available with capacities between 15 and 60 Ah (at the C rate). Actual performance depends on the design trade off between energy and power.

The number of batteries installed in the aircraft depends on the system architecture. When more than one battery is required, it is often specified for logistic reasons (reduced number of parts, interchangeability, etc.) that they must be of the same type. In smaller aircraft, there are commonly two batteries. One is used for starting and, in order to maximize power availability, its voltage

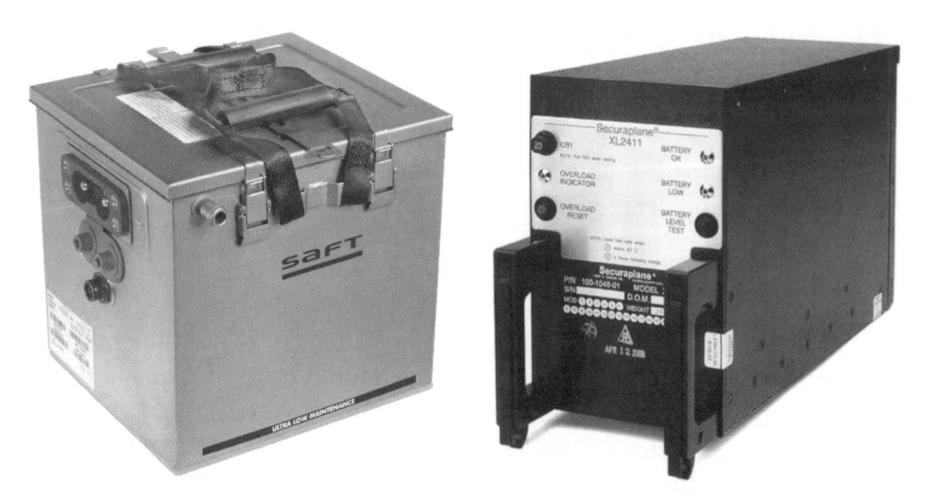

Figure 4.53. Left: 24 V, 50-Ah Ni–Cd main aircraft battery; right: 24 V, 10.5 Ah lead-acid emergency battery.

may be allowed to decrease to 12–14 V. During the start, the second battery supports equipment that would not work properly if the supply voltage fell below 18–20 V. During emergency conditions, the batteries may be connected in parallel to provide more energy. At the other end of the spectrum, the Airbus A380 has a complex architecture requiring three 24 V, 50 Ah Ni–Cd batteries (see Figure 4.53). A fourth 50 Ah battery is dedicated to APU starting. The total weight of the batteries is about 210 kg.

The life duration of an aircraft battery depends on various factors such as number of operating hours, ambient temperature, start frequency and on-board charge. It is therefore difficult to determine in advance how long the expected life of a battery will be in the real situation. Typically, the life of Ni–Cd batteries on long-range transport jets is 6–9 years, while in commuter aircraft is 5–7 years. On the other hand, in military trainers and fighters it is typically 4–6 years. By comparison, the life of lead-acid batteries is half to one-third that of Ni–Cd.

Because weight reduction leads to reduced fuel burn and/or increased payload, the aircraft industry is committed to minimizing weight and this applies to the equipment as well, including batteries and their supporting structures. Significant weight reduction is unlikely to come from traditional battery technologies, so Li-ion should be the solution.

Increasing equipment complexity is leading to a growth in power demand. Today's aircraft have four sources of power generation: mechanical, hydraulic, pneumatic and electric. Weight saving, maintainability, reliability and efficiency suggest increasing electric generation, for example by replacing hydraulic circuits and actuators with their electrical counterparts. These developments are leading to a reconsideration of the aircraft electrical systems, for example by doubling the voltage (115 V) of AC and DC networks or even moving to a 270 VDC network to which the battery would be directly connected. Again, it seems likely that a technology like Li-ion will be best suited to respond to the needs of such "more electric aircraft" systems, because of its lower weight and the reduced number of cells required to support a given voltage [41].

4.11.2. Planetary and Space Exploration Missions

Space exploration missions may be broadly classified into two categories: robotic and human. In addition, there are several science missions, for example deploying telescopes in space, which see a partial human intervention. Planetary missions with robotic exploration include (i) orbiters, (ii) fly-by and sample return missions (iii) landers, (iv) rovers, (v) probes, (vi) penetrators or impactors and (vii) miscellaneous science missions. Human exploration missions, on the other hand, include (i) crew exploration vehicle (CEV), a proposed replacement for the space shuttle; (ii) international space station (ISS); (iii) ascent and descent modules; (iv) extravehicular activities (EVAs); (v) surface exploration missions and (vi) miscellaneous science missions.

These missions are briefly described here, especially in terms of characteristics of the energy storage system [53]. Table 4.6 summarizes the most important missions and their batteries.

4.11.2.1. Robotic Space Exploration

Owing to advances in robotics, communication and power technologies, significant success has been achieved, especially by NASA, in exploring various planetary bodies in our solar system.

Orbiters

Orbiters are typically deployed for mapping planetary surfaces, to assist in studying atmospheres, and to monitor daily global weather. Orbiters, especially for Moon and Mars, use solar energy as their primary power source and require secondary batteries to provide power during eclipse periods. There have been several major Mars missions recently, that is, Global Surveyor, Odyssey, Express and Reconnaissance Orbiter, performing continuous monitoring of Mars. On the other hand, orbital missions to outer planets (i.e. Jupiter, Saturn and Neptune as well as their satellites) may preferentially use nuclear power sources, either with capacitors or long-life rechargeable batteries for load-levelling purposes. Examples of missions that used nuclear power source are Galileo around Jupiter and the on-going Cassini around Saturn.

Orbiters revolve around the planets multiple times, for example 12 times per day around Mars for the Mars Global Surveyor, from pole to pole for several years. The rechargeable batteries for these missions are therefore required to have excellent cycle life ($\sim$5000 cycles/year) combined with good active life (5–10 years). The batteries undergo rapid cycling, that is, discharge for $\sim$30 min followed by charge within the next 60 min to the original state of charge. At partial depths of discharge, some battery systems, for example Ni–Cd and Ni–H_2, have proved able to deliver thousands of cycles. High specific energy and energy density are desirable, but are not as critical as longevity. With the energy available from the solar arrays, the batteries are maintained above subzero temperatures, so that their low-temperature performance is not an issue. Orbiters to outer planets, especially Jupiter, encounter high intensities of radiation and the batteries are required to withstand them.

Nickel rechargeable battery systems, especially Ni–Cd and Ni–H_2, have been the preferred choice for this application, owing to their excellent cycle life characteristics. Li-ion batteries have shown impressive cycle life performance,

Table 4.6. Most relevant exploration missions and their battery systems (mission life at the beginning of 2008).

Exploration	Mission Type	Mission	Purpose	Year	Battery System	Mission Life
Robotic	Orbiter/fly-by	Lunar Orbiter	Earth's moon	1967	Ni-Cd	
		Mariner-4	Venus fly-by	1962	Ag-Zn	3 Years
		Mariner-9	Mars Mapping	1972	Ni-Cd	11 Months
		Viking-2	As above	1975	Ni-Cd	2 Years
		Mars Global Surveyor	As above	1996	Ni-H_2	11 Years
		Mars Climate Orbiter	As above	1998	Ni-H_2	Unsuccessful
		Mars Odissey	As above	2001	Ni-H_2	>6 Years. On-going
		Mars Express	As above	2003	Li-ion	>4 Years. On-going
		Mars Reconnaiss. Orbiter	As above	2005	Li-ion	>2 Years. On-going
	Lander	Viking-2	Mars Surface Exploration	1975	Ni–Cd	3.5 Years
		Mars Pathfinder	As above	1995	Ag–Zn	84 Days
		Mars Surveyor Lander	As above	2001	Li-ion	Suspended
		Phoenix Lander	As above	2007	Li-ion	
	Rover	Mars Sojourner	As above	1995	Li–$SOCl_2$	84 Days
		Mars Explor. Rovers	As above	2003	Li-ion	>5 Years. On-going

(*Continued*)

Table 4.6. (*Continued*)

Exploration	Mission Type	Mission	Purpose	Year	Battery System	Mission Life
		Mars Science Lab.	As above	2009	Li-ion	
	Probe	Galileo	Jupiter Environment	1992	Li–SO_2	>8 Years (Cruise)
		Cassini-Huygens	Saturn Environment	2004	Li–SO_2	As above
		Mars Microprobe-DS2	Mars Sub-surface	1998	Li–$SOCl_2$	Unsuccessful
	Sample Return	Stardust	Comet Wild-2	1998	Li–SO_2	>3 Years (Cruise)
		Genesis	Solar Particles	2000	Li–SO_2	3 Years
		Deep Impact	Comet	2005	Li–$SOCl_2$	>3 Years (Cruise)
	Other	Hubble Space Telescope	Space Studies	1990	Ni–H_2	>18 Years. On-going
Human	Human Lander	Apollo	Lunar Surface			
	Other	International Space Station (ISS)	Base for Different Studies	1998	Ni–H_2	>10 Years On-going
	Exploration Missions	Crew Exploration Vehicle	Transport of Astronauts to Space/Lunar Station	2012	Li-ion	
		EVA	Astronauts Suit and Tools	2012	Li-ion	

Source: From Ref. [53].

combined with reduction in mass and volume, and are ready to replace the nickel systems in future orbital missions.

Fly-by and sample-return missions

A fly-by is a flight manoeuvre in which an aircraft or spacecraft passes close enough to a specified target to make detailed observations without orbiting or landing. Fly-by missions may have science instruments for *in situ* analyses or may be equipped with a sample-return capsule to bring samples back to earth.

Landers

A lander's mission aims at studying the biology, chemical composition, meteorology, seismology, magnetic properties, etc., of the planetary surface and atmosphere.

The landed spacecraft generally experiences extreme temperatures, for example temperature swings are as wide as −120 to +30°C on Mars, while Moon surface temperatures range from −120 to 90°C. Therefore, it is desirable to have batteries able to work in a wide temperature range. Furthermore, landing imposes constraints on mass and volume, due to the impacts on the entry descent and landing system. A reduced battery volume also favours the accommodation of science equipment within the lander.

Governed by the needs of high specific energy and energy density, Ag–Zn batteries were used in the past. However, their limited cycle life became a significant disadvantage. The advent of Li-ion batteries, which have superior cycle and calendar life characteristics, with comparable specific energy and energy density, marked the end of Ag–Zn batteries for this type of missions.

Rovers

Rovers perform similar studies of the planetary surfaces, such as imaging and performing *in situ* analyses of the surface rocks. Unlike landers, rovers are equipped with an additional capability to rove around on the surface as commanded, and perform exploration at different sites. The energy storage needs of rovers are similar to those of landers, but rovers are generally even more constrained in mass and volume.

High-energy primary batteries, specifically $Li/SOCl_2$, were used in the first planetary rover mission (Mars Sojourner) with a limited life. Li-ion batteries have been used in subsequent missions and contributed to extend their lives.

Probes

Another class of planetary *in situ* missions includes planetary probes, such as Galileo, Cassini-Huygens and Mars Deep Space 2 (a microprobe). These missions are equipped with relatively small payload containing science instruments to probe the atmosphere and/or the surface of the planetary object. Being small, they are only equipped with small batteries, whose specific energy and energy density are important criteria because of size and mass constraints. In addition, the batteries must work at the extreme surface temperatures of the planets, for example at −80°C, because of limitations on thermal insulation. Another unique requirement of these batteries is their tolerance to high levels of impact due to severe g-forces during landing.

Li primary systems, especially with liquid cathodes, are the obvious choice for this application, because of their high specific energies and long durations. The Li/SO_2 system was adopted for planetary probes, while for microprobe missions requiring ultra low-temperature performance and high impact resistance, modified $Li/SOCl_2$ batteries were developed.

Impactors and penetrators

In these missions, the spacecraft examines the surface of planetary objects more closely, but not in a passive manner. An impactor of large mass, for example of 370 kg in the case of the recent Deep Impact mission, is made to impact with the planet. Consequently, a crater is expected to be produced, with the surface layers ejected from the crater revealing fresh material beneath. Likewise, the penetrator is expected to penetrate deep into the planetary surface to probe the compositions underneath. More useful information, neither available from the orbiter images, nor from the rover/lander surface studies, is thus expected from these missions probing the inner layers of the planetary surfaces. In terms of energy needs, the impactor and penetrator missions require primary battery systems.

Miscellaneous science missions

These missions include various investigations that involve optical or spectroscopic studies performed outside the earth's atmosphere, so to have reduced interference from our atmosphere and enhanced resolution. The spacecraft are stationed in an orbit around the earth, for example at ~6500 km for the HST. Since these missions are expected to perform scientific measurements over a few years, the power subsystem typically contains a solar array and a long-life rechargeable battery. HST has a $Ni–H_2$ battery, while upcoming NASA missions, such as Kepler (searching for earth's size and smaller planets), will utilize Li-ion batteries.

4.11.2.2. Human Exploration Missions

Human exploration missions were undertaken in the late 1960s under the Apollo program, which involved human landing on moon. Recent human exploration missions have been focused on the International Space Station (ISS).

Space shuttle

The shuttle has been used to transport material to the ISS. It may carry a flight crew of up to eight persons for a mission duration of seven days in space. Its main power source is represented by alkaline fuel cells having a power of 12 kW.

Crew exploration vehicle

The Crew Exploration Vehicle mission is the next-generation shuttle in NASA's future exploration missions to Moon, Mars and beyond. It will probably be powered by Li-ion batteries of 5–10 kWh and with an average power of 4.5 kW. The most critical driver for this application is safety; in addition, the battery system must have high specific energy and energy density. Long cycle life and calendar life are desirable but not critical, while the ability to survive relatively high temperatures is requested.

Planetary ascent stage and descent stage modules

Ascent stage and descent stage modules are intended to facilitate human exploration of Moon surface in the near future and of Mars surface in later years. The estimated average power for these modules would be 4.5 kW and the energy storage is expected to be about 13.5 kWh, both coming mainly from Li-ion batteries. Safety, high specific energy, high temperature resilience, and long calendar and cycle life are requested.

Extra-vehicular activities (EVA)

Extra-vehicular activities are carried out in the following areas:

- Field science, which requires surface mobility for each crew member
- Emplacement science, which includes geophysical and space monitoring
- Technology demonstrations, for example testing of astronaut suit and human-carrying rovers.

The energy needs for these various applications are expected to be less than 1 kWh. In addition to safety, batteries must possess high specific energies and wide operating temperature range.

4.11.2.3. General Characteristics of Space Batteries

Aerospace batteries are developed to meet stringent requirements, so they must have unique characteristics especially in terms of reliability, robustness and safety. The focus on reliability and safety is justified by the high cost of space exploration and the need to ensure the safety of any humans involved. This invariably imposes significant constraints on the chemistry and design of the battery systems. In addition to safety and reliability, space batteries are characterized by the following features [53]:

- Ability to operate under vacuum, as encountered in space, without the out-gassing of any undesirable material that would contaminate the spacecraft instruments or compromise the spacecraft.
- Ability to operate under extreme temperatures.
- Ability to survive high-intensity radiations: Jupiter, for example, has high levels of radiation, $\sim$4 Mrad, which can potentially degrade polymeric materials.
- Ability to operate in any orientation and in microgravity.
- Ability to survive high levels of vibration and acceleration environments.

In addition to these generic requirements, aerospace batteries must meet the "form fit and functionality" needs of the particular mission. As shown in Table 4.6, both primary and rechargeable batteries have been used; however, long-lasting missions can only rely on rechargeable batteries.

Ag–Zn batteries are suitable only for short-term applications, for example, launch vehicles, because of their limited cycle and shelf life.

Ni–Cd batteries, on the other hand, have good cycle life and calendar life characteristics, and demonstrated many years of operation in orbiters. Additionally, super Ni–Cd batteries have outstanding radiation resistance. However, this chemistry does not comply with the requirements of reduced mass and volume for surface missions (landers, rovers and probes).

Like Ni–Cd, Ni–H_2 batteries have demonstrated impressive cycle life performance (>50 000 cycles at 30–40% DOD) and calendar life (>15 years of GEO operational life), superior to any other rechargeable battery system. However, their unfavourable energy densities and specific energies make them suitable only for orbital missions, where cycle life is the main factor.

The use of Li-ion batteries in planetary missions is rapidly increasing, mainly due to their higher gravimetric and volumetric energy and good low-temperature behaviour compared to the nickel systems. These advantages are particularly significant in planetary surface missions. The on-going improvements and demonstrations in both cycle life and calendar life will result in their use in orbital applications as well. Figure 4.54 gives a snapshot of the relative performance of Ni–Cd, Ni-MH and Li-ion batteries in space applications.

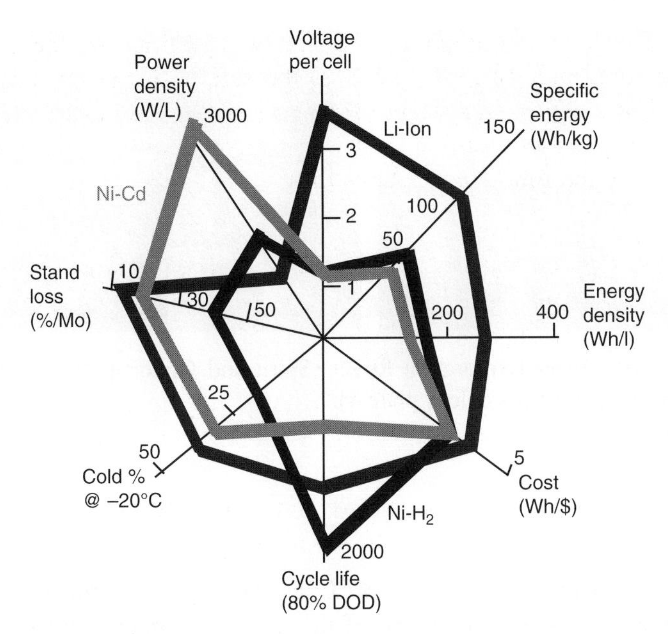

Figure 4.54. Comparison of Ni–Cd, Ni-MH and Li-ion cells for space applications. The stand loss for Li-ion is in %/year.
Source: From Ref. [53].

4.11.2.4. Examples of Missions

All relevant missions are reported in Table 4.6. Hereafter, only few examples of recent missions will be given.

Missions to Mars

Two twin rovers, Spirit and Opportunity, were launched toward Mars in June 2003 and landed in January 2004. These two "robotic geologists" were aimed at examining the surface of Mars and analysing the core of its rocks to detect past water activity. Since their landing, these rovers (see Figure 4.55 (left)) have successfully completed several exploration phases and were still active by March 2008.

The energy conversion system on the rovers is comprised of solar arrays with triple-junction GaInP/GaAs/Ge cells. The energy storage system is based on Li-ion batteries, used for the first time in such planetary exploration missions by NASA [54]. A distinct feature of the rovers' batteries is their ability to operate over a wide temperature range, being enabled by a low-temperature electrolyte developed at Jet Propulsion Laboratory (JPL). Specifically, the

Figure 4.55. Left: Mars Exploration Rover (Spirit and Opportunity are twin rovers); Right: Phoenix Lander (See colour plate 4).
Source: From Ref. [55].

batteries could operate down to −30°C as well as at ambient temperatures (30°C), this reducing mass and size of the thermal management system. They were designed to provide about 200 Wh during launch, about 160 Wh during cruise for supporting anomalies during trajectory control manoeuvres (TCM), about 280 Wh for surface operations, and to provide energy to fire three simultaneous pyros (each with a load of 7 A) several times during the entry, descent and landing (EDL) sequence. Both Spirit and Opportunity rovers have two parallel Li-ion batteries, each with eight 10-Ah cells in series. The 28 V, 20 Ah batteries were fabricated by Yardney (Lithion). In addition to good low-temperature performance, this chemistry showed excellent calendar and cycle life.

Several battery parameters were followed via telemetry during the mission. Measurements were obtained for all individual cell and battery voltages, for the current passing through each battery, and for the battery temperature. The four temperature probes are located on two end cells, on a middle cell and on the battery casing.

Another successful mission to Mars is the Phoenix Mars Mission. Launched in August 2007 and landed in May 2008, it is the first in NASA's "Scout Program". Scouts are designed to be highly innovative and relatively low-cost complements to major missions being planned as part of the agency's Mars Exploration Program. Phoenix is specifically designed to measure volatile molecules (especially water) and complex organic molecules in the arctic plains of Mars, where the Mars Odyssey orbiter has discovered evidence of ice-rich soil very near the surface. Phoenix is a fixed lander (Figure 4.55 (right)), using a robotic arm to dig the surface and collect samples to be analysed with a suite of sophisticated on-deck scientific instruments.

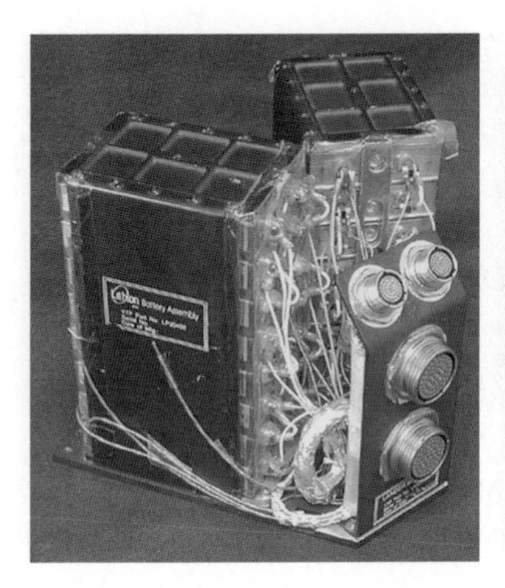

Figure 4.56. Left: Li-ion battery unit for the Phoenix Lander. The unit is formed by 8 cells in series connected in parallel; 28 V, 50 Ah (*source*: courtesy of Yardney); right: $Ni–H_2$ battery unit on board the International Space Station; basic characteristics: voltage 62.3–38 V, energy 4 kW, weight ~820 kg, dimensions 94 × 104 × 48 cm (*source*: From Ref. [57]).

The lander uses Li-ion batteries (Figure 4.56 (left)) similar to the ones used for Mars Surveyor Lander and the forthcoming Mars Science Laboratory [53].

Hubble Space Telescope

The Hubble Space Telescope (HST) is a cooperative programme of ESA and NASA to operate a long-lived space-based observatory. HST is a 2.4-m reflecting telescope, which was deployed in low-earth orbit (600 km) by the crew of the space shuttle Discovery on April 1990. Because of HST's location above the earth's atmosphere, its scientific instruments can produce high-resolution images of astronomical objects. HST's current instruments include three cameras, one spectrograph, and fine guidance sensors (primarily used for accurate pointing, but also for astrometric observations).

In addition to the solar array providing 176 A for battery charging and load support, the power system of the HST (Figure 4.57 (left)) still relays on the six batteries put on board at the time of launch. These are connected in parallel, and contain two modules each of three 88-Ah $Ni-H_2$ batteries. Each of the IPV (individual pressure vessel – see Section 2.4.1.1) batteries, manufactured by Eagle Picher/Lockheed Martin, comprises 23 cells in series, but only 22 cells are electrically connected. The battery's performance has been impressive through all these years in flight [53, 56].

Figure 4.57. Left: Hubble Space Telescope (HST); dimensions: length 13.2 m, diameter 4.2 m; weight 11.1 tons (at the time of launch); right: International Space Station (ISS); dimensions: length (max) 52.7 m, height (max) 71.6 m, span of solar panels 79.4 m; weight 217 tons.

International Space Station

Led by the United States, the ISS) (Figure 4.57 (right)) draws upon the scientific and technological resources of 16 nations: Canada, Japan, Russia, 11 nations of the European Space Agency and Brazil. NASA scientists have gathered vital information on the station that will help with future long-duration missions, as the station has a unique microgravity environment that cannot be duplicated on earth. ISS, the largest and most complex international scientific project in history, is in an orbit (altitude: 400 km, inclination: 51.6°) allowing the station to be reached by the launch vehicles of all the international partners for the delivery of crews and supplies. The orbit also provides excellent earth observations with coverage of 85% of the globe.

In addition to the 100-kW solar arrays, the ISS electrical power system has a Ni–H_2 battery. 38 IPV Ni–H_2 cells (81 Ah) are series-connected and packaged in an orbital replacement unit (ORU, see Figure 4.56 (right)) [57]. The ORU has been designed for about 38 000 LEO cycles over a period of 6.5 years. The batteries, formed by two ORUs in series, should cycle at 35% DOD, but currently operate at approximately 15–25% DOD. The power subsystem has been successfully maintaining power for all on-board loads, thus meeting or exceeding all ISS requirements.

Stardust

Stardust is the first US space mission dedicated solely to the exploration of a comet, and the first robotic mission designed to return extraterrestrial material from outside the orbit of the Moon. The Stardust spacecraft, shown in Figure 4.58 (left), was launched on February 1999, with the primary goal to collect dust from the Wild-2 Comet. Comets are believed to be the oldest, most primitive bodies in the solar system, possibly comprised of some of the basic building blocks of life. They contain the remains of materials used in the formation of stars and planets, holding

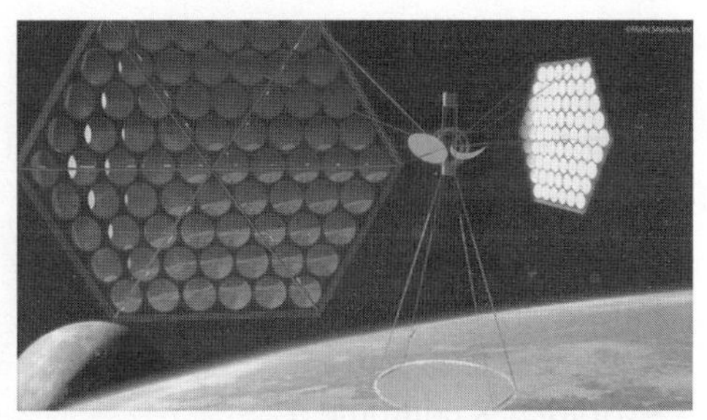

Figure 4.58. Artist's rendition of the Stardust spacecraft (left) and Space Power Solar Station (right).

volatile, carbon-based rich elements that are likely to provide clues about the nature of our solar system. Stardust spacecraft had a rendezvous with the comet in January 2004, after nearly 4 years of space travel. In January 2006, the sample return capsule (SRC) of Stardust successfully returned to earth, bringing the comet dust for further detailed analysis.

Stardust was powered by two solar arrays even when it was nearly three times farther from the sun than is the earth. One Ni–H_2 battery containing eleven 16-Ah CPV (common pressure vessel) cell pairs provided power during eclipses and for peak power operations [58]. The SRC was aided by Li/SO_2 batteries during its re-entry and parachute-supported landing. Two batteries with four cells each were connected in parallel for redundancy purpose. The batteries exhibited a shelf life of ~7 years during this mission.

Space Solar Power Station (SSPS)

The future solar power station (Figure 4.58 (right)), to be operative by about 2015, will be a space-based gigawatt system able to provide electrical power by converting the sun's energy and beaming it to the earth's surface. According to the Electric Power Research Institute (EPRI), photovoltaic arrays in a GEO would receive, on average, eight times as much sunlight as they would on earth's surface. Such arrays would be unaffected by cloud cover, atmospheric dust or by the earth's day–night cycle. Certainly, launching thousands of tons of solar arrays into space will be expensive, but the arrays' area may be reduced by concentrating sunlight with large mirrors or lenses.

Once the solar energy is captured in space, it could be converted into microwave radiation and beamed down to a rectifier-antenna combination located in an isolated area. According to some estimates, an efficient microwave transmitter should have a diameter of 1.5 km, and the receiving station should have an elliptic area of 14×10 km. The microwave energy would then be converted back to DC power. Batteries could be used to store energy, just as is currently done with terrestrial PV arrays.

4.12. Military Applications

Military applications powered by batteries can be broadly divided into the following categories:

- Mobile Communications
- Underwater warfare
- Soldier's equipment
- Autonomous underwater vehicles
- Missile and Ammunitions Systems
- Aviation
- All-electric ships
- Autonomous aerial vehicles

A more specific list of applications includes [59]:

- Night vision equipment
- Data terminals
- GPS
- Radios
- Gas masks
- Loudspeakers
- Counter measures
- Jammers
- Radiation detector
- Mini ocean buoy system
- Undersea mines
- Robotics
- Rescue radio/beacon
- Chemical agent monitors
- Antennas
- Scramblers
- Radars
- Range finders
- Weather instruments
- Cooling systems
- Handheld metal detector
- Laser-based battle simulation
- Thermal imaging
- Portable electronics

A few applications are described in the following.

4.12.1. Ammunitions

A projectile or missile may be endowed with electronic components, for example to reach and hit a target with the utmost precision. The electronics is contained in the so-called fuse and obviously needs a battery to work. In more detail, a fuse: (1) senses the presence of a target and initiates function of the ammunition; (2) increases effectiveness of the ammunition and (3) provides for safe handling and separation of projectile. According to the activation mode, fuses may be divided into four groups: proximity (within a selected distance from target), time (at a selected time from launch), point-detonate (PD, at the instant of impact) and delay (following a short delay after impact) [60]. An example of multifunction fuse is reported in Figure 4.59.

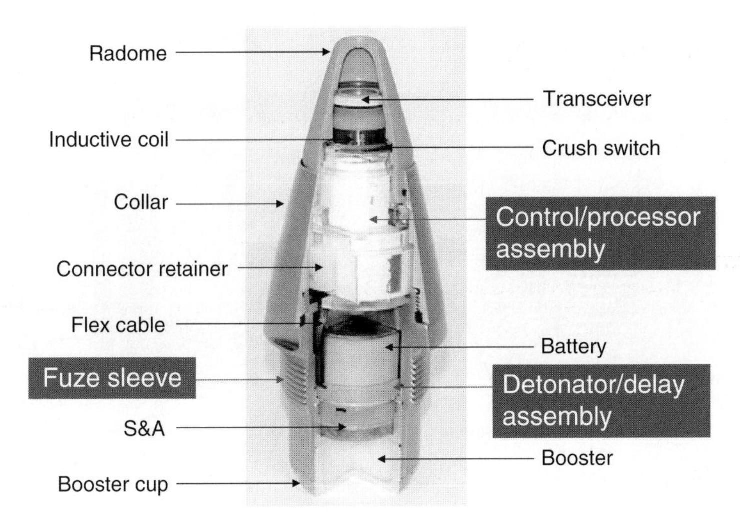

Figure 4.59. M 782 fuse used in artillery ammunitions. It has four-function modes: time, proximity, point detonate and delay. Dimensions: length, 15 cm; diameter 4.4 cm. Weight: 746 g.

Batteries powering the fuses belong to the class of reserve batteries. They are kept inactive even for a long time, and activated just prior to launch by making the electrolyte available to the electrode stack.

Two types of activation are known: (a) by breaking a reservoir containing a liquid electrolyte and (b) by heating a solid electrolyte so to melt it.

In Figure 4.60, a liquid electrolyte reserve cell and a thermal cell are shown. In terms of chemistry, there are three main reserve batteries using liquid electrolytes: lead-acid, $Li/SOCl_2$ and Ag–Zn. All these batteries are spin-activated: artillery projectiles possess high-spin forces that are exploited to cause rotation of the blades and, so, cutting of the electrolyte reservoir and battery activation [61].

The lead-acid battery does not contain H_2SO_4 as an electrolyte, but HBF_4 for a better low-temperature performance. It usually consists of a stack of bipolar electrodes, if high voltage is needed, or a set of alternate anodes and cathodes connected in parallel to give a high current, or a combination of the two.

The $Li/SOCl_2$ reserve battery is used, for instance, in communications jammers and for projectiles and sub-munitions that must operate while being slowed by parachute. One advantage of the reserve construction for lithium cells is that no passivating layer forms on the anode surface during storage, thus allowing very high rates of discharge with no voltage delay.

Automatically activated Ag–Zn reserve batteries are suitable for discharges of 20 s or longer. The cells are assembled with the KOH electrolyte stored in a separate compartment. One type of Ag–Zn single-cell reserve battery has a built-in DC/DC converter in order to raise the low voltage of the cell to a more useful,

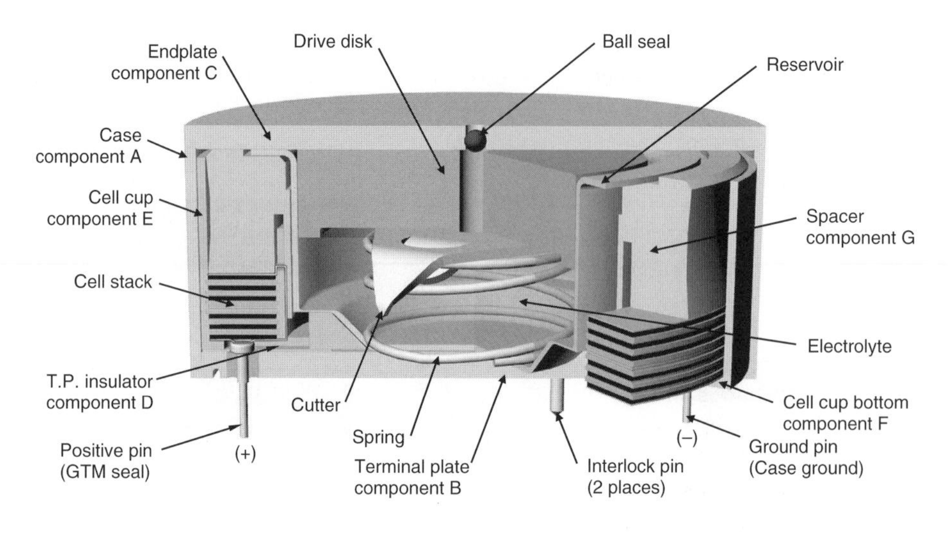

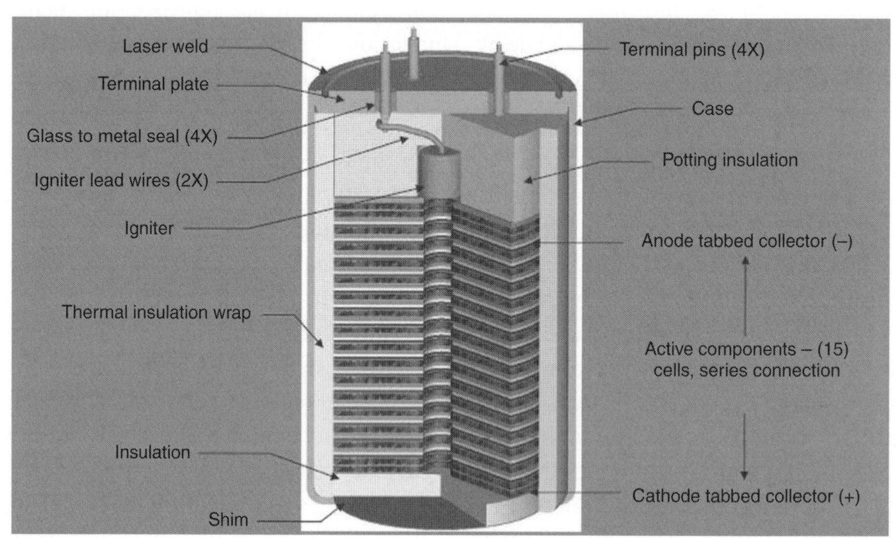

Figure 4.60. Up: reserve battery activated with a liquid electrolyte; the electrolyte is contained in a reservoir cut by blades at the moment of activation; down: thermal battery: the solid electrolyte in the electrode stack is melted to activate the battery. *Source*: From Ref. [60].

stabilized level. These reserve batteries can deliver many kW for several minutes and have been used for combat torpedoes and for many types of guided missiles.

Thermal batteries are activated by an energy impulse from an external source that ignites pyrotechnic materials within the battery to melt the electrolyte, which becomes conductive. The exact lifespan of individual thermal batteries depends on both design and application. A common thermal battery for military applications uses Li as an anode, eutectic mixture, for example LiCl and KCl (45:55 by weight) as an electrolyte, and a cathode chosen in a wide variety of materials ($CaCrO_4$, $K_2Cr_2O_7$, $PbCrO_4$, V_2O_5, FeS_2, etc.) [62].

Liquid-electrolyte batteries are generally chosen for rotating ammunitions (rifled artillery), while thermal batteries are the preferred choice for rockets and missiles [60]. Thermal batteries feature higher power and shorter mission times vs those with liquid electrolytes.

General requirements of reserve batteries include the following [60]:

- Current pulses >1 A
- Working voltage (3–20 V) reached in a few ms after activation
- Operating temperatures: −45 to +100°C
- Storage temperatures: −50 to +110°C
- >20-year maintenance-free storage.

4.12.2. Unmanned Air Systems

An unmanned air vehicle (UAV) is a pilotless, small aircraft with an on-board computer, or microprocessor, together with control, sensor and communication electronics. Any aerial application, in which the payload weighs less than an average adult male (~85 kg), could be performed more safely, less expensively and in a more environmentally friendly way by using an UAV. UAVs afford missions that are impossible for manned aircraft. These latter can fly at a maximum altitude of 20 km and for a maximum endurance of 10 h; an UAV can fly at an altitude of 30 km with an endurance of up to 60 h.

According to their weight and power, unmanned vehicles can be divided into [63]:

- Micro air vehicle (MAV): <<1 kg, 1–20 W
- Small UAV: 2–10 kg, 50–300 W
- Mid UAV: 20–100 kg, 1–10 kW
- Unmanned combat air vehicle (UCAV): >>2000 kg, 100 kW.

Unmanned aircraft already exploit more forms of propulsion than do manned aircraft, from traditional gas turbines and reciprocating engines to batteries and solar power, and are exploring, for example, fuel cells and even nuclear isotopes. Micro and small UAV use batteries for their propulsion. Examples are reported in Figure 4.61. Obviously, larger UAVs make use of batteries for the same tasks previously mentioned about manned aircraft, for example avionics.

Different batteries for UAVs are mentioned in the literature: Zn-air, Ni-MH, Ni–Cd, Li-ion and the emerging Li/S. The high energy of Zn-air is exploited in the Dragon Eye of Figure 4.61, while the Pointer uses a Li-ion battery.

Figure 4.61. Examples of battery-powered UAVs. (*Top*) BATCAM: weight 0.38 kg, length 0.61 m, payload 40 g, endurance 18 min; (*middle*) Dragon Eye: weight 2.0 kg, length 0.74 m, payload 0.45 kg, endurance 1 h: (*bottom*) Pointer: weight 3.7 kg, length 1.86 m, payload 0.45 kg, endurance 2 h.
Source: From Ref. [64].

4.12.3. Soldier Equipment

A modern soldier's equipment includes a series of battery-powered devices. In Figure 4.62, a dismounted soldier wearing a complete set of devices is shown (details can be found in the caption) [65].

The complete equipment is rather heavy, and reducing it as much as possible is mandatory. This action may start from the battery packs, for instance reducing the multiple packs to two [66]. Non-critical devices should operate from a common rechargeable or primary pack, while critical devices should operate from the common pack and an embedded rechargeable battery for backup or extended runtime. Tight integration of an embedded

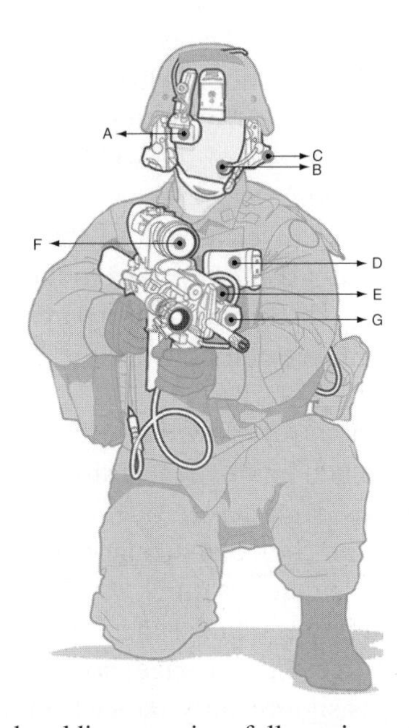

Figure 4.62. A dismounted soldier carrying full equipment as specified. 1. Helmet subsystem: the helmet weighs approximately 1.5 kg. An eyepiece *(A)* bolted to the helmet acts as a full-colour computer interface, displaying maps and images from the camera on the weapon. A microphone *(B)* and headphones *(C)* allow communications. 2. Soldier control unit: a computer "mouse" *(D)* is strapped to the chest and used to toggle among screens in the eyepiece; it is also used to key the radio and send data to other soldiers. 3. Weapon subsystem: a laser rangefinder *(E)* mounted on the rifle pinpoints enemy positions and sends their coordinates across the network. The thermal sight *(F)* provides night vision, and a digital camera *(G)* lets soldiers send video clips to commanders. 4. Controller: a toggle switch (hidden by the soldier's left hand) is used to control the weapon's laser and video sights; it can also be used to operate the radio.
Source: From Ref. [65].

battery pack and host equipment is a critical issue. In the Land Warrior System, a program initiated by the US Army in 1996 and now stopped, the total battery (1.1 kg) could provide 8–24 h of power for the devices and computer. Primary batteries lasted 4–12 h, while the rechargeable ones lasted 8–10 h.

Batteries for the soldier must meet the following environmental requirements [66, 67]:

- *Extreme temperatures.* Military equipment must operate in both temperature extremes from −40 to +60°C
- *Water submersion.* The equipment has to withstand temporary immersion in several meters of water. Watertight and hermetic seals are achieved with ultrasonic welding of plastic enclosure pieces
- *Excessive shock and vibration.* The equipment must withstand a drop test of at least 1 m onto a concrete floor. Ultrasonic welding of the battery pack enclosure maintains pack integrity during such tests as the resultant joint strength can match the strength of the welded material
- *Minimal electro magnetic interference (EMI).* Minimizing EMI and radiation reduces any negative impact on the electronics of the portable device. A conductive paint can be applied to the inside surface of the pack enclosure. Additionally, this protection ensures that portable devices are undetectable by surveillance equipment.

From a performance point of view, requirements for military-grade portable power sources are much more demanding than commercial batteries. Military equipment requires high-power rate, safety and lightweight. These requirements pose limits to the choice of primary batteries in terms of chemistry and design. Combat readiness requires quick response, that is immediate availability of full and consistent power dictates the use of primary batteries that have no voltage delays, even after long storage periods. The implementation of modern lithium batteries has significantly increased power density and reduced weight, when compared to earlier alkaline and carbon-zinc technology. The Li/SO_2 battery technology is still widely used. However, apart from possible voltage delays in some conditions, a major concern is the pressurized cylinder containing SO_2, which can explode if punctured by enemy fire or physical abuse. Modern portable electronic equipment is now moving towards the Li/MnO_2 chemistry, which has higher energy density and higher safety than Li/SO_2.

In recent years, several armies have shifted to the use of rechargeable batteries for peacetime, training and garrison operations, as well as specific combat uses, to save in cost and transportation. Li-ion is the preferred system for the advantages pointed out repeatedly in the previous sections.

4.12.4. Miscellaneous Naval Applications

Naval warfare makes use of various devices that can be powered by, or anyway take advantage of, special batteries.

4.12.4.1. Torpedo

A marine torpedo is a self-propelled explosive projectile, which explodes on contact with, or in proximity to, a target. It may be launched by, for example submarines, surface ships or naval mines. These last have a submerged sensor platform that releases a torpedo when something hostile is detected. Torpedoes may have various dimensions, but can be roughly classified as light (diameter: 32 cm) and heavy (diameter: 53 cm). Different propulsion technologies are possible, and one of them uses batteries. Several chemistries may suit reserve Ag–Zn, sea-water activated, Al/AgO and Li-ion.

4.12.4.2. Submarines and all Electric Ships

Nuclear and conventional submarines may use batteries for energy storage, lighting and electronics power supplies, and for tactical mobility in combat situation. The all-electric ship is a very recent concept and it will take some time before realization. In this ship, batteries will be used not only for propulsion but also to power auxiliary systems using, at present, steam, hydraulic or pneumatic power.

Today, most military vessels use mechanical drive. These systems convert the engine's high-speed revolutions to low-speed revolutions using a set of gears. These ships actually have two sets of engines: one is used for ship propulsion, and a second and separate set, connected to generators, is used to create on-board electricity. With an electric drive, a generator converts the engine's high speed into electricity. Ships with such a system can be designed so that a single set of engines produces a common pool of electricity for propulsion and non-propulsion use.

Ideal batteries for these naval resources are large Li-ion batteries, for example the Dauphin modules developed by SAFT (3.5 V, 9 kWh, 120 kg).

4.12.4.3. Unmanned Undersea Vehicles

These vehicles are often similar to the gliders described in Section 4.8.4. A typical military mission for an unmanned undersea vehicle (UUV) is to map an area to determine if there are any mines. With a convenient sensor equipment (sonar, magnetometer, etc.), other missions are possible, that is anti-submarine warfare, navigation aid, payload delivery, time critical strike, etc.

For these demanding applications, Ag–Zn, primary Li (Li/SO_2 and Li/MnO_2) and Li-ion are all suitable.

4.12.4.4. Sea Mines

There are countless types of offensive or defensive naval mines, which can undertake different missions. All of them require a low-magnetic signature to avoid malfunctions. Primary Li/MnO_2 batteries are ideal for this application.

4.13. Robotics

According to the definition of the Robotics Institute of America, a robot is a reprogrammable, multifunctional manipulator designed to move material, parts, tools or specialized devices through variable programmed motions for the performance of a variety of tasks [68].

Robots generate specific motion of the robot joints, simultaneously allowing tooling or sensors to perform certain functions, when the arm is moving or at specific operational configurations. The arm and attached tooling may perform the operations themselves (such as painting) or carry parts to other devices which perform the operations.

Robots are used in almost any industry where repetitive tasks are involved, or the task is manually difficult or dangerous, such as [68]:

- welding, painting or surface finishing in the aerospace or automotive industries
- electronics and consumer products assembly and inspection
- inspection of parts by robot-assisted sensors
- underwater and space exploration
- hazardous waste remediation in government labs, nuclear facilities and medical labs.

A more complete list of industrial activities includes:

Welding	Material handling	Other
Arc welding	Pick and place	Grinding
Spot welding	Dispensing	Drilling
Mig welding	Palletising	Thermal spray
Tig welding	Part transfer	Bonding/sealing
Laser welding	Machine loading	Painting
Resistance welding	Machine tending	Metal finishing
Plasma cutting	Order picking	Coating
Flux cored welding	Packaging	Assembly
Electron beam		Material removal

Basically, a robot consists of [69]:

- A mechanical device, such as a wheeled platform, arm, or other construction, capable of interacting with its environment.
- Sensors on or around the device that are able to sense the environment and give useful feedback to the device.
- Systems that process sensory input in the context of the device's current situation and instruct the device to perform actions in response to the situation.

For a machine to qualify as a robot, five basic parts are to be present: microcontroller, arm, drive, end effector and sensor [70].

Microcontroller. Every robot contains a microcontroller, which keeps the pieces of the arm working together. The controller also allows the robot to be networked to other systems, so that it may work together with other machines, processes, or robots.

Arm. The arm, that may have any shape and size, is the part of the robot that positions the end-effector and sensors to do their pre-programmed work. The number of robot's joints determines the degree of freedom. Each joint is said to give the robot one degree of freedom. Therefore, a simple robot arm with three degrees of freedom could move in three ways: up and down, left and right, forward and backward. Most robots have six degrees of freedom: in this way, a robot can reach any possible point in space within its working range.

Drive. The drive is the engine driving the links, that is the sections between the joints, into their position. Most drives are powered by air, pressure or electricity.

End-effector. This is the hand connected to the robot's arm. It could be a gripper, a vacuum pump, tweezers, scalpel, etc. Some robots can change end-effectors, and be reprogrammed for a different set of tasks.

Sensor. Sensors provide the robot with a variety of information about its surrounding. For instance, sensors enable the robot to see in the dark, detect radiation, etc. This information is sent back to a controller that may act accordingly.

Most robots are controlled, operated and programmed by using a device called teach pendant. Some modern pendants are battery-powered and can connect wirelessly to the robot's control unit by using the Wi-Fi technology (IEEE 802.11a/b/g – see Section 4.18). An example of wireless pendant is shown in Figure 4.63: it has an operating range of up to 100 m and autonomy of 6 h.

Figure 4.63. A wireless teach pendant to keep robots under control using the Wi-Fi technology.
Source: Courtesy of Comau.

The types of technologies that are often integrated with a robot are [68]:

- vision systems
- end-of-arm tooling and special compliance/manipulation devices
- welding technologies
- optical devices such as lasers
- sensors such as acoustical or other proximity sensors
- wrist sensors capable of measuring wrist forces and torques
- control software and hardware, such as AC or DC motors, encoders, tachometers, amplifiers
- part delivery systems such as conveyors, part feeders
- application software, interface software
- real-time operating systems, programming languages
- communication protocol/networks
- I/O devices such as programmable logic controllers (PLCs), either discrete or analog.

The advantages of using robots may be summarized as:

- Greater flexibility, reprogrammability, adjustable kinematic ability
- Quicker response to inputs than humans
- Improved product quality

- Equipment use in multiple work shifts
- Reduction of hazardous exposure for human workers
- Automation less susceptible to work stops.

In contrast, there are disadvantages, such as:

- Replacement of human labour (more unemployment)
- Hidden costs because of the associated technology that must be purchased and integrated into a functioning section.

4.13.1. Details on the Robot's Hardware

4.13.1.1. Motors

A variety of electric motors provide power to robots, allowing them to move materials, tools, etc. Motors currently being used in robots are: DC, AC, stepper and servomotor [69]. Servomotors are useful in many kinds of smaller robots, because they are compact and quite inexpensive. They have built-in motor, gearbox, position-feedback mechanisms and controlling electronics. Standard radio-controlled servomotors, which are used in model airplanes, cars and boats, are useful for making arms, legs and other mechanical appendages, which move back and forth rather than rotating in circles.

4.13.1.2. Driving Mechanisms

Common driving mechanisms are used in robots: gears and chains, pulleys and belts, gearboxes [69]. All of them can transmit motion, thus allowing the robots to walk and/or move their arm(s).

4.13.1.3. Power Supplies

The main sources of electrical power for robots are batteries; solar cells are used in open-air robots, and in this case batteries are used for backup. The robot platform runs off two separate battery packs. In this way, the motor uses one power source while the electronics can run off the other.

Given the enormous variety of robots, several batteries different in chemistry and size can be used. Small robots often use primary batteries: the alkaline ones are cheaper and convenient for many uses, but primary Li have a better performance, especially at low temperatures, and longer shelf life.

However, even extremely small robots may use rechargeable batteries. Seiko Epson's flying robot (Figure 4.64) is powered by a Li-polymer battery.

Figure 4.64. A "micro" flying robot developed by Seiko Epson. Weight: 12.3 g (battery included); height: 8.5 cm; diameter: 13.6 cm; battery weight: 3.7 g; flight time: ~3 min; power consumption: 3.5 W.
Source: Courtesy of Seiko Epson.

This helicopter has two tiny ultrasonic motors driving two propellers in opposite directions for lift. It has an on-board camera able to beam images wirelessly. It could be used for surveillance and for searching dangerous and narrow places. Specifications can be found in the caption.

Lead-acid, Ni–Cd, Ni-MH and Li-ion rechargeable batteries have all been used. More details will be given in the examples of Section 4.13.3.

4.13.1.4. Electronic Control

There are two major hardware platforms in a robot: mechanical and electronic (5 V signals). These two platforms need to be bridged in order for digital logic to control mechanical systems. The classic component for this is a bridge relay. A control signal generates a magnetic field in the relay's coil that physically closes a switch. MOSFETs, for example, are highly efficient switches that can operate as a solid-state relay to control the mechanical systems.

4.13.1.5. Sensors

Examples include the following [69].

- *Logical sensor.* It is a unit of sensing or module that supplies a particular perception. It consists of the signal processing, from the physical sensor, and the software processing needed to extract the perception.

- *Proximity sensor.* It measures the relative distance between the sensor and objects in the environment.
- *IR sensor.* It is another type of proximity sensor. It emits near-IR energy and measures whether any significant amount of the IR light is returned.
- *Bump and feeler sensor.* Robot's tactile sensing is done with a bump and feeler sensor. Feelers are constructed from sturdy wires; a bump sensor is usually a protruding ring around the robot.

4.13.2. Examples of Mobile Autonomous Robots

Robotic platforms can be built to autonomously patrol and secure the interior of, for example warehouses. A robot with these characteristics, developed by the US Department of Defence, also accomplishes intruder detection with a module employing microwave and passive-IR motion sensors and a controllable video zoom camera. An RFID module tracks stored inventory and reports missing, moved and found items.

This mobile robot (Figure 4.65 (left)) is powered by a 24-V battery (e.g. lead-acid). The RF transceiver (interrogator) for bi-directional communication with interactive tags has a range of up to 45 m and is powered by a 6-V, 600 mAh Li primary battery.

The robot of Figure 4.65 (right) has been developed at the Dartmouth College (UK) to deploy instruments in hostile areas, for example Antarctica. It has solar panels as a primary energy source and three Li-ion batteries for power storage. Potential missions include deploying arrays of magnetometers, seismometers, radio receivers and meteorological instruments. In this way, such information as ionosphere disturbances, presence of crevasses and glaciological

Figure 4.65. Left: a mobile robot for inspection of interiors; right: an instrument-deploying robot in Antarctica (empty weight, max: 75 kg; payload, max: 15 kg; travel range: up to 500 km; max dimensions: $1.4 \times 1.5 \times 1$ m; speed, max: 0.8 m/s).

profiles can be obtained. Robot arrays could also provide high-bandwidth communications links and mobile power systems for field scientists. A suitable Li-ion battery (by Ultralife, UK) has a voltage of 14.4 V and a capacity of 12 Ah. It may provide a continuous current of 12 A, and a pulse current of 36 A for 5 s.

A human-like robot able to act autonomously and perform uninterrupted service at home, in office or industry has been developed by Honda (Figure 4.66). Honda's work on intelligence technologies has enabled the latest model of this robot ASIMO to operate in an environment with people and other ASIMOs, so that its practical use in a real world requiring coexistence with people does not seem too far [71].

Multiple ASIMOs can now share tasks by adjusting to the situation and work together in coordination. For example, if one ASIMO is idle while recharging, other ASIMO robots will step in and perform assigned tasks. ASIMO identifies oncoming people through its eye camera, calculates travelling direction and speed, predicts forthcoming people's movements, and chooses the most appropriate path so that it will not block the movement of others.

When the battery state of charge falls below a certain limit, the robot automatically identifies, and walks to, the closest available battery charging station and recharges while standing (see figure).

ASIMO is powered by a 51.8-V Li-ion battery (in substitution of a previous Ni-MH battery) granting 1-h operation. The battery weighs 6 kg and

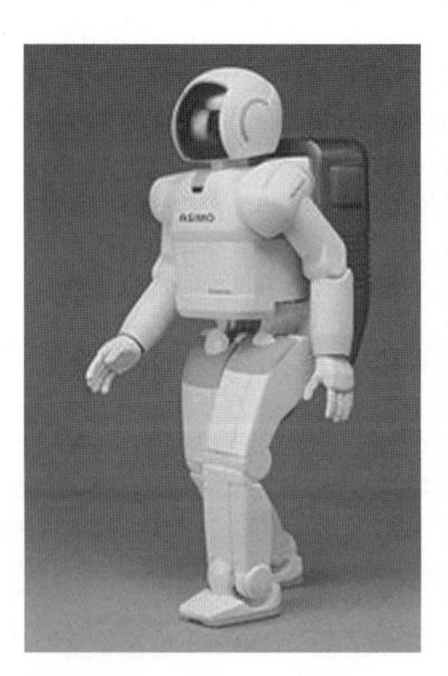

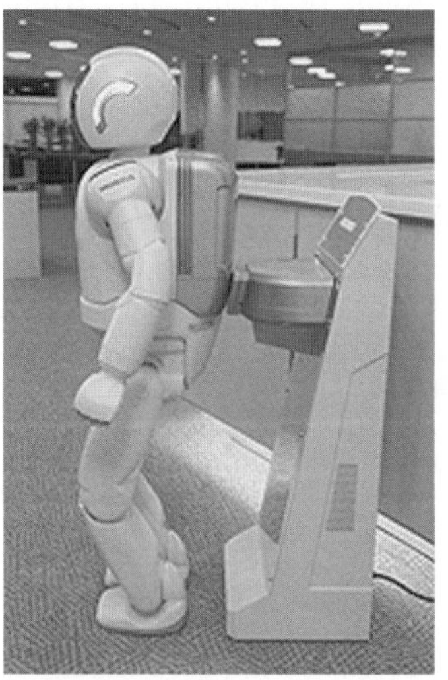

Figure 4.66. Honda's robot ASIMO (the acronym stands for: advanced step in innovative mobility). On the right, the way it recharges itself at a charging station is shown. The robot stands 130 cm and weighs 54 kg; it can run at 6 km/h and has a normal walking speed of 2.7 km/h; operating time: 1 h; degrees of freedom (DOF): 3 for the head, 4 for the hands, 14 for the arms and 12 for the legs.

is located in the robot's backpack. The use of a relatively large battery is justified by the complexity of the robot, as also evidenced by the presence of 34 separate servomotors.

4.13.2.1. Mobile Microrobots

Mobile autonomous microrobots are defined as millimeter-sized mobile robots with on board power and control. These robots offer numerous advantages due to their size and low-power requirements. For example, they could be used to add mobility to sensors in large-scale sensor networks as the size of those integrated sensors shrinks (see Section 4.4).

There is a general tendency to classify small robots as "microrobots". A more precise classification might be the following: minirobots (volume, $\sim 1\,dm^3$), microrobots ($\sim 1\,cm^3$), nanorobots ($\sim 1\,mm^3$) [72]. With this classification, the "micro" flying robot of Figure 4.64 should be more correctly defined as minirobot (its volume is $\sim 1.2\,dm^3$).

Another concept to be clarified is autonomy. When the robot has an onboard power supply or a generator (batteries or solar panels) it is energetically autonomous. In addition, when it is able to decide how to react to the environment it has an autonomous behaviour. A really autonomous robot possesses both energetic and computational autonomy [72].

Just like larger robots, a microrobot consists of an electronic section, a mechanical section, sensors and power source. Miniaturization is a challenge for all components, but is perhaps more critical for the power source, which has to be simultaneously small enough and capable of providing the robot with the power it needs to move and act.

The tasks expected from mobile microrobots (MMRs) may be summarized as navigation, localization, mapping, multirobots cooperation, research and education. Sending one or more MMRs into a small environment will make it possible exploring and modifying that environment. In other words, sending intelligent and autonomous machines into a micro-world will allow them to work autonomously and perform useful jobs without controlling human beings behind them.

Different microrobots have been developed by various laboratories since about 1995. Figure 4.67 shows some of them, while in Table 4.7 their main characteristics are listed. With the exception of the microrobot developed at the MIT, all have dimensions within one cubic inch. The runtime is normally between 10 and 30 min, but an impressive 600 min is reported for the microrobot developed at the Lausanne University. The batteries are commercially available. Some microrobots are reprogrammable and allow modularity, that is accept extension modules (sensors, tools).

Actuators are one of the major problems in designing miniature robots, as those used in large systems are very difficult to build in small scale. One

Figure 4.67. Examples of microrobots. See Table 4.7 for description.
Source: From Ref. [72].

Table 4.7. Specifications of autonomous mobile microrobots.

Robot	Volume (cm^3)	Battery	Motor	Run Time (min)	Speed (cm/s)
1. Inchy by LAMI	16	Ni-MH	Smoovy 5[a]	30	30
2. New MARV by Sandia	4	watch	Smoovy 3[a]	10	1
3. EMRoS by Seiko Epson	~1	watch	watch mov.	10[b]	10
4. Ants by MIT	36	Ni-Cd	3 DC motors	20	15
5. Old MARS by Fukuda Lab.	16	lithium	step motors	10[b]	4
6. Kity by KAIST	16	watch	DC motors	10	5[b]
7. Alice 2002, Lausanne Univ.	8	Ni-MH	watch mov.	600	4

[a] DC micromotor.
[b] Estimated values.
The numbers in the first column correspond to those in Figure 4.67.
Source: From Ref. [72].

actuator already available in these dimensions is the electromagnetic motor. Some motors may be optimized for high power, others for low cost or for good efficiency.

In autonomous mobile robots, sensors for environment perception are of the utmost importance (Table 4.8). Generally, one can distinguish between passive and active sensors. Passive sensors (e.g. camera, microphone) do not irradiate energy into the environment whereas active sensors send out some sort of signals which support the measurement. The power consumption of passive sensors is dominated by the signal conversion and processing which barely change with the robot's size. In contrast, active distance sensors, such as sonar or IR proximity sensors, emit energy and use the reflected beams dispersed by the measured object. Their power consumption may be not acceptable by MMRs; therefore, passive sensors or very simple active sensors are the right choice for small systems.

The controller of any robot has to process information and generate adequate actions. This task and its complexity might not change much with the size of the robot. However, smaller systems tend to be used in simpler and more restricted spaces and, thus, have less complex environments to deal with.

Communication in MMRs takes place between different units or between the robot and the user. The communication can be unidirectional or bidirectional, involving a receiver, a transmitter or both on the robot. The dimensions of communication devices often depend less on the communication distance but more on

Table 4.8. Possible sensors for MMRs.

Sensor	Principle	Output	Comment/Complexity
Bumpers	Contact	On-off	Theoretically feasible
Compass	Magnetic	Angle	Feasible or existing
Inclinometer	Inertia	Angle	Feasible
Barometer	Pressure	Bar	Feasible
Temperature	Heat	°C	Easy
Microphone	Sound	Hz	Feasible
Photodiode	Light	Lux	Very easy
Camera	Light, Position and Color	Value per Pixel	Much data to process but very promising
Sonar	Time of Flight	Proximity	Not for short ranges
Infrared	Reflected Light	Proximity	Feasible and existing; depend on surface reflection
Optical Triangulation	Triangulation	Distance	To be integrated
Laser Ranging	Time of Flight	Distance	Difficult and complex

The first eight are passive sensors, while the last four are active. The comment refers to the feasibility in small size.
Source: From Ref. [72].

the precision, the conversion technique and the communication speed. Communication can be established through IR waves, visual signalling, sound or radio.

If the microrobot Alice (Table 4.7) is taken as a reference, the necessary components making it a standalone mobile robot can be summarized as microcontroller, rechargeable battery, two-watch motors with aluminium wheels and rubber tyres, extension bus (connector), IR remote control receiver and four IR proximity sensors. In Figure 4.67 (7′) it is equipped with a CMOS colour camera module [72].

This section may be concluded with a few words on nanorobots. Strictly speaking, these machines would range in size from 0.1 to 10 μm.

At this time, they remain a hypothetical concept, as no artificial nanorobots have been created. If realized, they would work at the atomic, molecular and cellular level to perform tasks in the medical and industrial fields. A recent paper has reviewed the available nanotechnology for building nanorobots, and has proposed the CMOS VSLI technology using deep ultraviolet lithography to obtain nanorobots assembly with high levels of resolution [73]. For so small devices, a major issue is the choice of a suitable power source. Among the different proposals, one considers using integrated circuits allied with Li-ion batteries. In standby mode, the power consumption would be ~1 μW; once activated, the nanorobot could work at low powers, with a maximum of ~1 mW for RF communications [73].

4.14. Micro Electromechanical Systems (MEMS)

MEMS are small integrated devices or systems that combine electrical and mechanical components. They range in size from the sub-micron level to the millimeter level, and some systems may contain millions of them. In MEMS, the fabrication techniques developed for the integrated circuit industry are extended to add mechanical elements, such as beams, gears, diaphragms and springs [74].

These systems can sense, control and activate mechanical processes on the micro scale and function individually or in arrays to generate effects on the macro scale. The micro fabrication technology allows obtaining large arrays of devices, which individually perform simple tasks, but in combination can accomplish complicated functions.

The three characteristic features of MEMS fabrication technology are miniaturization, multiplicity and microelectronics. Miniaturization enables the production of compact, quick-response devices; multiplicity refers to the batch fabrication typical of semiconductor processing, which allows easy fabrication of millions of components; microelectronics provides the intelligence to MEMS and allows the merger of sensors, actuators and logic to build smart components and systems.

MEMS are not only identified as the miniaturization process of mechanical systems, but also represent a new model for designing mechanical devices and systems [74].

A number of materials can be used to build MEMS. Silicon is the most attractive material for a wide variety of applications. The basic techniques for producing all-Si based MEMS devices are deposition of material layers, followed by photolithography and then etching to produce the required shapes. As crystalline Si is still a complex and relatively expensive material, polymers may be used in substitution as they are cheap and have a broad range of material characteristics; for polymers, the injection moulding process may be applied. Metals can also be used, although they do not have the mechanical properties of Si. Examples of metals are: Au, Ni, Al, Ti, Pt and Ag. They can be deposited, for example by electroplating, evaporation or sputtering.

MEMS may be used in a variety of applications, some of them briefly mentioned in the previous sections. Examples of available, or currently studied, applications include:

- GPS sensors for constant tracking
- Microengines
- Optical switching devices that can switch light signals over different paths at 20-ns switching speeds
- Sensor-driven heating and cooling systems that significantly improve energy savings
- Tyre pressure monitoring

- Accelerometers for vehicle airbags
- Fluid pumps
- Miniature robots
- "Smart dust".

The last possible application requires some comments. "Smart dust" is a hypothetical network of tiny wireless MEMS sensors that can detect everything from light to vibrations. These dust devices could eventually have the size of a grain of sand, though each would contain sensors, computing circuits, bidirectional wireless communications technology and a power supply. The "grains" would gather data, run computations and communicate that information using two-way band radio between "grains" at distances approaching 300 m [75]. A device of this type can be represented as done in Figure 4.68.

Potential commercial applications are varied, ranging from catching manufacturing defects by sensing out-of-range vibrations in industrial equipment to tracking patient movements in a hospital room. Military applications are also possible, for example in spying activities.

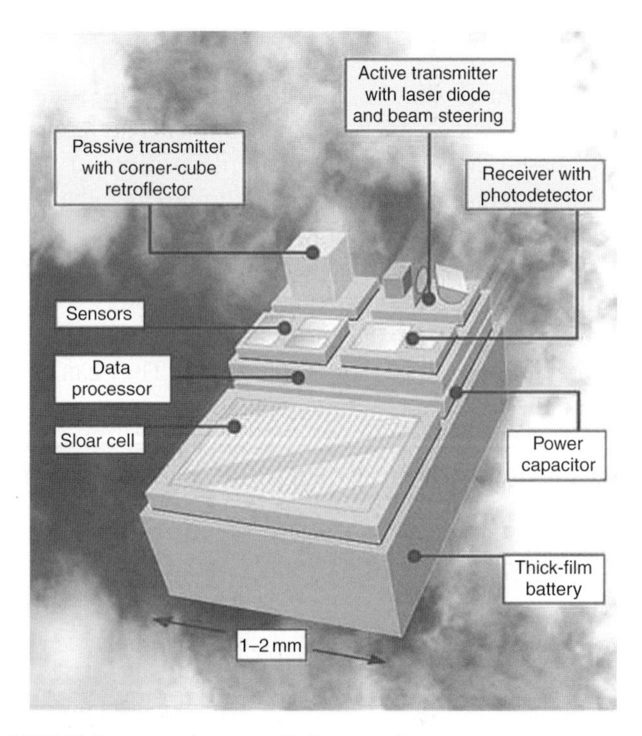

Figure 4.68. MEMS forming the so-called smart dust.
Source: From Ref. [75].

However, there are some hurdles on the way to commercial production, the main one being the difficulty of having MEMS and electronics onto a single chip [75]. In particular, an autonomous device with the dimensions shown in Figure 4.68 has to face the problem of having an extremely small power source.

This issue has been considered by several researchers, whose proposed solutions are rather diverse, for example solar power, capacitors, nuclear power, batteries. As with microrobots, batteries have to be simultaneously small and powerful enough to provide the MEMS with the necessary runtime.

The batteries proposed thus far can be briefly described as follows. A 2D microbattery based on the Ni–Zn couple in an alkaline electrolyte, constructed with the same technique used for MEMS, is reported to have a high-power capability [18, 77]. To comply with the power requirement of ~5 mW during transmission and with the dimensions of the microsensors, the battery should have an area of ~0.1 cm^2. This means a very high specific power of ~50 mW/cm^2. In Ref. [18], this battery is described in detail in association with a solar panel also mounted on the substrate hosting MEMS and battery.

Another bidimensional battery is based on Li anode, a solid electrolyte – for example lithium phosphorous oxynitride (Lipon) – and a cathode like $LiCoO_2$, $LiMn_2O_4$ or V_2O_5 [78]. Thin-film cells with these components can be fabricated in various shapes and on different substrates. For instance, they can be added to integrated circuits or to individual circuit components. Their thickness may be as low as 15 μm.

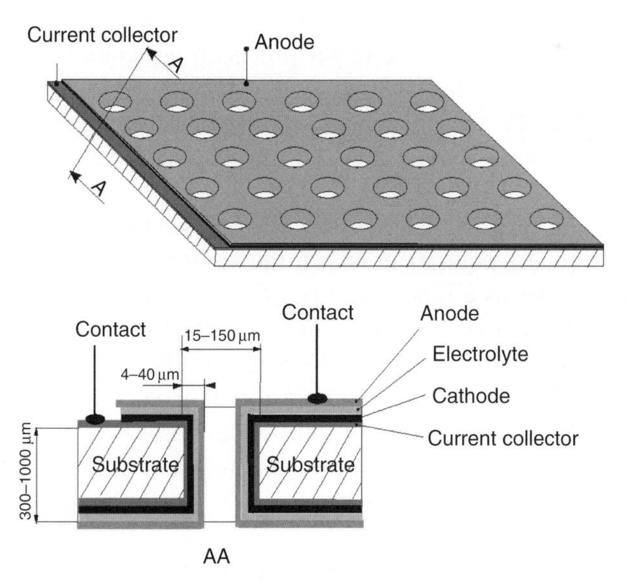

Figure 4.69. 3D "nanobattery" for MEMS.
Source: From Ref. [79].

3D "nanobatteries" have also been proposed and claimed to have substantially better performance than 2D cells. The proposed chemistry is: Li_xC/HPE/Mo_yS_z, where HPE is a hybrid polymeric electrolyte [79].

An experimental cell of this type, with a 1-μm-thick cathode, shows rates up to 2 C and can be cycled for 200 cycles with ~100% efficiency. Its claimed specific capacity, 2 mAh/cm^2, is 30 times higher than that of 2D cells with the same area and the same cathode thickness.

Details on the construction of this battery are evidenced in Figure 4.69. Using a silicon or glass substrate, a team from Tel Aviv University has created a matrix of tiny holes each 50 μm in diameter and 500 μm deep. Each of these holes functions as an independent micro battery or microchannel with an output power of 8–10 μW. The power of a 3D battery having an area of 1 cm^2 is 150–200 mW. Because each battery is comprised of thousands of small batteries, even if one of them has a short circuit and fails, the entire battery can keep functioning.

4.15. Farming Applications

Modern agriculture needs tools and technologies that can improve production efficiency, product quality, post-harvest operations, and reduce their environmental impact. In this respect, some of the advanced technologies described in this chapter are needed to make the activities more efficient, productive and sustainable.

Automation in agriculture brings about a fundamental contribution to what is now known as precision farming (or precision agriculture). A definition of precise farming may be the following: the technique of applying the right amount of input (water, fertilizer, pesticide, etc.) at the right location and at the right time to enhance production and improve quality, while protecting the environment [80].

Application of precision agriculture is based on two requirements: knowledge of where the farm equipment is as it moves across a field, and the value of one or more variables as a function of position within the field. Using the GPS system allows to automatically guide farm machines to stay along a track with only centimeter-scale deviations; furthermore, if the machine carries a mass flow sensor, yield data in a given field portion may be obtained (see later). The second requirement is fulfilled with a field-wide sensor network able to monitor relevant parameters, for example, soil moisture and air temperature, and to transmit data wirelessly to the farmer location, so that appropriate measures can de adopted [81].

The GPS system is used in a configuration like the one reported in Figure 4.70 (up), which allows equipment positioning. More precisely, this system is referred to as differential GPS (DGPS); it allows more accurate measurements,

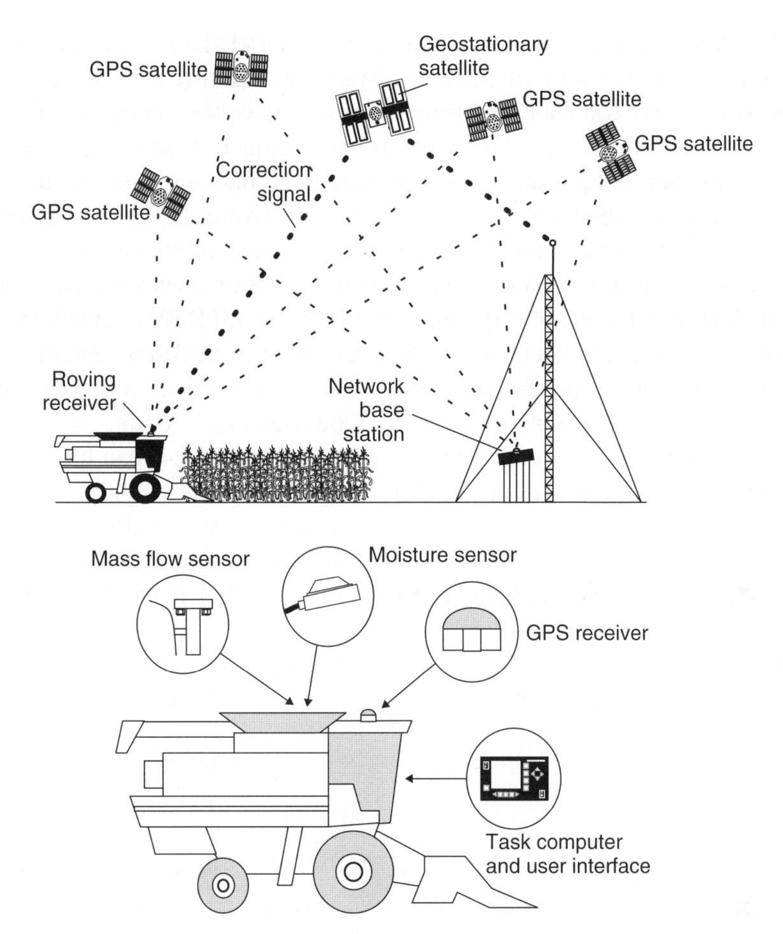

Figure 4.70. Up: GPS system for agricultural equipment positioning; down: components of a grain yield monitoring system on a typical combine harvester. *Source*: From Ref. [82].

due to corrections made possible by the simultaneous utilization of GPS and GEO satellites [82].

The yield monitor (Figure 4.70 (down)) is intended to give the user an accurate assessment of how grain yields vary within a field. Yield monitors typically include several different sensors and a task computer. The sensors measure the mass or the volume of grain flow, separator speed, ground speed, grain moisture, and combine's header height. The task computer serves several functions: it integrates and calibrates the sensors; converts their output signals into data for storage, display, and later use; contains the DGPS receiver interface, and controls the interaction of these devices. Yield is determined from the various parameters being sensed [82].

Another issue of particular relevance in large fields is irrigation. Electronic devices are needed to monitor watering systems, as depicted in the example of Figure 4.71. Field controllers operate pumps and valves; being such systems often located away from a farm, control and data transmission occur via a wireless modem. In complex watering systems, like the one of the figure, large and powerful batteries are to be used to activate the closing/opening of valves, in outdoor environmental conditions that are often rather harsh (with temperatures between −20 and +50°C). Primary lithium packs (often $Li/SOCl_2$) made with small (1/2R6) or large powerful cells (spiral R20) are preferred. The quiescent current is rather low (<150 μA), but the peak current may reach 500 mA for automatic valves, and a runtime of 5–10 years is requested. In less demanding systems, alkaline batteries may be used [1].

Wireless sensor networks (WSNs, see also Section 4.4) can help monitoring fields, vineyards and orchards, thus helping farmers to prevent damages to their crops. In particular, frost conditions need to be signalled in real time, so that countermeasures may be taken. In a technology developed at the Washington State University, a low-power (~10 mW) network operating in

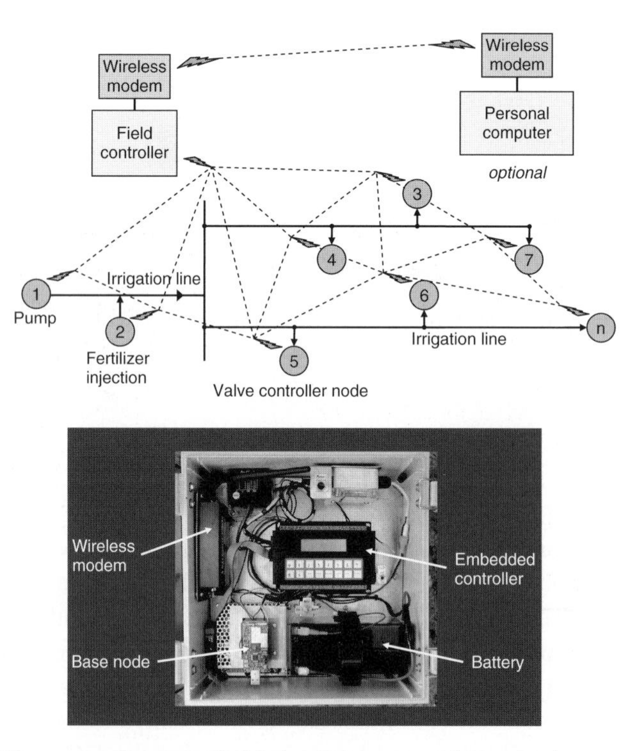

Figure 4.71. Up: remotely controlled irrigation system; down: basic components of the field controller.
Source: From Ref. [83].

the 900 MHz frequency band has proven efficient in monitoring and transmitting data to the farmer in a range of environments, from bare fields to orchards, and in variable weather conditions [84].

The low-power radio transceiver used in this application is programmed to operate as a base or remote station (see Figure 4.72 (up-left)). One radio can function as a gateway interfaced to a portable computer where data can be

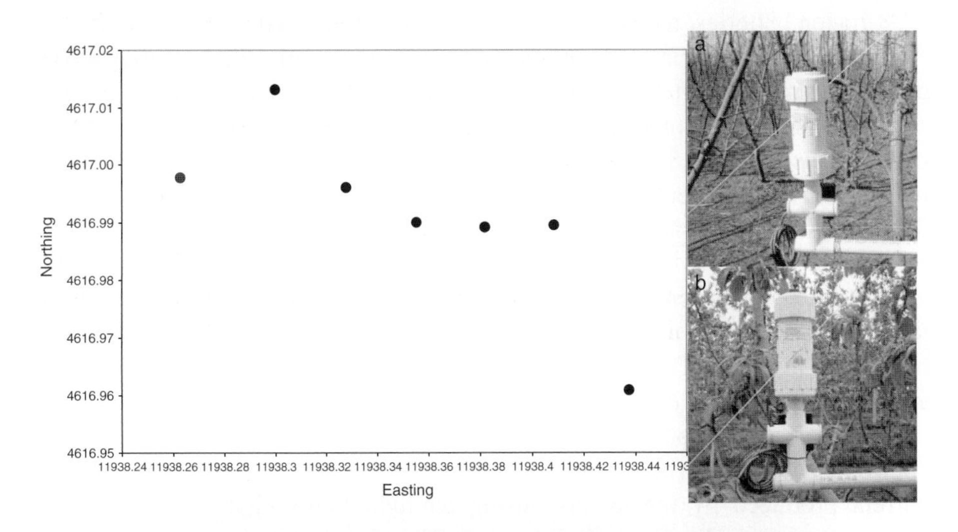

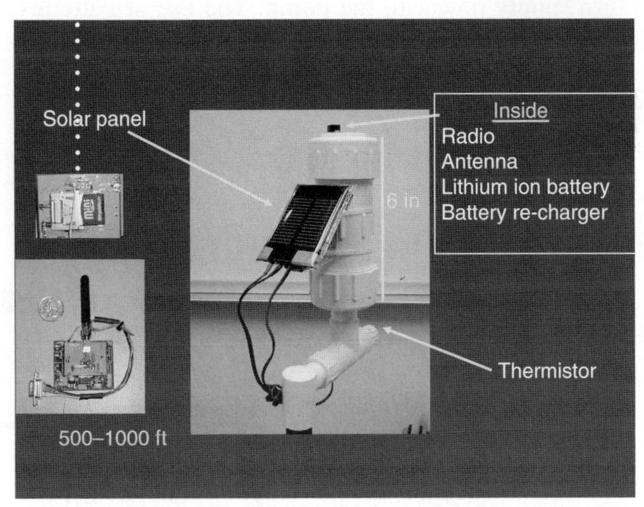

Figure 4.72. Up: wireless sensor network in an apple orchard: the remote stations transmit data, e.g. temperature, to the base station (the leftmost circle); on the right, a remote station in March (a) and April (b) is shown (*source*: From Ref. [84]); down: components of the station; the board in the low-left corner is the data-logger portion also controlling the radio transceiver circuitry (*source*: From Ref. [80]).

stored. Each radio is fitted with a precise thermistor for temperature measurements, as shown in Figure 4.72 (down).

A radio network of this kind is only moderately affected may the changes that a crop may experience during the seasons. For instance, in an apple orchard (Figure 4.72a and b) passsing from dormant to full canopy conditions, which means significant variations in line of sight, the radio signal maintains sufficient strength and quality at a distance of 150–300 m.

Li-ion batteries, recharged by solar panels through a dedicated charger, are suitable for this application. Primary batteries, for example alkaline, may also be used if the installation of solar panels is not judged convenient.

WSNs can be implemented in other applications demonstrating the wide range of uses for WSN in agriculture. Indeed, this technology, using radio transceivers of appropriate power and ad-hoc software, can be used to monitor, for example soil moisture and temperature, and weather conditions over a region-wide area.

A watering system for livestock kept in an area away from a stream needs to extract water from a well. If AC electric power is not available, solar pumps have to be installed. There are two basic types of solar-powered water pumping systems, battery- and direct-coupled [85]. Only the first is described here.

It consists of photovoltaic panels, charge control regulator, batteries, pump controller, pressure switch, pressure tank and DC water pump (Figure 4.73). The current produced by the panels during daylight hours charges the batteries, which in turn supply power to the pump. The use of batteries allows pumping over a longer period by providing a steady operating voltage to the pump's

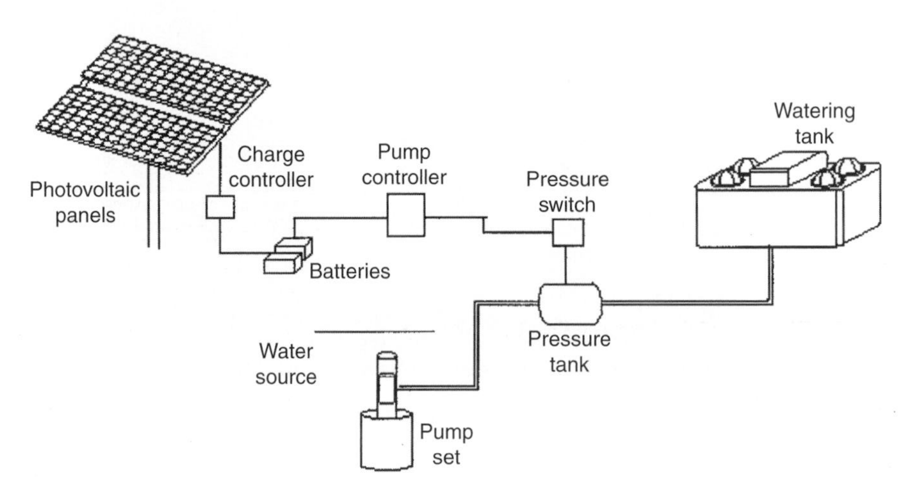

Figure 4.73. Battery-coupled solar water pumping system for livestock watering.
Source: From Ref. [85].

motor. Thus, during the night and low light periods, the system can still deliver a constant source of water for livestock.

However, in this kind of system, a pump controller is necessary. Indeed, the batteries can reduce the efficiency of the overall system as the operating voltage is dictated by them and not the PV panels. Depending on their temperature and state of charge, the voltage supplied by the batteries can be one to four volts lower than the voltage produced by the panels during maximum sunlight conditions. The pump controller boosts the battery voltage supplied to the pump [85].

The batteries most commonly used in this application are deep-cycle lead-acid batteries. They are easily maintained and can withstand discharges of up 80% of their capacity. Ni–Cd batteries are also used in spite of their higher cost. Indeed, they are more tolerant to extreme environmental conditions and to complete discharge, thus reducing the life-cycle cost in these conditions.

Above, few examples of battery-aided automation in farming have been given, but other technologies described in the previous sections of this chapter, for example robotics, animal identification and tracking, and UAV, are increasingly being used.

4.16. Energy-Related Stationary Applications

The use of batteries in conjunction with electricity produced by traditional or non-conventional (wind or photovoltaic) power plants covers a wide application range: energy storage, load levelling, UPS, energy quality. Different aspects may be found in each of these areas, as is detailed in the following discussion.

4.16.1. Load Levelling, Power Quality and UPS

Energy storage technologies do not generate electricity but can deliver stored electricity to the electric grid or to end-users. They also improve power quality by correcting voltage sags, flickers, surges or frequency imbalances. Storage devices are also used as UPS that deliver electricity during short utility outages.

In Figure 4.74, the benefits brought about by energy storage and management, along the complete line from fuel to customer premises, is shown for a traditional power plant. In particular, the volatility of fossil fuel prices can be tackled, supply and demand are balanced through load-levelling actions, power can be bought and stored at low prices and sold back at higher prices (power arbitrage), and the quality of power available to the final user is kept to a satisfactory value.

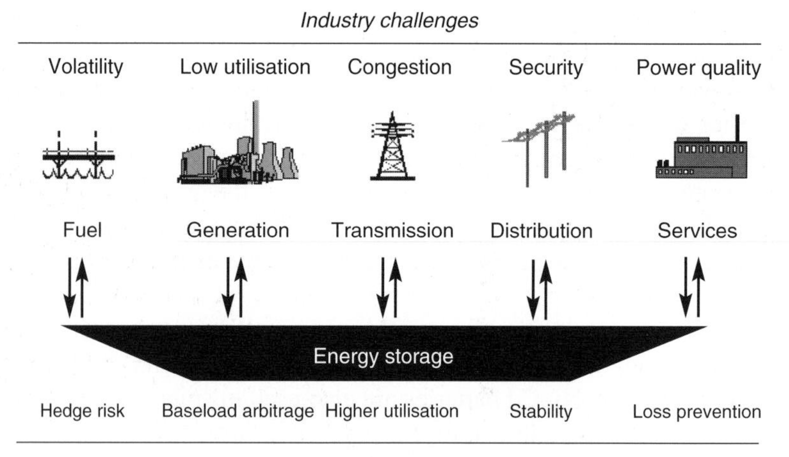

Figure 4.74. Advantages of energy storage for the electric power industry, exemplified for energy production based on fossil fuels.
Source: Courtesy of VRB Power Systems.

Large-scale stationary applications of electric energy storage can be divided in three major functional categories:

- *Bridging power*: stored energy is used for seconds to minutes to assure continuity of service when switching from one source of energy generation to another (for example from the main generator to a spinning reserve system).
- *Energy management*: storage, in these applications, is used to decouple the timing of generation and consumption of electric energy. A typical application is load levelling, that is storage of energy when its cost is low and utilization during peak hours.
- *Power quality*: stored energy is only applied for seconds or less to assure power continuity.

There are several operating technologies allowing energy storage [86, 87]:

- *Pumped hydroelectric energy storage (PHES) systems*: it represents the largest energy storage technology and is commercially available. Conventional systems comprise reversible pump-turbine generator-motors and two large reservoirs – one located at a low level and the other at a higher level. The reservoirs are connected to each other by water conduits via a pump-turbine. During off-peak periods at night, water is pumped from the lower to the upper reservoir where potential energy is stored. During peak hours at daytime, the water is released back to the lower reservoir, thus generating electrical power.

- *Compressed air energy storage (CAES) Systems*: these systems use pressurized air as the energy storage medium. An electric motor-driven compressor is used to pressurize the storage reservoir using off-peak energy, and air is released from the reservoir through a turbine during peak hours to produce energy. Aquifers or caverns are ideal storage locations, while small systems can use pressurized tanks.
- *Superconducting magnetic energy storage (SMES) systems*: these systems store energy in the field of a large magnetic coil (a dipole solenoid or a toroid) with direct current flowing. DC can be converted back to AC when needed. Low-temperature SMES cooled by liquid helium are commercially available.
- *Electric double layer capacitors (EDLC)*: electrostatic energy E_c that is stored in an electric double layer capacitor (a type of electrochemical capacitor or supercapacitor) with capacitance C_{edlc} at voltage V is equal to

$$E_c = \frac{1}{2} C_{edlc} V^2$$

Although the maximum voltage is as low as $\sim$2.7 V, C_{edlc} is very large and the energy density of EDLC is several hundreds times higher than that of electrolytic capacitors, although it is lower than that of secondary batteries. EDLC is suitable for fast discharge and charge applications as compared to secondary batteries.

- *Flywheel*: a flywheel is an electromechanical device that couples a motor generator with a rotating mass to store energy for short durations. Conventional flywheels are "charged" and "discharged" via an integral motor/generator, which draws power provided by the grid to spin the rotor of the flywheel. During a power outage, voltage sag, or other disturbance, the kinetic energy stored in the rotor is transformed into DC electric energy by the generator, and the energy is delivered at a constant frequency and voltage through an inverter and a control system.
- *Batteries*: they can be used in any of the three functional categories mentioned above, and will be treated more extensively in the following.

In Table 4.9, a comparison of the lifetime of several storage systems is reported [87].

The maturity level and the storable power of different systems are presented in Figure 4.75. The technically mature and secure systems are those based on compressed air, pumped hydro, and battery systems using lead-acid,

Table 4.9. Lifetime of energy storage systems.

System	Na/S	VRB	Ni-Cd	Pb-acid	Li-ion	Pumped Hydro	CAES	EDLC	Fly-wheel
Years	15	>10	15–20	15	10	>40	>20	20	–
Cycles	2250–4500	>12 000	Up to 10 000	4500	3500	–	–	70 000	150 000

Source: From Ref. [87].

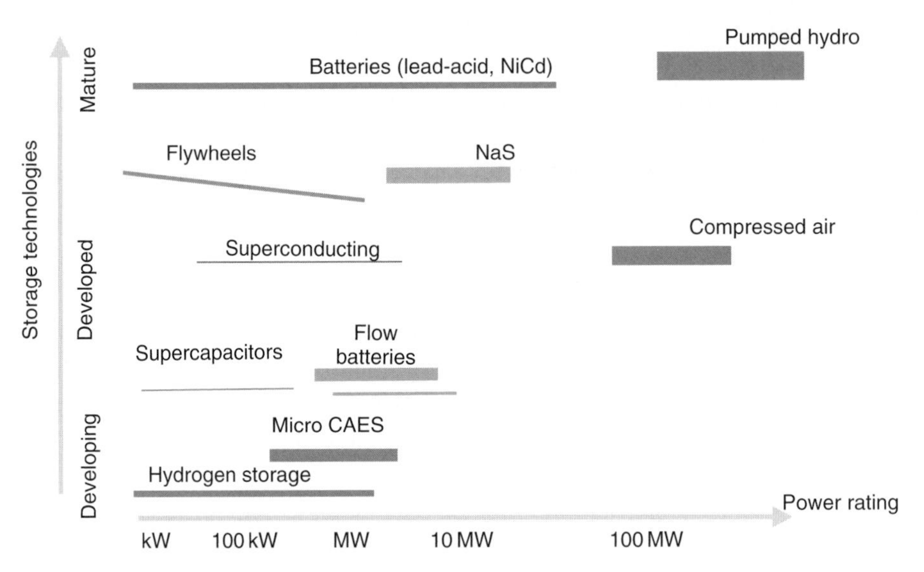

Figure 4.75. Development status and power capability of several energy storage technologies. The width of the bar indicates storage capacity.
Source: Courtesy of the New Zealand Electricity Commission.

Ni–Cd and Na/S batteries. Pumped hydro is the most technically secure large-scale storage technology. Battery systems have low technical risk, but are generally regarded as secure technology in the size range below 50 MW. Flow batteries are progressing, but they are not yet considered sufficiently mature. Li-ion systems provide technically mature solutions, but are generally considered only for applications below 1-MW power.

A comparison in terms of costs is presented in Figure 4.76. The abscissa reports the cost per kW, while the ordinate indicates the cost per kWh, and in this latter calculation the capacity and efficiency of each system are considered. High-power electrochemical capacitors have a low cost per unit power, but a high cost per unit energy. At the other extreme, long-duration flywheels have a

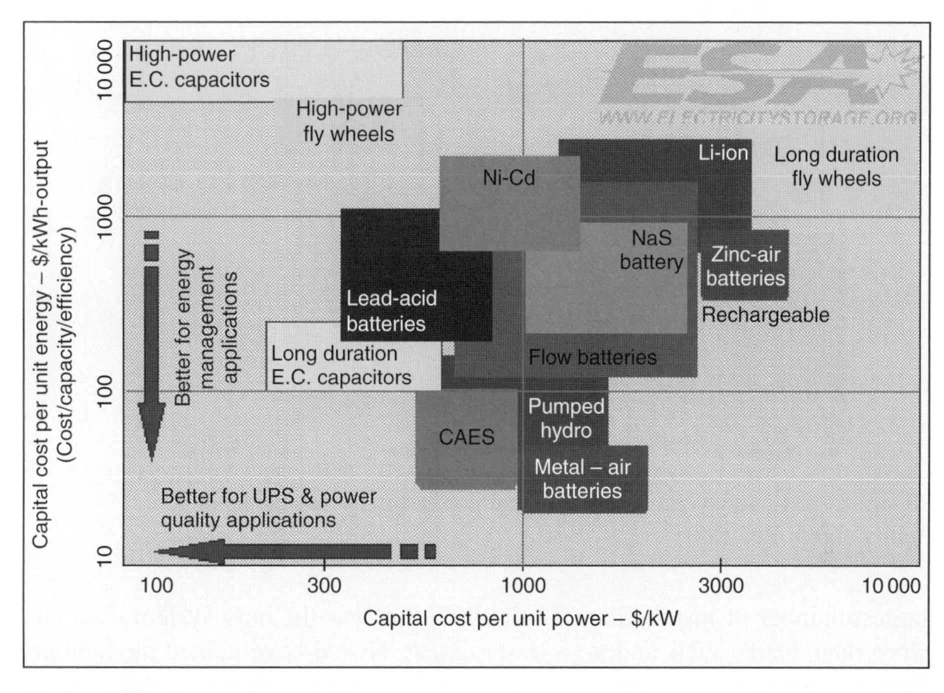

Figure 4.76. Comparison of the costs per unit power and energy of various energy storage systems. Metal-air batteries, including zinc-air, have not yet been significantly developed.
Source: Courtesy of ESA, Electricity Storage Association.

moderate cost per unit energy and a high cost per unit power. The battery systems are grouped in the middle, with Li-ion tending to manifest the higher costs in terms of both energy and power.

Figure 4.77 depicts how load levelling smoothes the power profile curve when batteries are used. The surplus energy produced overnight is used to charge large batteries. The increased power demand during daytime is partly met by battery discharging.

In Table 4.10, a comparison of battery systems is made. In the last column, the costs calculated for a 20-year operation time at the specified regime are reported. This kind of calculation produces a scale of merit where the systems able to be long cycled have the lower costs. Therefore, it is obvious that any extension of the battery life is an important factor to reduce the overall costs. This depends not only on the battery design but also on the application of appropriate charge/discharge strategies. Especially important are the impact from the battery DOD and the charge voltage limit.

Since 1986, several energy storage systems based on batteries have been installed worldwide. Lead-acid is still the dominating technology with the

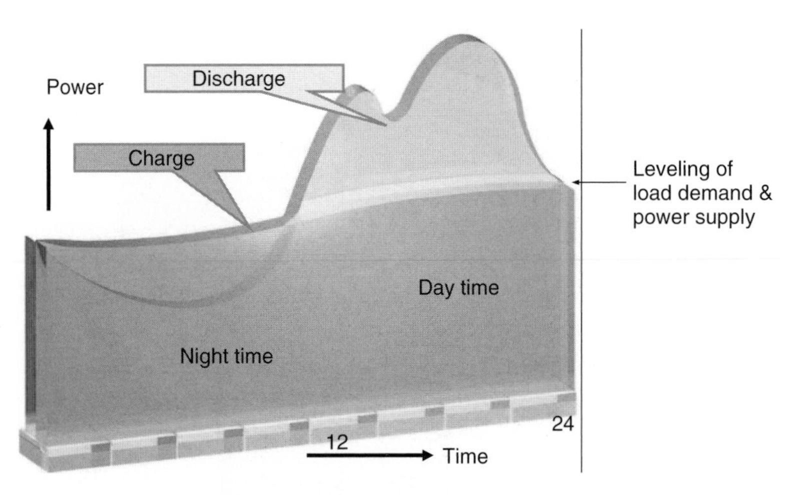

Figure 4.77. Load levelling to equilibrate power supply and demand.
Source: From Ref. [88].

largest number of installations. Until 1997, this was the only system used but, since then, Na/S, VRB and, to a lesser extent, Ni–Cd have gained momentum. These installations have often power capabilities in the MW range. An example is presented in Figure 4.78 [87], where a 2-MW Na/S battery is used to store the energy generated by a photovoltaic (PV) system.

The Na/S battery is characterized by high-energy density and high-charge/discharge efficiency. To take advantage of these characteristics, a storage facility based on this battery is especially applied to daily load levelling (see later). The gross efficiency of the battery of Figure 4.78 is 88% at the AC terminal, and the gross overall system efficiency is 75% during the year. The efficiency increases with deeper discharge operations.

The Na/S battery can also provide the power needed for short-term discharges typical of power quality applications, that is brief voltage sags. Furthermore, this battery, as well as other batteries with similar characteristics, can act as an emergency power supply, that is can deliver backup power for several hours.

Another emergent battery technology, VRB, will be discussed with special reference to storage of solar and wind energy.

While storage is now a convenient option for energy produced by the most diffused power plants, that is those burning oil, it becomes even more important when energy from renewable sources, for example wind and sun, is produced. Indeed, batteries can outbalance the intermittence typical of these power sources.

Solar and wind energy production has been growing in recent years at the rate of ~30%/year. According to the International Energy Agency forecasts, 15–20% of the globe electricity demand will be met by PV systems in 2040.

Table 4.10. Year 2007 comparison of battery systems for energy storage. 1-MW systems operating 8 h/day are assumed.

System/ Features	Cycle Life	Efficiency Round Trip AC to AC (%)	First-Time Capital Costs (\$/kWh)	Environm. Risk	Usual DOD (%)	Deep Discharge Capability (>70%)	20-Year Capital and Operating Costs (\$/kWh)[a]
Pb-acid[b]	2000	45	1550	Medium	25	No	11 769
Ni–Cd	10 800	65	2000	Medium	30	No	4644
Li-ion[c]	20 000	85	2700	Very little	40	Yes	2541
Na/S	3000	58	500	High	33	Yes	5003
Zn/Br_2	2500	60	1200	High	33	Yes	7828
VRB	10 000+	72	630	Very little	55	Yes	3078

[a] At a cost of $ 0.05/kWh in the first year escalating at 3% per year with 8 h per day of operation and a 20-year life of 10 800 cycles.
[b] Industry standard.
[c] Small systems; the cycle number is not in agreement with the value of Table 4.9, possibly due to a difference in DOD.
Source: Courtesy of VRB Power Systems.

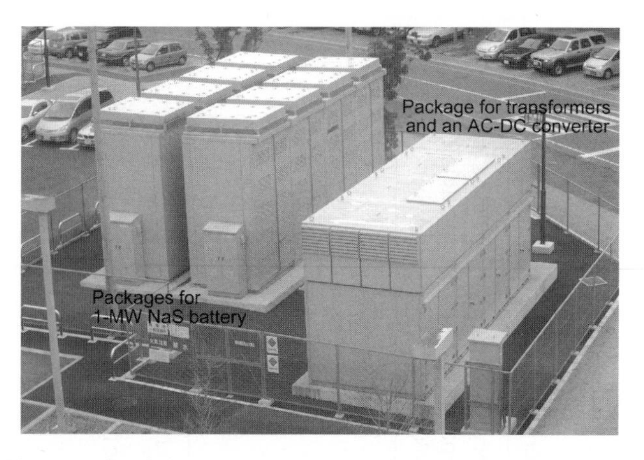

Figure 4.78. Photograph of the 2 MW_{ac}, 7.2 h Na/S battery system installed in 2004 at AIST Tsukuba, Ibaraki, Japan. There are two packages (10.3 m (W) × 2.3 m (D) × 5.2 m (H), 86 tons each) for a 1-MW battery and a package (9.2 m (W) × 3.3 m (D) × 4 m (H), 31 tons) for transformers and a 2400-kVA AC-DC converter. Each 1-MW Na/S battery comprises twenty 50-kW modules (52.6 kW_{dc} output/57 kW_{dc} input) in series. Each module consists of 384 cells (12 parallel and 32 series connections of 2-V, 1.5-kW cells). *Source*: From Ref. [87].

The advantages of this clean, abundant power sources are obvious, but their development was slowed down until some years ago by cost considerations. Fortunately, costs have been decreasing in the past few years, as shown in Figure 4.79 for wind generation, thus making these technologies competitive.

In what follows, the fundamentals of wind and solar energy generation, and the need of storage means will be discussed.

PV devices can be made from various types of semiconductor materials, deposited or arranged in various structures, to produce solar cells with optimal performance. Polycrystalline thin films are especially made of copper indium diselenide (CIS), cadmium telluride (CdTe) or Si; single-crystalline thin films are mainly based on gallium arsenide (GaAs) [89].

Silicon, the most popular material for PV devices, can be used in various forms, including single-crystalline, multicrystalline and amorphous. Crystalline Si cells exhibit outstanding longevity. Cells developed almost 40 years ago are still operating. Single crystal and polycrystalline cells perform similarly, but the efficiency of the former is generally higher. Amorphous Si absorbs solar radiation 40 times more efficiently than does single-crystal Si, so a film only ~1 μm thick can absorb 90% of the light energy shining on it. Therefore, amorphous Si could reduce the cost of PV systems. As it can be produced at low temperature and can be deposited on low-cost substrates, is increasingly utilized in thin-film PV devices [89].

The components of a typical photovoltaic cell are shown in Figure 4.80. The cell consists of a glass or plastic cover, a transparent adhesive, an

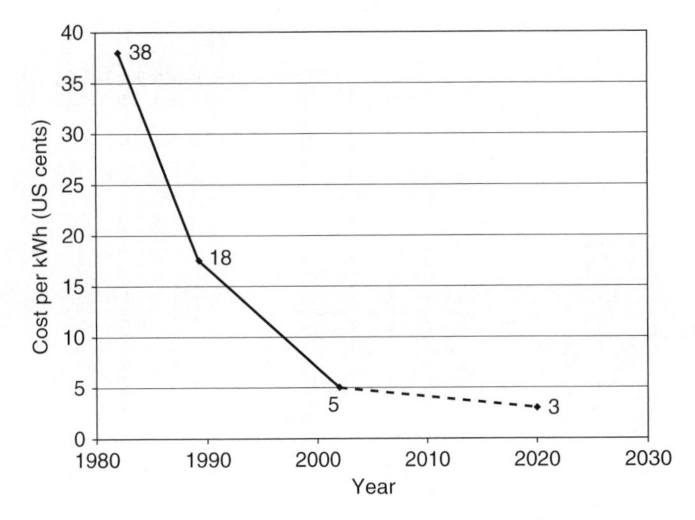

Figure 4.79. Average cost per kWh of wind-generated electricity for the period 1982–2002, with projection to 2020.

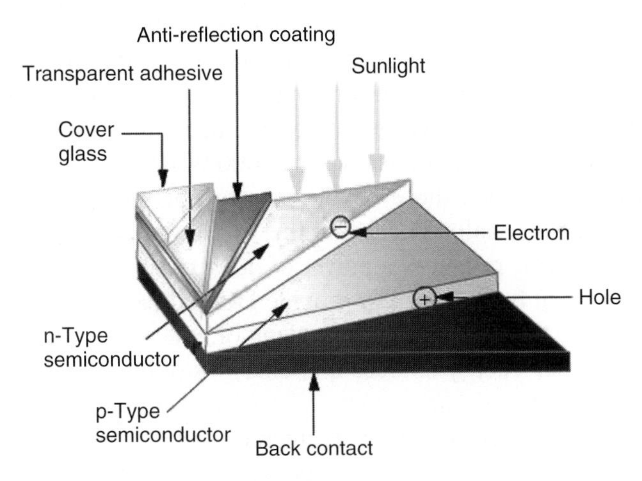

Figure 4.80. Scheme of a photovoltaic cell.
Source: From Ref. [89].

anti-reflective layer a front contact (not evidenced in the figure), n-type and p-type semiconductor layers and a back contact.

Photovoltaic cells are connected in series and/or parallel to form modules. The modules are sealed in an environmentally protective laminate, and are the building blocks of PV systems. Modules assembled as a pre-wired, field-installable unit, form the panels. A photovoltaic array is the complete power-generating unit consisting of several PV modules (see Figure 4.81).

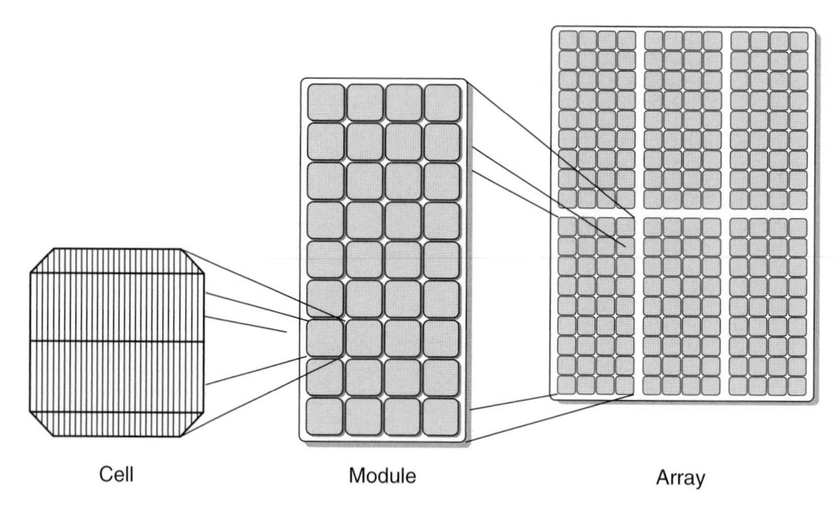

Figure 4.81. PV cell, module, panel and array.
Source: Courtesy of Florida Solar Energy Center.

PV modules are safe and reliable products, with rare failure reports and expected lifetimes of 20–30 years. The performance of PV modules and arrays are generally rated according to their maximum DC power output under standard test conditions (STCs). STCs are defined at an operating temperature of 25°C, and incident solar radiation level of 1000 W/m^2 and under air mass 1.5 spectral distribution. Since these conditions are not always met in the field, actual performance is usually 85–90% of the STC rating. A 36-cell module, like the one of Figure 4.81, produces a power of 40 W.

PV systems can be designed to provide DC and/or AC power service, can operate interconnected with, or independent of, the utility grid, and can be connected with other energy sources and energy storage systems. This is illustrated in Figure 4.82.

In PV systems where the array is interconnected with the electric utility grid (Figure 4.82a), the primary component is the inverter, or power conditioning unit (PCU). The PCU acts a DC–AC converter and makes the AC power consistent with the voltage and power quality requirements of the utility grid. Through a distribution panel, the AC power produced by the system can either supply on-site electrical loads or back-feed the grid when the PV system output is greater than the on-site load demand.

Standalone PV systems are generally designed to supply certain DC and/or AC electrical loads. These types of systems may be powered by a PV array only, or may use an auxiliary power source (wind or engine generator) in a PV-hybrid system. The simplest type of standalone system is a direct-coupled system, where the DC output of a PV module or array is directly connected to a DC

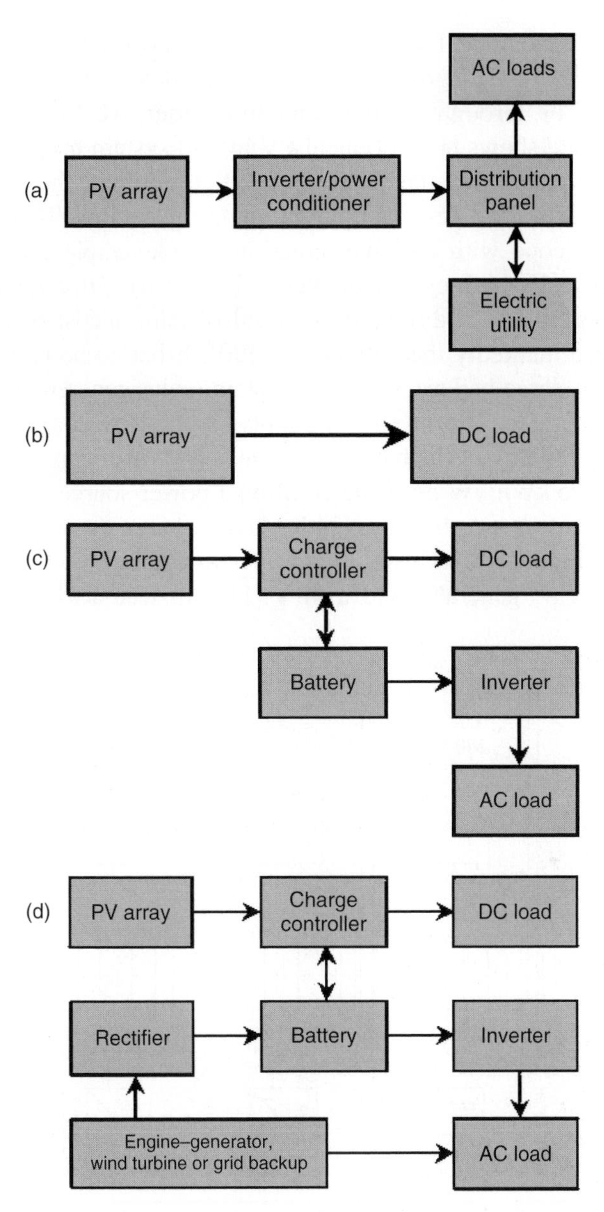

Figure 4.82. (a) PV array interconnected with the electric grid; (b) PV array directly connected to a load; (c) stand-alone PV array with storage battery; (d) hybrid system having an auxiliary power source.
Source: Courtesy of Florida Solar Energy Center.

load (Figure 4.82b). Without batteries for energy storage, the system only operates during sunlight hours, making these designs suitable for common applications such as solar thermal water heating systems.

In many standalone PV systems, batteries are used for energy storage. Figure 4.82c shows a diagram of a typical standalone PV system powering either DC loads or, through a battery and an inverter, AC loads.

Figure 4.82d shows how a typical PV hybrid system might be configured. The only difference with the previous configuration is the presence of an additional power source, for example an engine generator. This power source is necessary to cope with extreme situations, for example long bad weather periods or unusually high-energy demands. It improves, in any case, the reliability of the system and, additionally, it can also result in cost savings because it allows reducing markedly the battery size [90]. It has to be realized that even with a relatively large battery, the basic system (solar generator battery) cannot give a 100% guarantee for the energy supply. Systems with a ratio between the available stored energy of the battery and the peak power capability of the solar generator of <25 kWh/kW need the additional power source.

A scheme of photovoltaic system backed by batteries and engine generator is presented in Figure 4.83. This is a large system with a 1.8 kW_p solar generator, a 5 kW gas engine generator and a 48 V, 300 Ah lead-acid battery [90]. The

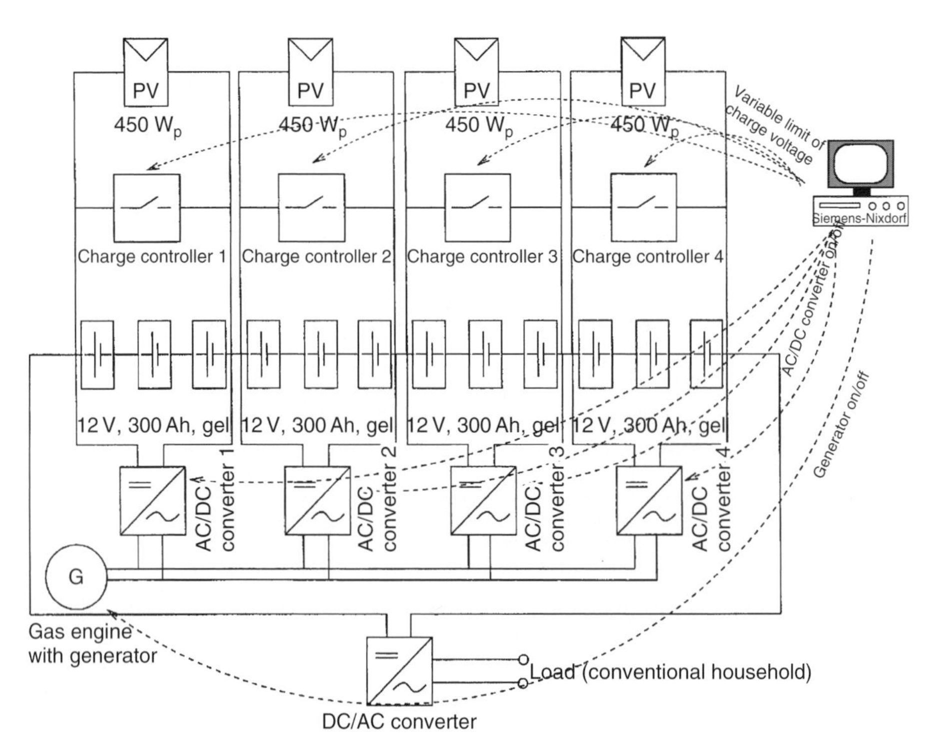

Figure 4.83. Schematic representation of a solar power system with a 48 V, 300 Ah lead-acid battery (VRLA with gel design) and a gas engine generator.
Source: From Ref. [90].

system is divided into four parts; each has a 12 V battery, a 450 W_p solar generator, a charge controller and a rectifier for the gas engine. A control system monitors the charge controllers, the AC/DC converters and the gas engine.

The importance of having battery storage in PV systems, with or without an additional power source, is emphasized when considering the so-called critical loads, for example refrigerators, water pumps, lighting, etc. Under normal circumstances, these loads use the energy generated by the PV array or that of the auxiliary power. In the event the latter becomes de-energized (e.g. a failure in the grid), and the PV array is not illuminated, the battery supplies power to the critical loads. Furthermore, using batteries as a storage system, output fluctuations due to day–night cycles and sunny–cloudy conditions are smoothed.

The issue of fluctuating power output is even more important in wind generators, as the output is proportional to the cube of the wind speed. Adding a battery to a generator removes fluctuations and can provide a constant supply to the utility. Where multiple generators are connected, as in a wind farm, the fluctuation in total power output is reduced. Under ideal conditions the variations of power output will drop with $1/n$, where n is the number of generators [89]. A high wind penetration will reduce the overall power fluctuations, but power stability concerns still remain.

Very large wind turbines and wind farms may be built, as shown in Figure 4.84. The large battery installation of the figure uses the vanadium redox-flow technology to smooth the output fluctuations of the 32-MW wind farm.

Figure 4.84. 32 MW wind farm in Sapporo, Japan. The building houses a Vanadium Redox Battery (VRB, see Section 2.4.1.6) of 4 MW, 1.5 h, 6 MW peak; the battery can smooth power output fluctuations due to wind variations (See colour plate 5).
Source: From Ref. [91].

Wind-generated power is often intended to be used in a grid, for realizing the so-called distributed generation (DG). If one considers the scheme: Generation → Transmission → Distribution → Customers, an efficient DG placed between distribution and customers would address the issues of [91]:

- Power system losses and power factor
- Voltage variations or regulation based upon load variations – especially light loading over weekends or nights
- Frequency control and system stability
- Reliability and power quality.

The devices allowing continuous power supply represent another important aspect of battery utilization in power distribution. The power provided by UPS devices is often referred to as emergency power. The different technologies available for UPS are discussed in the following.

These systems are usually divided into standby and online UPS, but this does not correctly describe many of the UPS systems available. A more correct classification that takes into account the most common design approaches is [92]:

- Standby
- Line interactive
- Standby-ferro
- Double conversion on-line
- Delta conversion on-line.

These designs are depicted in Figure 4.85 and will be briefly explained here.

4.16.1.1. Standby UPS

This is the most common type used for personal computers. In the block diagram of Figure 4.85a, the transfer switch is set to choose the AC input as the primary power source (solid line), and switches to the battery/inverter (dashed line) if the primary source fails. High efficiency, small size and low cost are the main benefits of this design.

4.16.1.2. Line Interactive UPS

This UPS, illustrated in Figure 4.85b, is the most common design used for small business, web and departmental servers. In this design, the inverter is always connected to the output of the UPS. When the input AC power is normal, the inverter provides battery charging. When the input power fails, the transfer

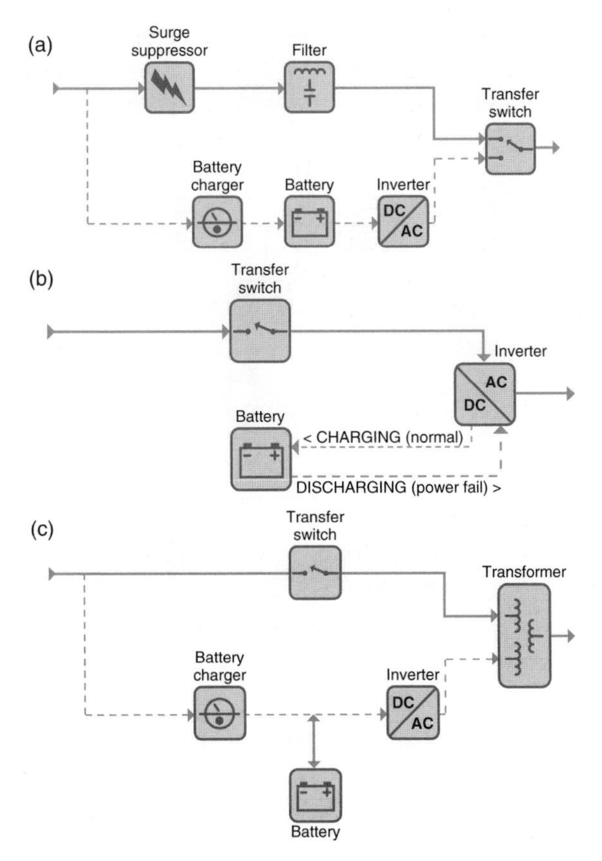

Figure 4.85. The different types of UPS: (a) standby; (b) line interactive; (c) standby-ferro.
Source: From Ref. [92].

switch opens and the power flows from the battery to the UPS output. With the inverter always on and connected to the output, this design provides additional filtering and yields reduced switching transients when compared with the standby UPS.

High efficiency, small size, low cost and high reliability coupled with the ability to correct low or high line voltage conditions make this the dominant type of UPS in the 0.5–5 kW power range.

4.16.1.3. Standby-Ferro UPS

This was once the dominant form of UPS in the 3–15 kW range. This design (Figure 4.85c) depends on a special saturating transformer that has three windings

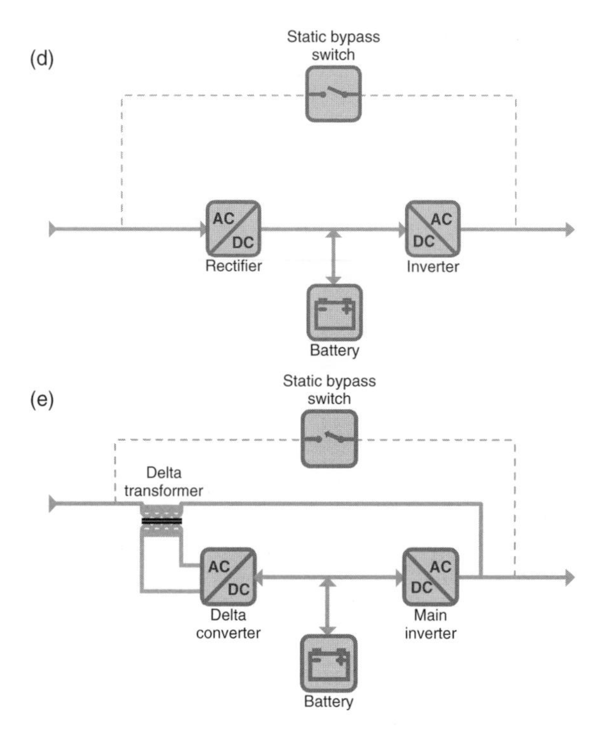

Figure 4.85. (Continued) (d) double conversion on-line; (e) delta conversion on-line. *Source*: From Ref. [92].

(power connections). In the case of a power failure, the transfer switch is opened, and the inverter, which is normally in standby, picks up the output load.

The transformer has a special "ferro-resonant" capability, which provides limited voltage regulation and output waveform "shaping". The isolation from AC power transients provided by the ferro transformer is as good as, or better than, any filter available; however, the ferro transformer itself creates severe output voltage distortion and transients, which can be worse than a poor AC connection.

High reliability and excellent line filtering are this design's strengths. However, it has very low efficiency combined with instability when used with some generators and newer power-factor corrected computers, so that the popularity of this design has remarkably declined.

4.16.1.4. Double Conversion On-Line UPS

This is the most common type of UPS above 10 kW. In this design (Figure 4.85d), failure of the input AC does not cause activation of the transfer switch, because the AC is charging the battery that provides power to the inverter. Therefore, during an AC power failure, on-line operation results in no transfer time.

Both the battery charger (rectifier) and the inverter convert the entire load power in this design, resulting in reduced efficiency with its associated increased heat generation.

With no transfer time, this UPS (also called "true" UPS) provides nearly ideal electrical output performance. However, the power components are constantly under load, this reducing reliability and efficiency.

4.16.1.5. Delta Conversion On-Line UPS

This UPS design, illustrated in Figure 4.85e, is a relatively new technology introduced to eliminate the drawbacks of the double conversion on-line design and is available in sizes ranging from 5 kW to 1.6 MW.

The delta converter on-line UPS also has the inverter supplying the load voltage. However, the additional delta converter also contributes power to the inverter output. In other words, the converter and the inverter can both handle DC and AC current, so providing the two power sources for the UPS. Bypassing the battery for part of the power during normal operation reduces power consumption. Under conditions of AC failure or disturbances, this UPS behaves just like the double conversion on-line UPS.

Table 4.11 summarizes some of the characteristics of the various UPS types.

Thus far, three fundamental aspects of energy storage in connection with batteries have been presented:

- Load levelling, or load peak shaving (PS; timescale: hours)
- Power quality (PQ), that is delivery of stabilized power (timescale: milliseconds to seconds)
- Emergency power supply (EPS) or UPS (timescale: seconds to minutes, sometimes few hours).

It is to be remarked that some battery systems for these applications have huge dimensions, and a couple of examples will help understanding the power and energy values they may control.

In Figure 4.86, the 57.6-MWh Na/S battery built in Japan by NGK for the Hitachi's Automotive System Factory is shown. This system is on daily load levelling duty to significantly reduce electricity costs. The tall battery modules are flanked by the power conversion systems (PCS). Other technical specifications are reported in the figure caption.

The Golden Valley Electrical Association (GVEA) based in Fairbanks, Alaska, decided to build there a battery energy storage system (BESS). Its basic function would be covering the time from AC mains failure to diesel generators reaching generation speed (spinning reserve); in practice, it would also cover a number of different power schemes, including improved voltage regulation,

Table 4.11. Technical characteristics, costs, benefits and limitations of the five main designs of UPS systems.

	Power Range (kW)	Voltage Conditioning	Cost per W	Efficiency	Inverter Always Operating	Benefits	Limitations
Stand-by	0–0.5	Low	Low	Very high	No	Low cost, high efficiency, compact	Uses batteries during brownouts, impractical over 2 kW
Line interactive	0.5–5	Design dependent	Medium	Very high	Design dependent	High reliability and efficiency, good voltage conditioning	Impractical over 2 kW
Stand-by ferro	3–15	High	High	Low-medium	No	Excellent voltage conditioning, high reliability	Low efficiency, unstable with some loads and generators
Double conversion on-line	5–5000	High	Medium	Low-medium	Yes	Excellent voltage conditioning, ease of paralleling	Low efficiency, expensive under 5 kW
Delta conversion on-line	5–5000	High	Medium	High	Yes	Excellent voltage conditioning, high efficiency	Impractical under 5 kW

Source: From Ref. [92].

Figure 4.86. The world's largest battery system based on Na/S batteries. This system implements daily load levelling. Specifications include: 4 × 40 module units, 9.6 MW, 57.6 MWh, length: 51.2 m, width: 22.6 m, height: 5.2 m (for a total footprint of 1160 m^2). The Na/S battery operates in the temperature range 290–360°C (see also Section 2.4.2).
Source: From Ref. [88].

electricity loss reduction, load levelling and power quality. This system came into operation by the end of 2003.

The main requirement for the BESS is that it should be capable of providing 40 MW for 15 min [93]. The 15-min time would allow alternative generation to be brought on-line.

There are two primary components making up the BESS; these are as follows

- The nickel–cadmium battery, developed by Saft, which constitutes the energy storage medium
- The converter, designed and supplied by ABB, which changes the battery's DC power into AC power, ready for transmission over the grid.

The BESS battery consists of four parallel strings of 3440 high-performance pocket plate cells (920 Ah). Hence the storage capacity is 4 × 920 Ah = 3680 Ah and the total number of cells in the battery is 3440 × 4 = 13 760. In order to simplify handling, the cells were built into 10-cell modules, which were mounted into an industrial racking system, as shown in Figure 4.87.

The complete battery weighs some 1300 tons and the area in which it is located has a footprint of 3120 m^2. A 20–25 year life is expected without any

Figure 4.87. Part of BESS built in Golden Valley, Alaska. The Ni-Cd batteries can deliver up to 46 MW for 5 min. This system may implement several functions, but is especially used as an emergency power supply in case of mains failure.
Source: From Ref. [93].

loss of the beneficial characteristics of Ni–Cd batteries. The pocket plate cells can deliver 80% of their rated capacity in 20 min and can withstand repeated deep discharges with little effect on battery's life.

Although it is possible for the BESS to produce up to 46 MW for a short time, its principal role is acting as an UPS [93]. Its ability to prevent power outages is shown in Table 4.12.

On a smaller scale, another example of battery installation that may accomplish multiple energy management functions is presented in Figure 4.88. The system, based on VRLA batteries, is installed in a Pb recycling plant and provides power quality protection for the plant's pollution-control equipment, preventing an environmental release in the event of a loss of power. The system carries the critical plant loads while an orderly shutdown occurs, and operates a daily load levelling, reducing the plant's energy costs [94].

Table 4.12. Outages prevented to June 2007 by the Alaska's BESS.

Year	Total Outages Prevented	Total Outage Time (min)	Number of Consumers
2003	3	24	11 122
2004	56	529	286 598
2005	34	416	226 052
2006	82	725	311 000
2007, to June 30	28	232	125 309

Source: Courtesy of GVEA.

Figure 4.88. A 5-MW, 3.5-MWh VRLA battery for peak shaving (load levelling) and power quality. This system is installed at a lead recycling plant in the Los Angeles area. *Source*: From Ref. [94].

Although lead-acid batteries have many disadvantages such as low specific power and energy, short cycle life, high maintenance requirements and environmental hazards, they were accepted as the default choice for energy storage because of their low cost and ready availability. Continuous improvements in design and techniques have mitigated many of these disadvantages, and lead-acid batteries remain the most popular energy storage system for most large-scale applications [87].

Newly designed VRLA batteries have a long cycle life. These batteries have excellent characteristics of 4500 cycles, 15-year lifetime with 70% DOD and 87% cycle efficiency [87].

Advanced flowing electrolyte batteries offer the promise of longer lifetimes and easier scalability to large, multi-MW systems (see Figure 4.84). Indeed their capacity and energy depend only on the amount of electrolyte available, that is on the size of the electrolyte reservoirs.

New systems, for example Li-ion and Ni-MH, are limited at present to installations managing powers of some kW [87].

Batteries for these applications may be inactive for long periods, submitted to trickle charge from time to time, or, as in telecommunications installations (see the next section), they may be on float charge to have them fully charged at all times. By their nature, these batteries must be always ready to work, and their management especially points at knowing the status of the battery and whether it can be relied upon to support its load during an outage. To this end, the SOH (state of health, see Section 3.5) and the SOC of the battery need to be accurately measured [95]. In a very simple way, the SOH could be determined by completely discharging a fully charged battery and measuring its capacity. By comparing this value with the nominal capacity of the fresh battery, the SOH can be derived. Such testing is very inconvenient, as for a large installation it could take several hours to discharge the battery and even more to recharge it. During this time, the installation would be without emergency power.

A better way to measure true battery SOC and SOH, is by real-time impedance tests, as described in Section 3.5.3.1 [95]. These tests can be carried out without discharging the battery, and the monitoring device can remain in place providing a permanent on line measurement. Impedance tests allow an up to date assessment of the battery condition, so that any deterioration in cell performance can be detected and future problems can be predicted. This technique identifies the problem area as well as the single cell involved [96].

Monitoring and management is of the utmost importance in energy storage, as is often ascertained that these batteries exhibit shorter lifetimes than those observed in traditional applications. An optimized battery management system (BMS) has recently been proposed to improve the battery runtime in renewable energy systems [97].

The basic idea behind this BMS was to split the battery array into several strings connected in parallel, as shown in Figure 4.89. The SOH and SOC of

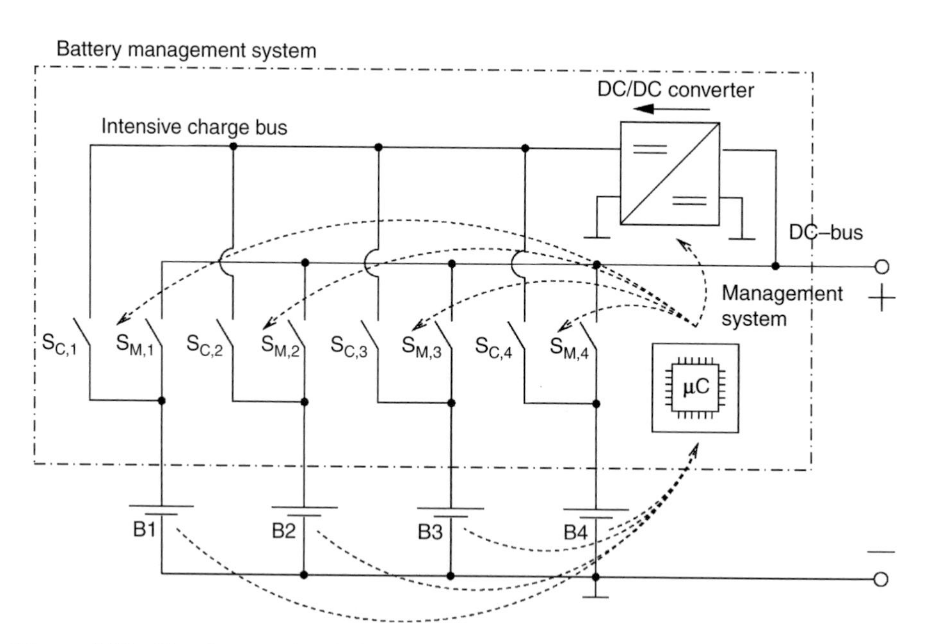

Figure 4.89. Circuit concept of the BMS in a renewable energy system. The internal management system (μC) controls 4 battery strings (B1-B4), four main switches ($S_{M,1}$–$S_{M,4}$), and 4 switches connected to the charge bus ($S_{C,1}$–$S_{C,4}$). The BMS comprises the DC-DC converter, also connected to the DC-bus.
Source: From Ref. [97].

each string are continuously measured, so that each string is individually treated in such a way as to prolong its life. The BMS is able to shorten discharges at low SOC (which are often detrimental, especially for the Lead-acid battery), increase the battery current rate if needed, and fully charge the string, via a DC-DC converter, during normal operating conditions. Furthermore, this BMS takes into account the ageing of each string, and can operate with different battery types or technologies.

The battery characteristics have been tested in a stand-alone system with and without the BMS for one year. Analysis of the results has indicated that this management strategy significantly extends lifetime and improves reliability of the battery system.

4.16.2. Telecommunications

Batteries for telecommunications have a backup function: they are kept in the full charge state to be always up to the task.

In Figure 4.90, the complexity of a telecommunication network is illustrated. The various infrastructures of the network can all be considered critical

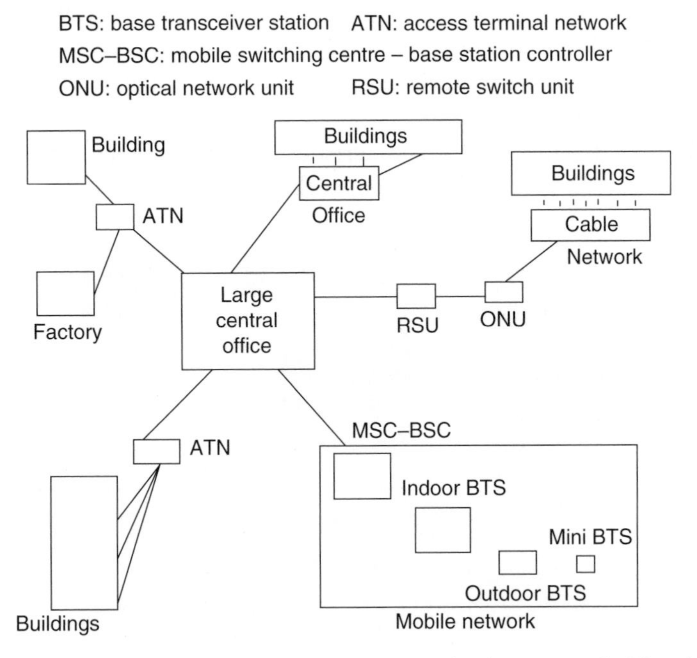

Figure 4.90. Block diagram of a typical telecommunication network. The sites can be identified with the legend in the upper part of the figure.

loads: power discontinuity must be avoided, and this may be accomplished with backup batteries.

Telecom equipment, including the power equipment that has traditionally resided in central offices, is increasingly shifted to outdoor cabinets. This evolution is equally true for wired and wireless systems. However, there is a great difference between power systems in offices and power systems in cabinet environments. Equipment located at a remote site must operate reliably across a wide range of temperatures and humidity levels, and must also cope with the possible presence of dust and dirt [98].

A typical outdoor cabinet is shown in Figure 4.91. An important cabinet design consideration is how the base station electronics are protected against AC blackouts and deep brownouts. In general, the local AC utility powers the rectifiers, which output 24 or 48 VDC. Connected to the DC bus is the backup battery, which floats at the bus voltage. If the AC utility suffers a blackout or brownout, the reserve battery continues supplying power to the DC bus.

This is a typical UPS behaviour and schemes like those of Figure 4.85 could be adopted to avoid power discontinuity. However, as the network moves from conventional systems to Internet protocol (IP) telephony, the standby power requirements change [99]. These new systems require higher power levels, but availability and reliability cannot be compromised if the network has to work when a failure of the public utility occurs. Many parts of these new networks are AC rather than DC powered with UPS systems for backup power. These networks have generally lower levels of reliability than DC systems and need to be designed with appropriate levels of redundancy to maintain a satisfactory reliability.

In Table 4.13, the main characteristics of conventional networks and IP-based networks are listed. The IP network, consisting of routers and servers

Figure 4.91. Typical outdoor cabinet installation in a telecom network. *Source*: From Ref. [93].

Table 4.13. Main features of fixed-line telephone and IP networks.

	Telephone	IP
Reliability	Guaranteed	Best effort
Bandwidth	Narrowband	Broadband
Information transfer	Voice/fax/slow data	Voice/fax/fast data/video
Cost/unit of data transmitted	High	Low
Power supply	DC 48 V	AC 240 V or higher

Source: From Ref. [99].

connected by a variety of broadband services, affords fast data transfer and video transfer at a lower cost. On the other hand, IP-based communications need to make further steps to approach the very high reliability of conventional networks.

The fact that many parts of these new networks are powered by AC power entails using a different architecture for UPS systems, i.e. both AC and DC loads must be powered. An example is shown in Figure 4.92, where a parallel processing (or line interactive) UPS system is adapted to feed the DC loads through a DC/DC converter, that could be replaced with a rectifier connected to the AC line; both AC and DC loads are supported by a single high-voltage battery. This offers the best compromise in terms of efficiency and cost compared to systems with additional power conversion or separate standby

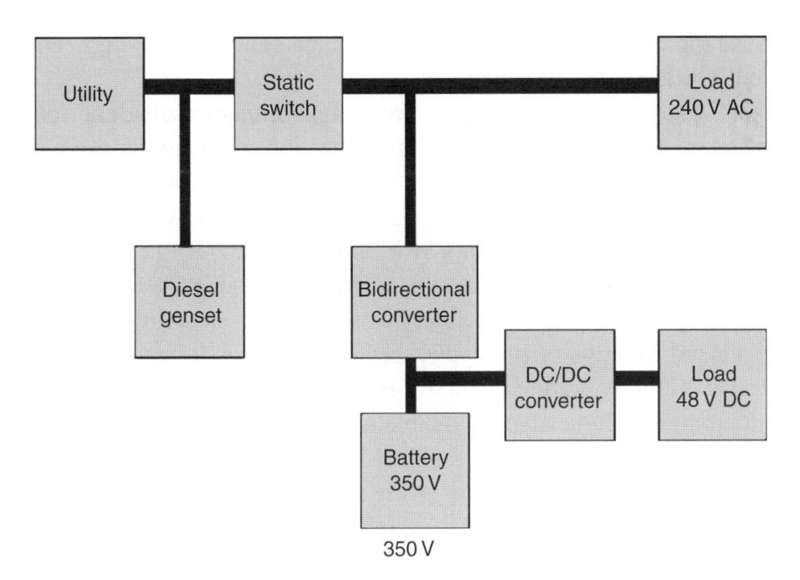

Figure 4.92. A DC/AC power supply system for the convergence of IP and telephone networks.
Source: From Ref. [99].

batteries [99]. The system also includes a diesel generator as a redundant backup power supply in case of prolonged blackout.

It should be noted that mobile networks have different power requirements: as they have a high density of nodes, power continuity can be achieved with relatively short periods of backup power.

The changes in telecom network architecture brought about by the IP technology are in turn changing the requirements of the backup batteries. Hereafter, the main battery systems used today as backup power in telecommunications are reviewed.

Lead-acid batteries (vented type, not VRLA) remain important for large central office applications. The types more frequently used are: (a) pasted plate, (b) tubular plate, and (c) large round batteries [99–101]. The lifetimes of these three types are ~25, ~15 and ~40 years, respectively.

VRLA (or SLA) batteries exist in three main types: (a) tubular gel cells, (b) pasted plate cells with Pb–Ca–Sn grids, and (c) pasted plate cells with pure Pb or pure Pb–Sn grids (see also Section 2.3.4.1). These differ in construction and performance but all are sealed and maintenance-free [99, 100]. Their service lives are around 12 years, but batteries with pure Pb thick plates can live in excess of 25 years at room temperature.

The VRLA compactness constitutes a major requirement for an outdoor cabinet of the type shown in Figure 4.91. However, these batteries are inherently susceptible to thermal runaway. In the summertime, the metal cabinet's interior temperature may reach very high values. Adding the heat of the base station electronics, the temperature within the cabinet can become high enough to significantly reduce overall reliability.

In this respect, the Ni–Cd battery represents a viable substitute for VRLA. Ni–Cd batteries are deployed in telecommunication applications throughout the world, and, in recent years, thousands of strings of batteries, with sintered positive and plastic bonded negative electrodes, have been put into service. These batteries, used in outdoor cabinets, have shown lifetimes in excess of 15 years. This is possible if careful attention to charging and preparation prior to installation are carried out (plus watering at 5-year intervals). Under these conditions, provided the cost is evaluated in terms of life cycle, they can be a cost-effective solution [93]. Saft is leader in the marketing of these batteries.

Li-ion batteries have also been developed for telecom backup applications; their good performance at high rates can offset their higher cost. Both the conventional Li/$LiCoO_2$ chemistry (with or without Ni-doping) and the relatively new Li/$LiFePO_4$ chemistry have been implemented. Li-ion batteries are bound to loose capacity remarkably on float charge especially above room temperature.

Li-metal-polymer-electrolyte (LMP) batteries, commercialized by Avestor (now Bolloré group, see Section 2.4.2) have also been used in telecom

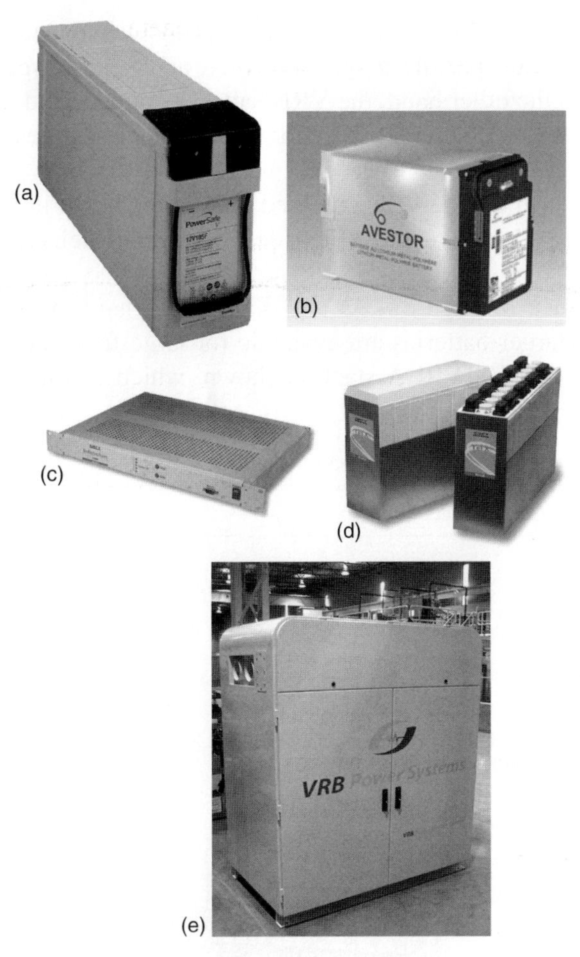

Figure 4.93. Batteries for telecommunications. (a) VRLA by Powersafe; (b) LMP by Avestor-Bolloré; (c) Li-ion Intensium Power 3000 by SAFT; (d) Ni-Cd Tel.X by SAFT; (e) VRB by VRB Power Systems.

applications. Single cells are connected to create the 24 or 48 V commonly requested by these applications. Heating element layers interspersed in the cell stack ensure a temperature range (normally 40–60°C) that is maintained with any external temperature thanks to the insulation. Tests carried out in operating conditions, that is during float and backup periods, have shown that the battery can be operated for at least 12 years (with projections to 20 years).

Another battery reported to be suitable for these application if the one using the vanadium redox chemistry. This battery (see also Section 2.4.1.6), developed for telecom backup by VRB Power Systems, has been compared with an equivalent 5-kW lead-acid battery and has revealed some advantages [102].

In particular, a better DC-DC efficiency (70% vs 45%), a longer lifetime (10 000 vs 1200 cycles), a deeper discharge capability, and reduced replacement costs are reported. On the other hand, the VRB battery is bulkier and heavier than the Lead-acid battery, this being a hurdle for use in a space-limited ambient, e.g. a cabinet.

Examples of the batteries mentioned above are shown in Figure 4.93.

In Table 4.14, the energy densities of various telecom batteries are reported.

The batteries of Figure 4.93 and Table 4.14 have power in the range 1–5 kW. Much larger batteries are available for large-footprint installations. In Figure 4.94, a 48-kWh VRLA stack is shown, which can deliver ~50 kW in

Table 4.14. Main battery types for telecom applications.

Battery Type	Producer	Energy Density (Wh/L)
VRLA	Various	65
Ni–Cd	SAFT	65 100[a]
Li-ion	SAFT[b]	140
LMP	Avestor[c]	170
VRB	VRB Power Systems	45

[a] Up to 100 Wh/L claimed by SAFT for the new (mid-2008) Tel.X battery.
[b] Both Intensium 1 and Intensium 3.
[c] Avestor acquired by Bolloré Group in 2007.

Figure 4.94. 48 V, 1000 Ah, ~50 kW (15 min) VRLA battery with OPzV, A600 type gel cells for telecom applications in a horizontal position.
Source: From Ref. [100].

15 min. This stack is made with gel-type no-spill Lead-acid cells, this allowing their horizontal positioning [100].

4.17. Real Time Clock and Memory Backup

Memories and real time clocks can be found in a very large number of electronic portable and industrial devices [1]. Some of them are listed here:

- automatic bank tellers, gaming and gambling machines
- industrial tools, for example computers, programmable logic controllers, numerical control units, test machines, timers, welding machines, scales, etc.
- instruments and terminals of all kinds
- metering systems (e.g. electricity meters)
- controllers (e.g. contactless ticket readers)
- office equipment (e.g. printers, copiers), telecom equipment, cash registers, smart keys
- on-board instruments (aircraft, boats and terrestrial vehicles)
- some memory cards.

As this matter regards applications of both Chapters 3 and 4, a brief treatment is given at the end of this chapter.

Backup for real time clocks (RTC) and volatile memories can be accomplished with different power sources, i.e. primary or secondary batteries, and capacitors. Consideration must be given to such factors as current consumption of the equipment in which the backup power sources are to be used, their expected lives, and temperature in the operating environment. Deterioration of batteries in long-term backup is significant at high temperatures.

In Table 4.15, the characteristics of different power sources used for backup are shown [103]. Primary lithium cells, of coin or cylindrical shape, are commonly used for RTC and memory backup if long operation times are expected. They are normally sized to function for the expected life of the product. Li/CF_x and Li/MnO_2 are mostly used, but $Li/SOCl_2$ cells are also suitable.

These cells cannot withstand a reverse (charging) current, therefore a diode and a protective resistor must be inserted in the circuit. This is depicted in the scheme of Figure 4.95 (up) [104]. The memory holding voltage, equal to: $V_B - V_F - I_F \bullet R$, must be greater than the IC's voltage.

The voltage characteristics are shown in Figure 4.95 (down): with the main power supply *off*, the battery voltage curve is above the load voltage curve throughout the backup time, thus making the memory non-volatile.

Table 4.15. Common backup supply sources and key selection criteria.

Technology	Operating Temperature (°C)	Self-Discharge Rate	Charging Circuit/ Cycles	Backup Time
Primary lithium	−30 to +80	Low		Long
Capacitor	−40 to +85	High	Simple/unlimited	Short
Rechargeable (Ni–Cd/Ni-MH)	0 to +40[a]	High	Simple/≈500	Short
Rechargeable ML (Li/MnO_2)	20 to +60	Low	V_{limit} controlled/ >1000	Medium[b]

[a] Ambient temperature during charge. The allowed ambient temperature during discharge may be higher.
[b] Total backup time is dependent upon the DOD between each charge cycle.
Source: From Ref. [103].

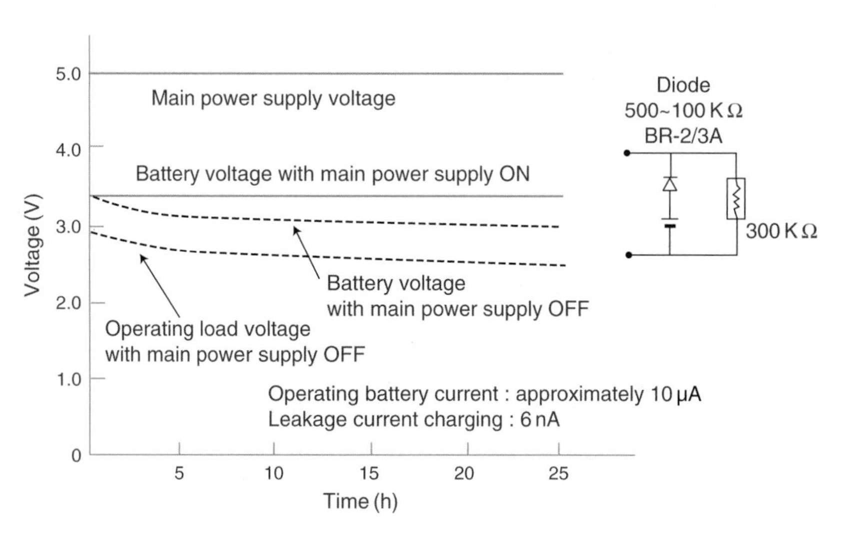

Figure 4.95. Up: circuit used for memory backup with a primary cell (2/3A cylindrical Li/CF_x). A silicon diode or Schottky diode with a low reverse current can be used; down: voltage characteristics of the above circuit.
Source: From Ref. [104].

Large low-leakage capacitors are sometimes used for backup. Unlike primary cells, capacitors can stand reverse currents. However, capacitors require a charging circuit, and provide backup operation for a relatively short time. Capacity may decrease with use, especially at higher operating temperatures [103].

Ni–Cd and Ni-MH can be used for short-time, high-current backup, and can be recharged. Their main disadvantage is the high self-discharge rate ($\sim$10 and $\sim$20% per month, respectively, at room temperature). The charge circuit has to be such to avoid life-shortening overcharges.

Rechargeable Li/MnO_2 cells (see Table 4.15), useful for medium backup times, require a regulated-voltage charging source. Indeed, the maximum voltage must be set to a precise value or permanent damage will occur, while too low a voltage results in incomplete charging.

A memory backup battery in a motherboard typically powers the RTC and associated memory function in a dedicated chip. The memory is often of the SRAM type with a limited density (128–512 kB). However, SRAM chips with larger densities, e.g. for routers and servers, may also be made non-volatile by battery backup. A type of SRAM targeted by this backup is the cache memory; a level 2 cache memory may have a size of 256 kB to 2 MB.

4.18. Wireless Connectivity

In Chapter 3 and in this chapter, several devices (for example, cell phones) and systems (for example, sensor networks) using wireless communication protocols have been mentioned. Here, the characteristics of protocols used in local networks, that is Bluetooth and ZigBee for wireless personal area networks (WPANs) and Wi-Fi for wireless local area networks (WLANs), are summarized. Bluetooth and Zigbee are intended for short-range communications between devices typically used by a single person, for example a wireless keyboard with a computer, or a mobile phone with a hands-free kit. Wi-Fi allows wireless Internet access and communication with other systems in a WLAN, for example computers or printers.

4.18.1. Bluetooth

The Bluetooth specification was developed in 1994 by Ericsson Mobile Platforms, and then formalized by the Bluetooth Special Interest Group (SIG, formally announced in 1999). This association was established by Ericsson, Sony Ericsson, IBM, Intel, Toshiba, and Nokia, and later joined by many other companies (70 by the end of 2007).

With this standard, small devices such as laptops, digital cameras and cell phones may be connected and communicate by radiofrequency if they are in a given range. This range and the transmission power consumption depend on the class:

Class	Maximum Permitted Power	Range (m)
1	100 mW (20 dBm)	~100
2	2.5 mW (4 dBm)	~10
3	1 mW (0 dBm)	~1

The maximum rate of data transmission is ~1 Mbit/s. Current (compatible) versions of Bluetooth are 1.1, 1.2, 2.0, and 2.1. In the future, the Bluetooth technology will be combined with the ultra wide band (UWB) radio technology to reach a data transfer of 480 Mbit/s. Another likely future application will be in the VoIP technology. In idle mode the power consumption is, and will remain, very low. In June 2007, the development of a technology enabling ultra low power consumption was announced (Nokia will have a role in this program).

A Bluetooth chip's block diagram is shown in Figure 4.96. In some devices, for example smartphones and PDAs, new generation Bluetooth units can coexist with WLAN units and enhance their data throughput (transmission rate).

Other relevant Bluetooth features not mentioned above include the following:

- This technology is geared towards voice and data applications
- It operates in the unlicensed 2.4 GHz spectrum
- It is able to penetrate solid objects
- It is omni-directional and does not require line-of-sight positioning of connected devices

4.18.2. ZigBee

The ZigBee standard is regulated by a group known as the ZigBee Alliance, with over 150 members worldwide. This technology has been developed to allow wireless networking of numerous low-power devices. It provides the ability to run for years on inexpensive batteries for a host of monitoring applications: lighting controls, AMR (Automatic Meter Reading), smoke and CO detectors, wireless telemetry, heating control, home security, environmental controls, etc.

While Bluetooth focuses on connectivity between large-packet devices, such as laptops, phones, and major peripherals, ZigBee is designed to provide

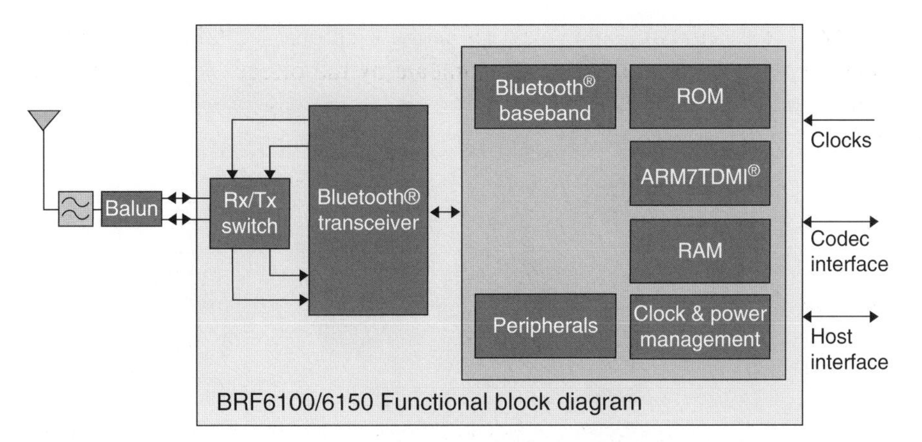

Figure 4.96. Block diagram of a Bluetooth chip. Rx: listening; Tx: transmitting; Balun: a passive electronic device that converts between balanced and unbalanced electric signals.
Source: From Ref. [105].

highly efficient connectivity between small-packet devices. As a result of its simplified operations, which are one to two full orders of magnitude less complex than a comparable Bluetooth device, pricing for ZigBee devices is extremely competitive, with full nodes available for a fraction of the cost of a Bluetooth node. In contrast, ZigBee devices have a maximum data rate of 250 kbps vs 1 Mbps of Bluetooth.

The ZigBee specification identifies three kinds of devices that incorporate ZigBee radios, all three found in a typical ZigBee network (Figure 4.97): a coordinator, which organizes the network and maintains routing tables; routers, which can talk to the coordinator, to other routers and to reduced-function end devices; reduced-function end devices, which can talk to routers and the coordinator, but not to each other. To minimize power consumption and promote long battery life, end devices can spend most of their time asleep, waking up only when they need to communicate and then going immediately back to sleep [106].

The ability to cover large areas with routers is one of the key features of ZigBee that helps differentiate itself from other technologies. Mesh networking can extend the range of the network through routing, while self healing increases the reliability of the network by re-routing a message in case of a node failure.

An example of home control by ZigBee technology is given in Figure 4.98. In the near future, more than 60 ZigBee devices may be found in an average home in developed countries, all communicating with one another freely and regulating common tasks seamlessly. The remote control of the figure has a

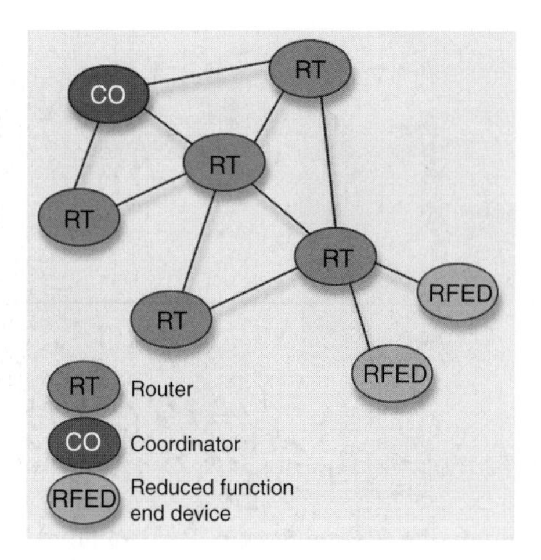

Figure 4.97. A ZigBee network incorporates coordinator (CO), routers (RT) and reduced-function end devices. The network of the figure corresponds to a mesh topology.
Source: From Ref. [106].

range of 30–100 m (indoor-outdoor) and 16 RF channels. It may be powered by AAA rechargeable batteries (Li-ion or Ni-MH).

In the industry, ZigBee is used for automated manufacturing, with small transmitters in all devices, so that they can communicate with a central computer.

4.18.3. Wi-Fi

Wi-Fi, commonly intended as the abbreviation of Wireless Fidelity, is a technology brand, owned by the Wi-Fi Alliance, aimed at improving the interoperability of wireless local area network products based on the IEEE 802.11 standards. The different versions of the 802.11 standards are:

Standard	Frequency/Data Rate	Range
IEEE 802.11	2.4 GHz/1–2 Mbps	Few meters
IEEE 802.11b (Wi-Fi)	2.4 GHz/5.5–11/22 Mbps	30–100 m
IEEE 802.11a (Wi-Fi 5)	5–40 GHz/Up to 54 Mbps	20–40 m
IEEE 802.11g	Up to 54 Mbps	50–80 m

The 802.11b version corresponds to what is commonly known as Wi-Fi. The frequency most often used is 2.4 GHz, just like for Bluetooth and ZigBee.

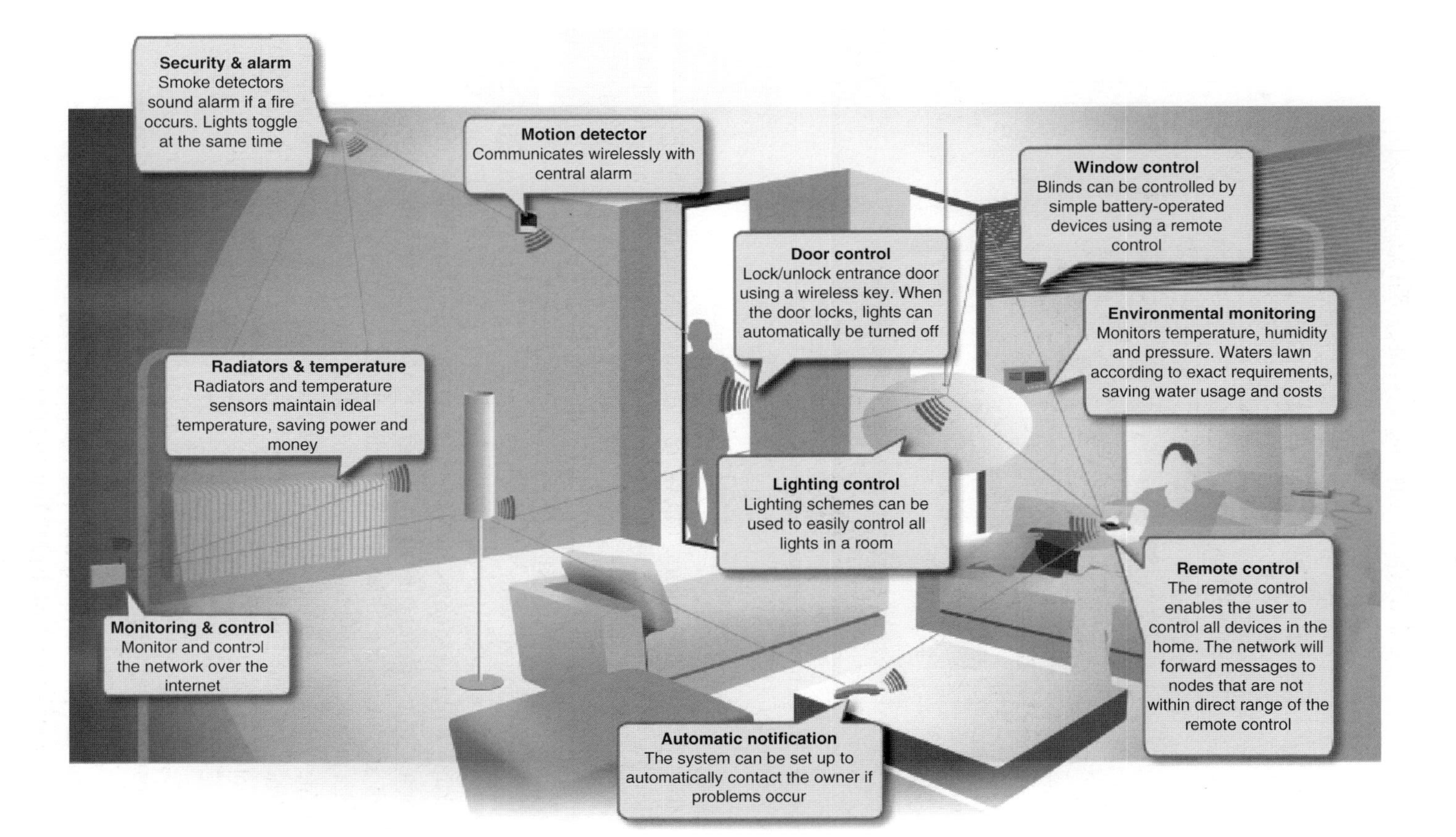

Figure 4.98. An example of home wireless networking using ZigBee technology (See colour plate 6). *Source*: From Ref. [107].

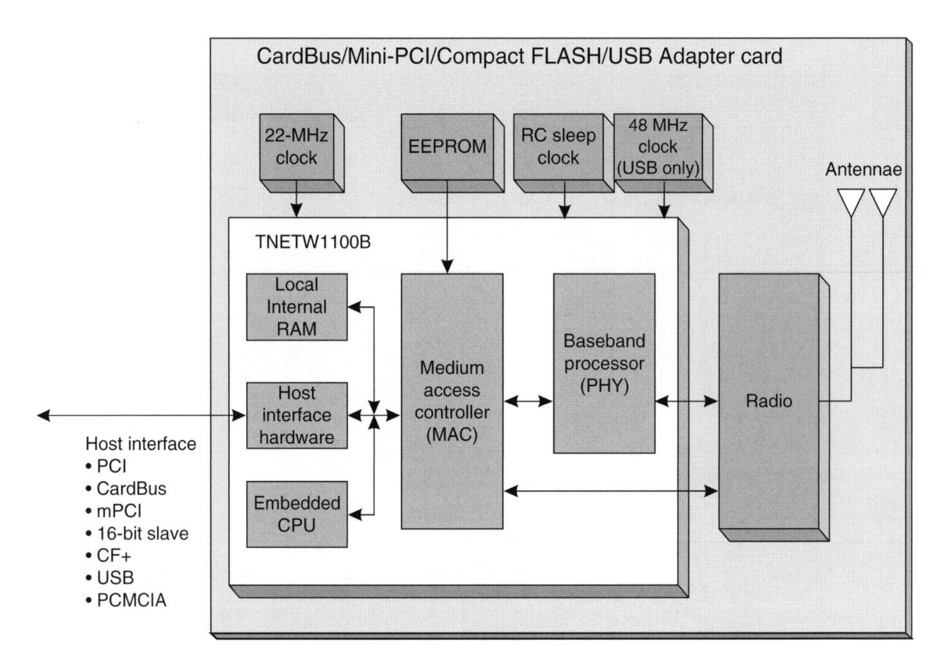

Figure 4.99. Wi-Fi block diagram.
Source: From Ref. [108].

Most modern computers, PDAs, or digital cameras have built-in electronics allowing WLAN connections. The block diagram of an embedded Wi-Fi chip (by Texas Instruments) is shown in Figure 4.99. This single chip (MAC and baseband processor) is fully compliant with the 802.11b standard and is characterized by a very low standby power consumption, 2 mW.

Devices endowed with Wi-Fi chips can be connected wirelessly using a specific router. An example of wireless router is presented in Figure 4.100. It allows using wireless signals (or Ethernet cables) to connect, e.g., computers to one another, to a printer and to the Internet. Most routers provide coverage of ~30 m in all directions, although walls and doors can block the signal. The area of Internet access created by the router is called a hotspot. Anyone with a Wi-Fi enabled device, e.g. a computer or cell phone, can use the hotspot to connect to the Internet. By the end of 2007, Wi-Fi was available in more than 250 000 public hotspots and millions of homes, offices and university campuses worldwide.

Power consumption of Wi-Fi chips is fairly high compared to Zigbee and Bluetooth, thus shortening the battery life. Wi-Fi performance also decreases exponentially as the distance increases. Finally, Wi-Fi is not as fast as Ethernet or other cable systems: its maximum data rate is 54 Mbps (see above), while cables can reach speeds of 1000 Mbps or more.

Figure 4.100. A wireless access point using the 802.11b/g (Wi-Fi) technology. It uses an antenna to send signals to wireless devices and a wire to send signals to the Internet. It is powered by an external power supply (12-V DC).
Source: Courtesy of Links.

The Wi-Fi router of Figure 4.100 is not powered by a battery; however, it is possible to realize a complete independence from mains with portable, battery-powered routers. Alkaline batteries or, for longer operation times, rechargeable batteries, for example Lead-acid or Li-ion, can be used. A sealed lead-acid battery of 12 V and 7 Ah provides the router with a runtime of 8–10 h.

A comparison of the Bluetooth, ZigBee and Wi-Fi technologies is done in Table 4.16. The longer battery life afforded by the ZigBee technology is stressed. ZigBee chips have a low power consumption and runtimes of 5–10 years. They can be powered by a single Li battery, most often a $Li/SOCl_2$ cell of small size (for example, 1/2R6); in some other cases, larger cells (for example, R20) prove necessary [1].

Owing to their higher power consumption, bluetooth and Wi-Fi chips require mainly rechargeable batteries (Li-ion or Ni-MH).

In addition to local area technologies for wireless connections, a wide area technology that is now emerging needs to be mentioned: worldwide interoperability for microwave access (Wi-MAX). It aims at providing wireless data over long distances in a variety of ways, from point-to-point links to full mobile cellular type access. It is based on the IEEE 802.16 standard. The Wi-MAX Forum was formed in June 2001 to promote conformance and interoperability of the standard. Unlike Wi-Fi, this technology allows wireless broadband connections over distances of many kilometers (up to 50) using any frequency below 66 GHz.

Table 4.16. Main characteristics of local-area wireless technologies.

	ZigBee 802.15.4	Wi-Fi 802.11b	Bluetooth 802.15.1
Transmission range (meters)	10–100	30 (indoor), 90 (outdoor)	1–100
Battery life (days)	100–1000	0.5–5.0	1–7
Network size (number of nodes)	>65 000	32	7
Maximum network Speed (kb/s)	250[a]	11 000	1000[b]
Bandwidth (MHz at 2.4 GHz frequency)	5	83.5	83.5
Advantage	Long battery life, low cost	High data rate, interoperability	Interoperability, cable replacement
Main application field	Remote control, battery-operated products, sensors	Internet browsing, PC networking, file transfer	Wireless USB, handset, headset
Power consumption during transmission	30 mA	400+ mA	40 mA

[a] At 2.4 GHz; at 915 MHz: 40 kb/s; at 868 MHz: 20 kb/s.
[b] 3000 kb/s with EDR (Enhanced Data Rate) Bluetooth.

References

1. M. Grimm, in *Industrial Applications of Batteries. From Cars to Aerospace and Energy Storage*, M. Broussely and G. Pistoia, Eds., Elsevier, Amsterdam, 2007.
2. M.-L. Chang, *J. Power Sources* 80 (1999) 273.
3. S. Jacobs, "Adapting Lithium Battery Technology to High Performance AMR Devices", *Metering International*, Issue 3, 2001.
4. Texas Instruments, "3D Metrology".
5. J. Flood, "Ultrasonic Flowmeter Basics", *Sensors*, October 1997.
6. S. Jacobs, "Advanced Lithium Battery Technology – Key to AMR Market", *REMOTE Site & Equipment Magazine*, April/May 2002.
7. G.O. Allgood, W.W. Manges and S.F. Smith, "It's Time for Sensors to Go Wireless; Part 1: Technological Underpinnings", *Sensors*, April 1, 1999.
8. Harvard University, School of Engineering and Applied Sciences, "CodeBlue: Wireless Sensor Network for Medical Care", September 2006.

9. T.R.F. Fulford-Jones, G.-Y. Wei and M. Welsh, "A Portable, Low-Power, Wireless Two-Lead EKG System", *Proceedings of the 26th Annual International Conference of the IEEE EMBS*, San Francisco, CA, September, 2004, p. 2141.
10. Mainwaring, J. Polastre, R. Szewczyk, D. Culler and J. Anderson, "Wireless Sensor Network for Habitat Monitoring", *WSNA '02*, Atlanta, GA, September 2002.
11. Bharathidasan and V.A.S. Ponduru, "Sensor Networks: An Overview", Department of Computer Science, University of California, Davis, CA, www.csif.cs.ucdavis.edu/~bharathi/ sensor/survey.
12. D. Culler, D. Estrin and M. Srivastava, *Computer* [*IEEE Computer Society*] 37 (2004) 41.
13. S.W. Arms, C.P. Townsend, D.L. Churchill, M.J. Hamel, J.H. Galbreath and S.W. Mundell, "Frequency Agile Wireless Sensor Networks", *Proceedings of the SPIE Symposium on Smart Structures & Materials/ NDE 2004*, San Diego, CA, March 2004.
14. N. Bulusu, D. Estrin, L. Girod and J. Heidemann, "Scalable Coordination for Wireless Sensor Networks: Self-Configuring Localization Systems", *Proceedings of the Sixth International Symposium on Communication Theory and Applications (ISCTA 2001),* Ambleside, UK, July 2001.
15. C. Ma and Y. Yang, *Mobile Networks Applications* 11 (2006) 757.
16. S. Jacobs, "Battery Power for Remote Wireless Sensors", *Sensors*, March 2004.
17. P.H. Humble and J.N. Harb, *Journal of Electrochemical Society* 150 (2003) A1182.
18. P.H. Humble, J.N. Harb and R. LaFollette, *J. Electrochem. Soc.* 148 (2001) A1357.
19. C. Park, K. Lahiri and A. Raghunathan, "Battery Discharge Characteristics of Wireless Sensor Nodes: An Experimental Analysis", *IEEE SECON 2005. Second Annual IEEE Communications Society Conference on Sensor and Ad Hoc Communications and Networks,* September 2005.
20. Texas Instruments, "Video Security over Internet Protocol (VSIP) Development Platform", 2003.
21. J.M. Major, "Ensuring the Health of Our Power Lines", *Technology Today*, Summer 2006.
22. H. Yamin, in *Industrial Applications of Batteries. From Cars to Aerospace and Energy Storage*, M. Broussely and G. Pistoia, Eds., Elsevier, Amsterdam, 2007.
23. Electrochem Pack Brochure, www.electrochempower.com.
24. D. Hensley, M. Milewits and W. Zhang, "The Evolution of Oilfield Batteries", *Oilfield Review,* Autumn 1998, p. 42.
25. J. Cohen, J.D. Rogers, E. Malcore and J. Estep, "The Quest for High Temperature MWD and LWD Tools", *Gas TIPS,* Fall 2002, p. 8.
26. R.A. Normann, "High Temperature Instrumentation", *Workshop on Extreme Environments Technologies for Space Exploration*, Pasadena, Texas, May 2003.
27. R.A. Guidotti, F.W. Reinhardt and J. Odinek, *J. Power Sources* 136 (2004) 257.
28. N. Brown, A.J. Fougere, A.L. Kun, "An Acoustic Current Meter Update (3D-ACM)", *Oceanology International,* Brighton, U.K., 1998.
29. A.J. Fougere and A.L. Kun, "Applications of an Acoustic Current Measurement Technique", *Ocean News and Technology*, pp. 36–37, September/October 1998.
30. Information on the SEAL Project provided by GFZ, Potsdam (Germany).
31. S. Jacobs, "Battery Technology for Remote Oceanographic Applications", *ST Power Systems Feature,* www.tadiranbat.com/articles/seatech
32. Information provided by IRD (Institut de Recherche pour le Développement, Paris, France).
33. R. Stephen, T. Pettigrew and R. Petitt, "SeisCORK Engineering Design Study", *Woods Hole Oceanographic Institution's Report WHOI*-2006-10.
34. C. Meinig, S.E. Stalin, A.I. Nakamura, F. Gonzalez and H.B. Milburn, "Technology Developments in Real-Time Tsunami Measuring, Monitoring and Forecasting", NOAA-PMEL Report, www.nctr.pmel.noaa.gov/Dart/Pdf/mein2836_final.

35. J. Sherman, R.E. Davis, W.B. Owens and J. Valdes, *IEEE Oceanic Engineering* 26 (2001) 437.
36. Information provided by Sirtrack, www.sirtrack.com.
37. D. Meldrum, "Developments in Satellite Communication Systems", *Scottish Association for Marine Science*, October 2004.
38. D. Katz, G. Ouellette, R. Gentile and G. Olivadoti, "Fast, Versatile Blackfin Processors Handle Advanced RFID Reader Applications", *Analog Dialogue* 40-09, September 2006.
39. Transcore (Trademark of TC IP), "Active/Passive and RF-Active/RF-Passive Tags". *Considerations for High Performance Toll Systems*', 2004.
40. "Electronic Toll Collection", *Wikipedia*, November 2007.
41. Y. Borthomieu and N. Thomas, in *Industrial Applications of Batteries. From Cars to Aerospace and Energy Storage*, M. Broussely and G. Pistoia, Eds., Elsevier, Amsterdam, 2007.
42. Philips Semiconductors, "Application of UART in GPS Navigation System", March 2005.
43. EXAR, "UARTs in GPS Applications", March 2007.
44. G. Naden, "Going the Distance. Power Supply Management in an Asset Tracking Device", *GPS World*, October 2003.
45. NASA, "Artificial Satellites", in *World Book Encyclopaedia*, 2007.
46. Boeing, "GOES-N – Mission Booklet".
47. NASA-NOAA, "NOAA GOES-N,O,P – The Next Generation", *Report* 2005.
48. Directory of U.S. Military Rockets and Missiles. Appendix 3: Space Vehicles, 2006.
49. Yardney, "Aerospace Batteries in Unprecedented Trifecta", March 2007.
50. T.R. Crompton, *Battery Reference Book*, 3rd edition, Newnes Pub., London, 2000.
51. A.O. Nilsson and C.A. Baker, in *Handbook of Batteries*, D. Linden and T.B. Reddy, Eds., Chapter 26, McGraw-Hill, New York, 2002.
52. J. Timmons, R. Kurian, A. Goodman and W.R. Johnson, *Journal Power Sources* 136 (2004) 372.
53. B.V. Ratnakumar and M.C. Smart, in *Industrial Applications of Batteries. From Cars to Aerospace and Energy Storage*, M. Broussely and G. Pistoia, Eds., Elsevier, Amsterdam, 2007.
54. B.V. Ratnakumar, M.C. Smart, A. Kindler, H. Frank, R. Ewell and S. Surampudi, *Journal Power Sources* 119 (2003) 906.
55. P.G. Backles, J.S. Norris, M.W. Powell and M.A. Vona, "Multi-mission Activity Planning for Mars Lander and Rover Missions", *IEEE Aerospace Conference*, Big Sky, Montana, March 2004, Paper 1498.
56. R. Hollandsworth and J.D. Armantrout, "Hubble Space Telescope Battery Capacity Trend Studies", *Proceedings of NASA Aerospace Battery Workshop*, Huntsville, AL, 2003.
57. F. Cohen and P. Dalton, "International Space Station Nickel-Hydrogen Battery Start-Up and Initial Performance", *Proceedings 36th Intersociety Energy Conversion Engineering Conference (IECEC),* Savannah, GA, July-August 2001.
58. C.P. Bankston, "Progress in Spacecraft Electric Power System Technologies for Deep Space Missions", *Proceedings 35th Aerospace Sciences Meeting and Exhibit,* Reno, NV, January 1997.
59. D. Linden and T.B. Reddy, in *Handbook of Batteries*, D. Linden and T.B. Reddy, Eds., Chapter 14, McGraw-Hill, New York, 2002.
60. J. Swank, "Reserve Batteries: The Power Sources That Make Munitions Tick", *Army Research Laboratory Report*, 2006.
61. A.A. Benderly, in *Handbook of Batteries*, D. Linden and T.B. Reddy, Eds., Chapter 19, McGraw-Hill, New York, 2002.
62. V. Klasons and C.M. Lamb, in *Handbook of Batteries*, D. Linden and T.B. Reddy, Eds., Chapter 21, McGraw-Hill, New York, 2002.

63. T. Reitz, "AFRL's Long Endurance Electrochemical Power Systems", *Air Force Research Laboratory Report*, May 2005.
64. Department of Defence, "Unmanned Aircraft Systems Roadmap", *Report* 2005.
65. D. Coburn, "Land Warrior System: Inside the Pentagon's New High-Tech Gear", *Popular Mechanics*, May 2007.
66. J. VanZwol, "Powering the Smart Soldier", *Cots Journal*, June 2005.
67. Micro Power, "Ruggedized and Mil-Spec Battery Packs", www.micro-power.com
68. E. Red, "Robotics Overview", http://research.et.byu.edu/eaal/html/RoboticsReview
69. D. Addison, "Introduction to Robotics Technology", *IBM's Developer Works*, September 2001.
70. TheTech, "Get a Grip on Robotics", www.thetech.org
71. Honda, "ASIMO – Frequently Asked Questions".
72. G. Caprari, "Autonomous Micro-Robots: Applications and Limitations", Thesis, École Polytechnique Federal de Lausanne, Switzerland, 2003.
73. A. Cavalcanti, W.W. Wood, L.C. Kretly and B. Shirinzadeh, "Computational Nanomechatronics: A Pathway for Control and Manufacturing Nanorobots", *IEEE CIMCA International Conference on Computational Intelligence for Modelling Control and Automation*, Sidney, Australia. November-December 2006.
74. S.A. Vittorio, "Microelectromechanical Systems (MEMS)", *CSA Discovery Guides*, October 2001.
75. T. Hoffman, "Smart Dust. Mighty Motes for Medicine, Manufacturing, the Military and More", *Computer World*, March 24, 2003.
76. E. Kolesar, "Introduction to Microelectromechanical Systems (MEMS). Lecture 13", Texas Christian University, Department of Engineering.
77. J.N. Harb, R.M. LaFollette, R.H. Selfridge and L.L. Howell, *Journal of Power Sources* 104 (2002) 46.
78. Infinite Power Solutions, www-mtl.mit.edu
79. M. Nathan, D. Golodnitsky, V. Yufit, E. Strauss, T. Ripenbein, I. Shechtman, S. Menkin and E. Peled, *J. Microelectromechanical Systems* 14 (2005) 879.
80. S.H. Upadhyaya, "Precision Agriculture. Specialty Crops", www.specialtycrop.info/finalagrisite/Sp_Crop_SKU.pdf
81. D. Rickman, J.C. Luvall, J. Shaw, P. Mask, D. Kissel and D. Sullivan, "Precision Agriculture: Changing the Face of Farming", *Geotimes*, November 2003.
82. S.A. Shearer, J.P. Fulton, S.G. McNeill, S.F. Higgins and T.G. Mueller, "Elements of Precision Agriculture: Basic Yield Monitor Installation and Operation", University of Kentucky, College of Agriculture, Cooperative Extension Service, 1999.
83. M. Delwiche, B. Coates and P. Brown, "Management of Water and Fertilizer", www.specialtycrop.info/finalagrisite/Sp_Crop_SKU.pdf.
84. P. Andrade-Sanchez, F.J. Pierce and T.V. Elliott, "Performance Assessment of Wireless Sensor Networks in Agricultural Settings", *ASABE Annual International Meeting*, Minneapolis, MN, June 2007.
85. M.J. Buschermohle and R.T. Burns, "Solar-Powered Livestock Watering Systems" www.utextension.utk.edu/publications/pbfiles/pb1640.pdf.
86. California Energy Commission, "Distributed Energy Resource Guide".
87. J. Kondoh, in *Industrial Applications of Batteries. From Cars to Aerospace and Energy Storage*, M. Broussely and G. Pistoia, Eds., Elsevier, Amsterdam, 2007.
88. Technology Insights, "Overview of NaS Battery for Load Management", February 2005. www.energy.ca.gov/pier/notices/2005-02-24_workshop/11%Mears.
89. U.S. Department of Energy – Energy Efficiency and Renewable Energy – Solar Energy Technologies Program.

90. R. Wagner, in *Industrial Applications of Batteries. From Cars to Aerospace and Energy Storage*, M. Broussely and G. Pistoia, Eds., Chapter 9, Elsevier, Amsterdam, 2007.
91. VRB Power Systems, "The Multiple Benefits of Integrating the VRB-ESS with Wind Energy – Case Studies in MWh Applications", *Report* March 2007.
92. N. Rasmussen, "The Different Types of UPS Systems", Revision 5, American Power Conversion, 2004.
93. A. Green, in *Industrial Applications of Batteries. From Cars to Aerospace and Energy Storage*, M. Broussely and G. Pistoia, Eds., Elsevier, Amsterdam, 2007.
94. U.S. Climate Change Technology Program – Technology Options for the Near and Long Term, August 2005. www.climatetechnology.gov/library/2005/tech-options/tor2005.
95. Y. Barsukov and B. Krafthofer, "Predicting Runtime for Battery Fuel Gauges", *Power Management DesignLine*, August 2005.
96. M.R. Laidig and J.W. Wurst, "Battery Failure Prediction", www.btechinc.com/docs/white-papers/BatteryFailurePrediction
97. R. Kaiser, *Journal of Power Sources* 168 (2007) 58.
98. B. Bonney, "Powering Wireless Telecom Basestations", *Telephony on Line*, March 1996.
99. G.J. May, *Journal of Power Sources* 158 (2006) 1117.
100. R. Wagner, Chapter 7, in *Industrial Applications of Batteries. From Cars to Aerospace and Energy Storage,* M. Broussely and G. Pistoia, Eds., Elsevier, Amsterdam, 2007.
101. W.D. Reeve, *Dc Power System Design for Telecommunications*, Wiley-IEEE Press, 2006.
102. VRB Power Systems, "The VRB-ESS System – A comparison with Lead Acid Batteries", September 2007.
103. Maxim, "Selecting a Backup Source for Real-Time Clocks", April 2006.
104. Panasonic, "Design for Memory Back-up Use", *in Lithium Handbook*, August 2005.
105. Texas Instruments, "Bluetooth Solutions", 2004.
106. T. Cutler, "Implementing ZigBee Wireless Mesh Networking", *Industrial Automation*, July 2005.
107. Texas Instruments, "ZigBee Wireless Communications Overview", *Report Slyb127*, 3Q 2006.
108. Texas Instruments, "WLAN Solutions: TNETW1100B Embedded Single-Chip MAC and Baseband Processor", 2002.

Chapter 5

VEHICLE APPLICATIONS: TRACTION AND CONTROL SYSTEMS

5.1. Introduction

Three fundamental approaches have been considered in the last decades to reduce oil consumption and CO_2 emissions in engine-driven vehicles. The first points to bio-fuels, for example bio-methane, bio-ethanol and bio-diesel; the second is the long sought, but not yet feasible, solution of hydrogen car (fuel cells or H_2-ICE); the third is based on the use of electric motors, either alone (electric vehicle, EV) or coupled with a traditional engine (hybrid electric vehicle, HEV).

The choice of an electric motor to substitute, or to be coupled with, an internal combustion engine (ICE) is suggested by the well-known higher efficiency of the former vs the latter. This is illustrated in Figure 5.1.

In a conventional engine, only ~15% of the energy generated by the fuel reaches the wheels to move the car and power accessories, such as air conditioner. The other ~85% of the energy content of the fuel is lost to engine and driveline inefficiencies, idling and braking [1]. In contrast, battery-powered electrical motors can be smaller, lighter, and up to 5–6 times as efficient as ICEs at converting energy to motion. The main loss in these motors is due to the resistance in the electric circuitry.

Pure EVs are still not widely commercialized, despite the many prototypes produced over the years. In contrast, HEVs have become commercially available in 1997, and are now gaining increasing market shares. The state-of-the-art of these vehicles and the role of batteries will be pointed out in this chapter.

Before describing EV and HEV systems, the organizations that have supported over the years research in this field must be mentioned:

- US Department of Energy (DOE)
- US Advanced Battery Consortium (USABC)
- FreedomCAR and Vehicle Technology (FCVT; USA)
- US Council for Automotive Research (USCAR)
- Partnership for a New Generation of Vehicles (PNGV; USA) (this project was accomplished from 1993 to 2001)
- New Energy and Industrial Technology Development Organization (NEDO; Japan)

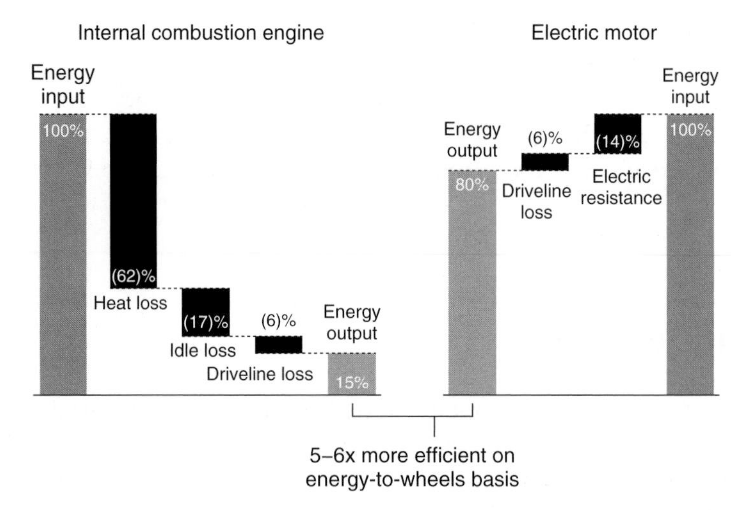

Figure 5.1. Comparison of an internal combustion engine with an electric motor on energy-to-wheels basis.
Source: From Ref. [1].

- Central Research Institute of Electric Power Industry (CRIEPI; Japan)
- Lithium Battery Energy Storage Technology Research Association (LIBES; Japan) (this project was accomplished from 1992 to 2001)
- Ministry of Economy Trade and Industry (METI; Japan)
- European Commission.

In the final part of this chapter, the role of batteries in the vehicle's control systems (braking, steering, transmission, control of electric devices, etc.) will be outlined.

5.2. Electric Vehicles

In EVs, only the electric motor powered by the battery transmits motive power to the wheels. They are also called zero emission vehicles (ZEV) as no chemical substances are released in the atmosphere.

To compete with ICE cars, the battery has to store an amount of energy granting a sufficient driving range, and has to be recharged in a relatively short time. However, a discharged battery needs few hours for a full charge, as the charging rate cannot exceed a given limit. A fast charge is possible for high-power batteries, but less feasible for the high-energy batteries needed to provide an acceptable driving range.

An EV battery is normally recharged overnight at home in 5–8 h. Additional partial recharges at medium rate can be made during the day, for example in parking lots with dedicated facilities. A typical battery for a medium EV car has energy of about 30 kWh (see Table 5.5). Fully charging this battery in 20–40 min, as recently claimed for Li-ion traction batteries, would therefore need power in the range 45–90 kW. This high-power charge cannot be carried out using home facilities and would require special structures with related cost issues.

As is obvious, the driving range with a full charge becomes the fundamental parameter, and this is directly related to battery size. Its volume is important for the car design and utilization, because it affects the available space. However, the battery weight is more critical: indeed, there is a maximum weight for a car of a given size that cannot be exceeded on the basis of (a) safety considerations: the car would be difficult to drive, and slowing down would take a long time; and (b) energy considerations: part of the energy would be wasted to move the battery, this reducing the theoretical driving range.

The influence of the battery's weight on the car range is illustrated in Figure 5.2, which describes the maximum range of a vehicle as a function of the total battery energy. The 800-kg vehicle (without the battery) is assumed to consume 135 Wh/ton/km. As the battery energy increases, the battery weight also increases and soon becomes intolerable for the vehicle.

Figure 5.3 displays the corresponding weights of several batteries of different specific energy. From this picture, it can be seen that a minimum of ~100 Wh/kg is needed to exceed a 200-km range. At present, only lithium-based batteries can achieve this goal (see later).

Based on the above discussion stressing the need of a battery with reduced weight and volume, the most important parameters for an EV's battery are energy density (Wh/L) and specific energy (Wh/kg).

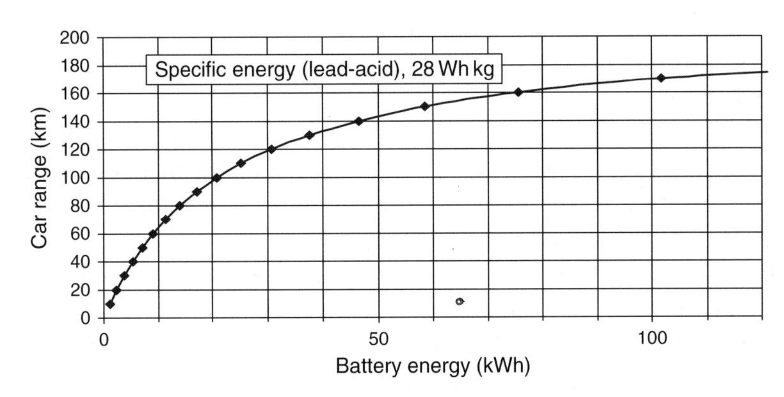

Figure 5.2. Example of a calculated mid-size EV range as a function of battery energy. *Source*: From Ref. [2].

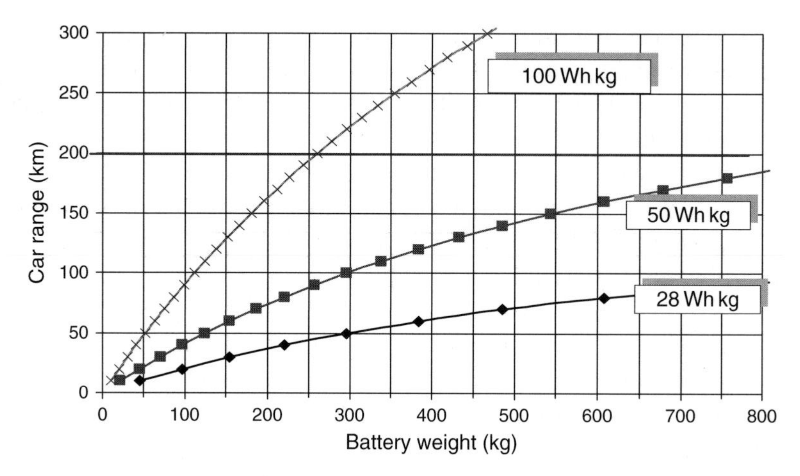

Figure 5.3. Calculated car range as a function of the weight of batteries with different specific energies.
Source: From Ref. [2].

As the total energy required is large (tens of kWh), the battery cost becomes a very sensitive issue. Ideally, the battery should not be replaced during the vehicle lifetime, that is at least 10 years. (The new goal of USABC is 15 years.) A cycle life exceeding 1000 cycles is therefore expected from the battery, and deep cycling capability is required, that is at depth of discharge (DOD) of ~80%.

Batteries for EVs need to meet the stringent requirements of the components of vehicles operating on the roads. They must not present safety hazards for people inside the vehicle or for the environment, even in abuse conditions. The potential hazards of an EV are proportional to the battery energy, and to the quantity and type of chemicals inside the battery.

From the above considerations, it can be concluded that the more challenging aspects for the realization of EVs are driving range (battery energy), cost and safety.

Lead-acid has long been the sole battery system used in EVs, but the limited car range, typically about 60 km, the considerable maintenance requirements and poor cycle life have severely limited its development. For about 20 years, Ni–Cd has been used in Europe, especially in France, extending the range to about 80 km, and with better life. Ni-MH has been developed for several demonstration programs, improving the energy density and car range (>100 km) with no battery maintenance, but has never actually been commercialized, partly due to high manufacturing costs. The new high-energy density technologies based on Li-ion have been studied for this application for more than 10 years, leading to demonstration fleets. They are now close to the commercialization phase, bringing the EV to a possible range of 200 km or more. Emerging technologies are being tested in new-generation vehicles: Zn-air, Zebra and Li-metal-polymer (LMP, see Chapter 2). Details on the use of these batteries will be reported later.

Tables 5.1 and 5.2 take as a reference a mid-size vehicle (weighing 1200 kg), and summarize the characteristics of different EV batteries (weighing 250 kg or having a volume of 200 L, respectively). The differences in the calculated driving range can be appreciated.

The vehicle's ability to accelerate and climb hills depends on the battery power, if not limited by the electric motor power. Here again, Li-ion batteries prove superior, as shown in Table 5.3.

The car range can be increased by a small combustion engine able to recharge the battery while driving or parking; this car is called EV with range extender. The low-power combustion engine can be used on either demand by the driver or when the battery charge level is below a certain limit. The battery is the same as for a pure EV of the same size, and has to fulfil the same requirements. This car can also be considered as a series hybrid vehicle (see below).

Table 5.1. Calculated range of a mid-size EV with different batteries weighing 250 kg.

Technology	Pb-Acid	Ni–Cd	Ni-MH	Li-Ion
Vehicle curb weight		1200 kg		
Battery weight allocation (typ.)		250 kg		
Battery structure, cooling, etc.		55 kg		
Module weight allocation		195 kg		
Energy density (module) (Wh/kg)	33	45	70	120
Onboard energy (kWh)	6.4	8.8	13.0	23.4
Range at 120 Wh/ton/km (km)	53	73	114	195

Source: From Ref. [3].

Table 5.2. Calculated range of a mid-size EV with different batteries of 200 L.

Technology	Pb-Acid	Ni–Cd	Ni-MH	Li-Ion
Battery volume allocation (typ.)		200 L		
Battery structure, cooling, etc.		70 L		
Module volume allocation		130 L		
Volumetric energy (module) (Wh/L)	75	80	160	190
Onboard energy (kWh)	9.8	10.4	20.8	24.7
Range at 120 Wh/ton/km (km)	81	87	173	206

Source: From Ref. [3].

Table 5.3. Power characteristics comparison for a 250 kg battery using different chemistries.

Technology	Pb-Acid	Ni–Cd	Ni-MH	Li-Ion
Vehicle curb weight		1200 kg		
Battery weight allocation (typ.)		250 kg		
Battery structure, cooling, etc.		55 kg		
Module weight allocation		195 kg		
Power density (module) (W/kg)	75	120	170	370
Battery power (kW)	15	24	33	72

Source: From Ref. [3].

5.2.1. New EV Proposals: Will They Succeed?

Since 1997, a number of EVs have been proposed; some of these cars were fully developed but never reached a significant level of commercialization, because of limited performance and high cost.

With the advent of new battery chemistries and with technologies able to tackle the safety issues, efforts have been renewed to make EVs more performing and affordable. Here, some recent proposals are briefly introduced, even though it is impossible to foresee their future.

The French group Dassault-Heuliez produces Cleanova II, an EV with a permanent magnet synchronous motor (35 kW) powered by Li-ion batteries developed by SAFT. This car is either fully electric or may be endowed with a 15-kW ICE using bio-ethanol, thus becoming an EV with range extender. The first version should allow a driving range of 150–200 km, and the batteries should last more than 300 000 km, that is the vehicle's lifetime.

About a thousand cars should be produced by 2008, based on two light commercial vehicles, Fiat Doblò and Renault Kangoo, The conversion into EV is expected to add around US$10 000 to the price of these vehicles.

The Dassault group has also introduced in 2005 Cleanova III, an EV based on Renault Scénic, with a driving urban range of 210 km (470 km with a range extender). This EV is also equipped with Li-ion batteries.

Nissan has presented in August 2007, in Europe, its compact concept EV, the Mixim, with two electric motors powered by Li-ion batteries. The two motors move both the front and the rear wheels, thus making the Mixim a four-wheel drive car. The Li-ion cells are of the laminate type for a better space utilization vs the cylindrical ones. Their negative electrode is carbon and the positive is Li–Mn oxide. Two batteries, each with a power of 50 kW, should allow reaching a speed of 180 km/h, while the driving range should be 250 km. Nissan claims a very short recharge time, that is between 20 and 40 min.

This city car (3.7 m long, weighing 950 kg) has a central seat for the driver, while two passengers' seats are located at both sides of the driver. Nissan has clearly stated that no mass production is guaranteed.

A peculiar EV, called Pivo 2, has been presented by Nissan at the Tokyo Motor Show (October 2007). This small car has, among other features, a 360° turning cabin, so that a reverse gear is not required, wheels that can rotate 90°, a "robotic agent" on the instrument panel that helps the driver (for example with directions), and electronically driven braking and steering. 15- kW electric motors powered by compact Li-ion batteries are in each wheel, and provide a driving range of 125 km. These batteries are based on thin prismatic cells instead of the cylindrical ones. Commercialization is expected by 2011.

The Think City car was produced by Ford until 2003 and then abandoned. The project has been resumed in 2006 by the Norwegian group Think Global. By the end of 2007, a new car has been produced, and by the end of 2008 there should be some 3500 cars on the road. The car should be equipped with Li-ion batteries, but at present Zebra thermal batteries are used, based on the $Na/NiCl_2$ couple (see Section 2.4.2).

This two-seater car should have a maximum speed of 100 km/h, and a driving range of 180 km with Li-ion batteries, but a lower reach with Zebra batteries. These latter, however, have several positive features, such as safety and resistance to overcharge and overdischarge. In optimal conditions, lifetime of a battery pack should be around 160 000 km.

In October 2007, Think Global has signed an agreement with the US battery company EnerDel to have Li-ion batteries by the end of 2008. These batteries should be based on the couple: $Li_4Ti_5O_{12}$ (as a negative electrode) and $LiMn_2O_4$ (as a positive) – see later for more details.

Owing to the batteries' cost, Think Global is planning to lease them at 120 euros/month. In contrast, the car should cost 25 000 euros. These prices raise obvious concern for its real commercialization, letting alone technical difficulties.

The joint venture between the French group Bolloré and the Italian design house Pininfarina should produce, by 2010, a four-seater electric car (Blue Car) with a driving range of 250 km. This EV will be powered by lithium-metal-polymer batteries that should grant a 250-km driving range and a maximum speed of 130 km/h. The LMP batteries, without liquid electrolyte, are characterized by enhanced safety, as they are thermally stable (combustion temperature above 200°C). Although no details have been released by Bolloré on the battery chemistry, it is likely that it will use the Li/LiV_3O_8 couple in a polymeric electrolyte. This is the chemistry used by Avestor (see Section 2.4.2), recently acquired by the Bolloré Group.

Finally, a sport car has been introduced by Tesla Motors (USA) in July 2006. The Roadster is equipped with Li-ion batteries and an electric motor of 185 kW. It has an impressive acceleration: 0–100 km in <4 s. Its driving range is 400 km and maximum speed 210 km/h. Tesla Motors will supply Think Global with the same Li-ion batteries.

In Figure 5.4, images of the cars described above (with the exception of Cleanova III) are presented.

Figure 5.4. Top left: Cleanova II; top right: Nissan Mixim; middle left: Nissan Pivo 2; middle right: Think City; bottom left: Blue Car (Bolloré); bottom right: Tesla Roadster (See color plate 7).

5.3. Basics of Hybrid Electric Vehicles

In a HEV, both combustion engine and electric motor are used. The electric energy is used to provide power for start and acceleration, when the conventional ICE would consume a large amount of fuel. The ICE can use natural gas, gasoline, diesel or alternative fuels, like the bio-fuels mentioned previously. In a particular case, a hybrid vehicle can have two electric motors, one powered by a battery, and the other powered by a fuel cell.

These different approaches are represented in Figure 5.5.

The goal is to have a vehicle with the same driving range of conventional ICE cars, with lower fuel consumption and lower emissions. The battery size

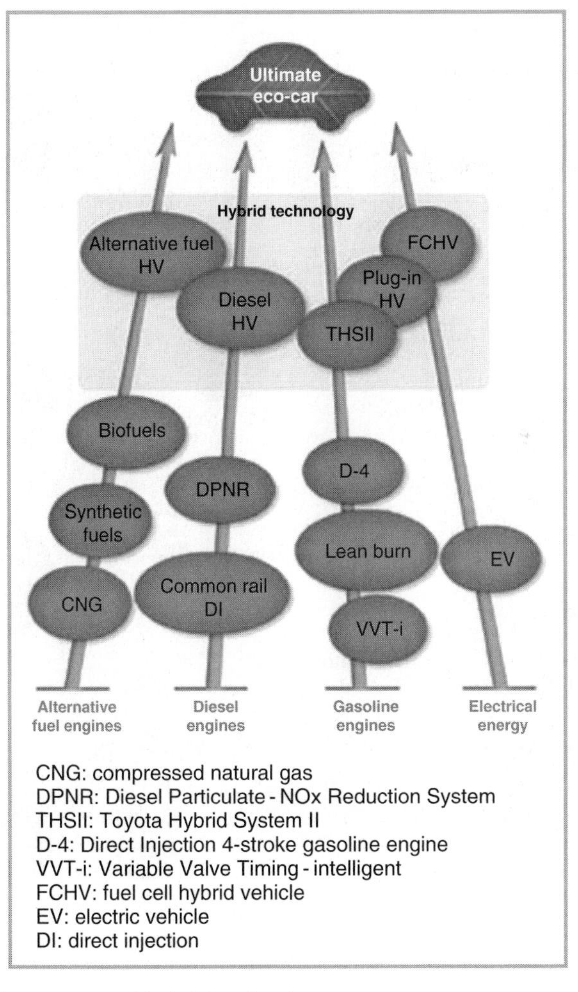

Figure 5.5. Various types of hybrid technology.
Source: From Ref. [4].

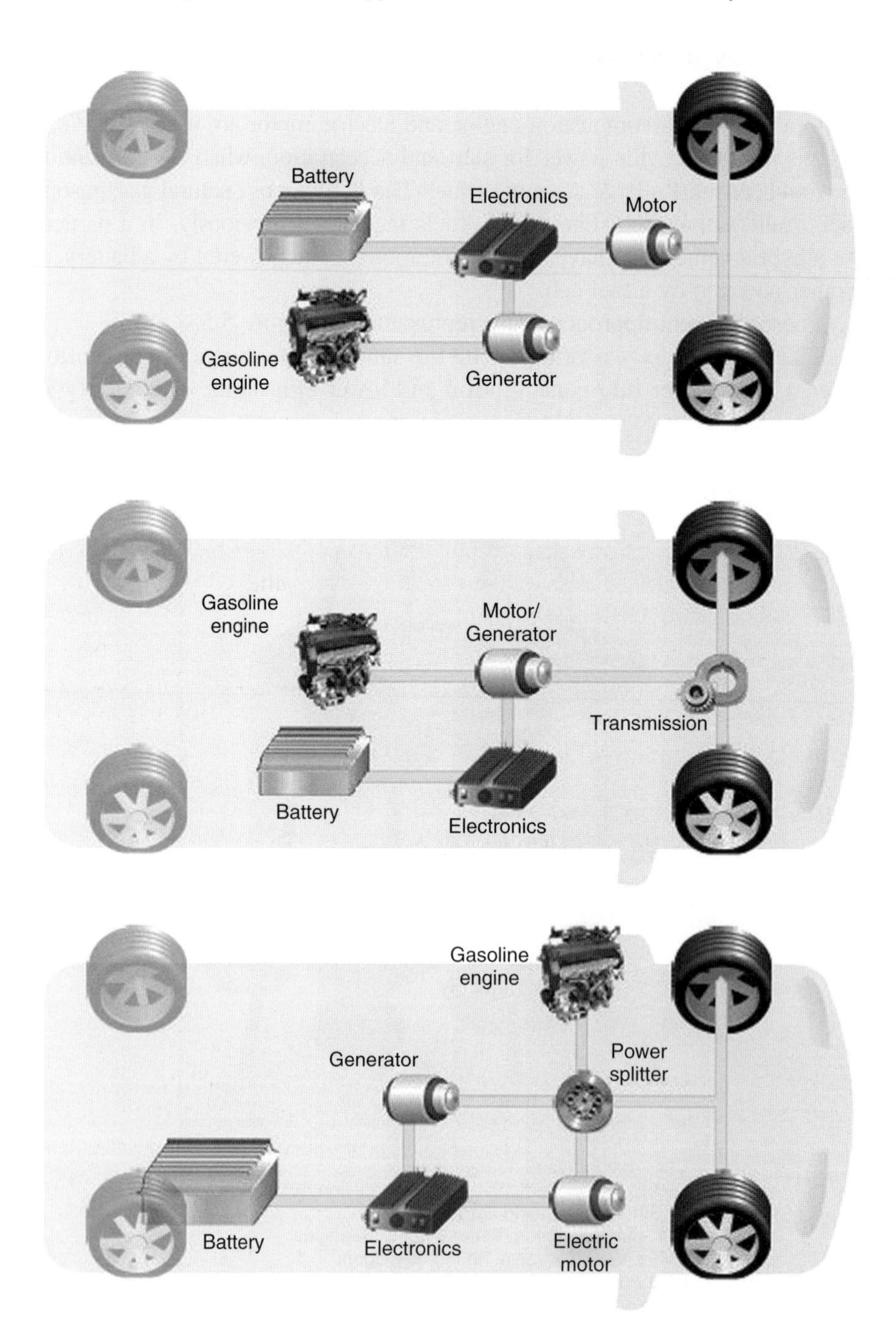

Figure 5.6. Drive train configurations in hybrid vehicles: series (top); parallel (middle); series/parallel (bottom).
Source: From Ref. [5].

of HEVs is much smaller than that of EVs, but the power has to be about the same.

To describe hybrid car technology, the drive train has to be introduced. The drive train of a vehicle is formed by the components that are responsible for transferring power to the drive wheels. There are three possible configurations: series, parallel and series/parallel (Figure 5.6) [5].

5.3.1. Series Drive Train

In a series hybrid, only the electric motor provides power to the wheels. The motor is powered by either the battery pack or a generator run by the ICE. A computer determines how much of the power comes from the battery or the engine/generator set. Both the engine/generator and regenerative braking (see later) recharge the battery pack. The engine is typically smaller in this configuration because it only has to deliver average power; instead, the battery pack is generally more powerful than the one in parallel hybrids in order to provide peak powers. This larger battery and motor, along with the generator, add to the cost, making series hybrids more expensive than parallel hybrids.

In a series hybrid, the ICE is not coupled to the wheels. This means the engine is not requested to satisfy the varying power demands experienced in stop-and-go (urban) driving and can instead operate in a narrow power range with a good efficiency. As series hybrids perform best in stop-and-go driving, they are the best choice for buses and other urban work vehicles.

It can be noted that the EV with range extender is typically a series hybrid configuration.

5.3.2. Parallel Drive Train

With this configuration, both the engine and the electric motor drive the wheels. The addition of electronics and a driving shaft (see Figure 5.6) allow these components to work together. This technology is implemented in the Insight, Civic and Accord hybrids by Honda, under the name of Integrated Motor Assist (IMA). Parallel hybrids can use a smaller battery pack and therefore rely mainly on regenerative braking to keep it recharged. However, when power demands are low, parallel hybrids also utilize the drive motor as a generator for supplemental recharging, much like an alternator in conventional cars.

Being the engine connected directly to the wheels, it eliminates the inefficiency of converting mechanical power to electricity and back, so these hybrids are quite efficient in extra-urban traffic. This configuration does reduce, but not eliminate, the city driving efficiency benefits of the series hybrid, that is the engine operates with low efficiency in stop-and-go driving because it is forced to meet varying power demands.

5.3.3. Series/Parallel Drive Train

This technology combines the advantages and complications of the parallel and series drive trains. Here, the engine can both drive the wheels directly (as in the parallel drive train) and be disconnected from the wheels so that only the electric motor powers the wheels (as in the series drive train). The Toyota Prius implements this concept, and a similar technology is also in the new Ford Escape Hybrid.

The engine operates at near-optimum efficiency more often. At lower speeds it operates more as a series vehicle, while at high speeds, where the series drive train is less efficient, the engine takes over and energy loss is minimized. This system is more expensive than a pure parallel hybrid since it needs a generator, a larger battery pack, and more computing power to control the dual system. However, the series/parallel drive train has the potential to perform better than either of the systems alone.

From the above, the characteristics of the hybrid systems may be summarized as follows:

- *Energy-loss reduction.* The system automatically stops the engine when the car stops (idling stop), thus reducing the energy waste.
- *Energy recovery and reuse.* The energy that would normally be wasted as heat during deceleration and braking is recovered as electrical energy.
- *Motor assist.* The electric motor assists the engine during acceleration.
- *High-efficiency operation control.* The system maximizes the vehicle's overall efficiency by using the electric motor to run the vehicle when the engine's efficiency is low and by generating electricity when the engine's efficiency is high.

These characteristics are not possessed to the same extent by the three technologies, as shown in Table 5.4. The series–parallel system stands out, as it can provide both superior fuel efficiency and driving performance.

Depending on the goals, a large variety of hybrid cars can be designed. The different categories are divided into several main "families" [2]. They are described below, and their requirements for battery systems are summarized in Table 5.5 which include EVs. It should be emphasized that these approximate values are given for comparative purposes, as they may vary significantly in each category depending on vehicle size.

Table 5.4. Comparison of the three hybrid technologies.

	Fuel Economy Improvement				Driving Performance	
	Idling stop	Energy recovery	High-efficiency operation control	Total efficiency	Acceleration	Continuous high output
Series	OO	OOO	OO	OO	O	O
Parallel	OO	OO	O	OO	OO	O
Series/ parallel	OOO	OOO	OOO	OOO	OO	OO

OOO, excellent; OO, good; O, unfavourable.
Source: From Ref. [4].

Table 5.5. Types of HEVs and their approximate energy/power requirements. Values for EVs are also enclosed for comparison.

Vehicle Type	Power Range (kW)	Energy Range (kWh)	Voltage Range (V)
Micro hybrids	2.5–5	0.5	12–36
Mild HEV	15–20	1	120–160
Full HEV	30–50	2–3	200–350
FCHEV	25–30	1–2	220
Plug-in HEV	30–100 (van)	5–15	200–350
EV	35–70 (van)	25–40	200–350

Source: From Ref. [2].

The different needs of electrical power and energy of the various hybrids have a strong influence on the battery choice. Each of the main battery systems (Pb-acid, Ni-MH, Li-ion) may find application, as will be described further. At present, lead-acid in micro hybrids, and Ni-MH in mild and full hybrids are considered satisfactory options. Li-ion is considered as the best future option for several hybrids.

5.3.4. Micro Hybrids

These vehicles can use the existing ICE platforms. This is a non-intrusive technology, as it only adds elements to the existing architecture of conventional cars. Generally, a simple electric machine, starter-alternator reversible system, is put in the power chain via a belt, powered by a suitable battery and electronic inverter.

In these cars, the engine is stopped as soon as the car stops. The starter-alternator is only used to start the engine; there is no regenerative energy recovered from braking. Power required is about 2.5 kW, with energy of a few hundred Wh.

Table 5.6. Typical idling stop rate in city drive (Kyoto).

Driving course	
Driving time/day (min)	140
Driving distance/day (km)	30
Average discharge current (A)	22
Number of stops	38
Accumulated stops duration (min)	35.2
Total idling stop rate (%)	25.1

Source: From Ref. [6].

Therefore, 12 V SLI (starting, lighting and ignition) lead-acid batteries of the largest size can supply the required power. The same battery provides the energy for on-board equipment and starting. AGM technology (see Section 2.3.4.1) is the most appropriate and offers the best cost/performance ratio. The battery is usually maintained at full SOC, but much more frequently discharged than the conventional SLI. Battery ageing is a real concern, as the projected lifetime is generally about 1–2 years less than SLI batteries in conventional cars. Because of the relatively low cost of this battery, the replacement cost may be compensated by increased fuel economy. In urban configuration, with many stops, fuel saving can be more than 15%.

Table 5.6 describes a typical utilization in urban conditions, recorded in Kyoto.

5.3.5. Soft Hybrids

Similarly to micro hybrids, this technique can be applied to existing ICE cars. In this case, the main difference is that some energy is saved by regenerative electric power produced on braking (50–60% braking energy is converted to electricity), and more electric energy is used on starting, because the car starts in an electric mode, while the ICE is being started. Compared to micro hybrids, the gain in fuel consumption can be almost doubled in the same driving conditions. It is expected that such vehicle will be rapidly expanding, and several developments are in progress.

Of course, the battery pack must be able to sustain high charge peak power and provide more power and energy during start. The average power required during car start, and average regeneration power, is about 6 kW. Because the battery provides more energy, the SOC will vary more than in the micro hybrid, and typical depth of cycling range is 20% DOD, between 100 and 80% SOC, for a battery of similar size to that of the micro hybrid.

The conventional lead-acid battery design with flat plates is not suitable for this application, and its lifetime would be reduced. A good technical solution is the advanced high-power technology of spirally wound design. The other solution considered today is an association of conventional SLI battery and supercapacitor. The capacitor may accept the high peak of regenerative power and help on starting, while the battery provides energy. Advanced lead-acid solutions are suited better when high energy level is required in upper class cars, causing deeper DOD cycling.

High-power Li-ion batteries possess both power and energy requirements and can be considered as the next solution.

5.3.6. Mild Hybrids

Mild hybrids represent a more intrusive technology in conventional car design with respect to the previous hybrids. The drive train must be completely redesigned, and manufacturing this type of car entails a deeper change in design and technology than the previous configurations. An electric motor is inserted in the drive axle, and the electric driving mode grows a step further. After starting, the car can stay in EV mode up to a speed of about 10–20 km/h before the ICE starts. On driving, a mild power boost is provided during accelerations, which allows some reduction of the ICE's size. Therefore, more power is required from the battery, which is more frequently and deeply discharged: as reported in Table 5.5, mild hybrids need power in the range of 15–20 kW.

Lead-acid, Ni-MH and Li-ion batteries have all been proposed for this type of hybrid. There is a trend to use a 12-V lead-acid (AGM VRLA) as the starter battery coupled with either Ni-MH (more used at present) or Li-ion batteries of higher voltage for traction and regenerative functions.

GM, BMW and Mercedes have announced mild hybrids with Li-ion batteries at the Geneva exhibition in February 2008. BMW's electrical architecture will be at 120 V, whereas the GM's system should work at about 100 V, with a battery pack 25% smaller and 33% more powerful than the currently used Ni-MH pack.

5.3.7. Full Hybrids or "Power Assist"

In this configuration, both the combustion engine and electric motor contribute to powering the car, with electric power being used for starting and acceleration. The car can be driven for a few kilometers in pure EV mode. The energy required to the battery is still small, but the power density has to be higher (see Table 5.5). The battery is permanently either on discharge or charge,

Table 5.7. Fuel efficiency as a function of electric to total power ratio in a conventional vehicle and in different HEVs.

	Electric to Total Power (%)	Fuel Economy Benefit (%)	Representative Model
Conventional vehicle	2	Baseline	
Weak hybrid[a]	5–10	5–20	GMC Sierra
Mild hybrid	10–30	20–50	Honda Civic
Full hybrid	30–50	20–80	Toyota Prius

[a] Micro and soft hybrids.
Source: From Ref. [1].

and the state of charge has to be maintained at an intermediate level in such a way that it can deliver peak power to the drive train, and accept power from the engine or regenerative braking. The main requirement of the battery is therefore its ability to sustain a very large number of high drain and shallow cycles. At present, Ni-MH batteries are almost exclusively used.

At this point, a comparison can be made about the fuel efficiency afforded by the HEVs presented thus far and a conventional vehicle. As shown in Table 5.7, the higher the electric to total power ratio, the higher the fuel economy. The values are in a rather wide range, especially for full hybrids: this depends on such factors as cycle drive (urban or extra-urban), driving style, environmental conditions (temperature, flat or hilly routes), etc. However, even the lowest economy for a full hybrids (~20%) is to be considered a remarkable saving.

5.3.8. Plug-in Hybrids (PHEV)

Unlike other hybrid cars, the battery of a plug-in hybrid (PHEV) can be recharged from outside, connected to the electricity network.

PHEV is a sort of intermediate between EV with range extender and full HEV. The car can run in pure electric mode (ZEV), as long as the state of charge of the battery is sufficient, and the speed of the car below a certain limit. The driving range in electric mode is much smaller than that of EVs and largely depends on the model. If higher speed is required, the combustion engine starts to operate. During this phase, the state of charge of the battery is depleting, and when the battery has reached a minimum SOC (for example, 30%), the car runs in full hybrid mode, and the state of charge of the battery is maintained.

Table 5.8. Cycle life requirements for electric drive vehicle (EDV) batteries, over a 10-year life.

EDV type	Deep Cycles		Shallow Cycles	
	Cycles at 80% DOD	Cumulative energy (MWh)	Cycles at 50 Wh	Cumulative energy (MWh)
Full HEV			200 000	10
PHEV-20[a]	2400	12	120 000	6
EV	1000	32		

[a] Plug-in vehicle endowed with a battery allowing a 20-mile continuous electric drive.
Source: From Ref. [3].

The battery size is about one-third to one-half that of a pure EV, and is recharged (plug-in) from an external source of electricity when the car is not used. The battery's energy density is more important than power density, as a long driving range is sought. The required cycle life is more demanding than in EVs, as strong cycling capability at deep DOD is needed. A cycle life requirement comparison is described in Table 5.8, for an expected 10-year life. PHEV is assumed to be driven ~6500 km/year in EV mode, and ~15 000 km/year in HEV mode.

At present, this type of car is still at the prototype level, but will probably have a bright future. Plug-ins are expected to be ~50% more fuel efficient than standard HEVs, thanks to their longer driving range in electric mode. The Electric Power Research Institute has calculated the range afforded by 1 L of fuel for a gasoline car, a HEV and a future PHEV able to run for 20 miles (32 km) with the electric motor only: the values are 10.5, 12.7 and 25.4 km/L, respectively. The present limitations lie in battery range, recharge time, life and cost. Li-ion batteries appear promising to cope with these limitations, and could also provide satisfactory power outputs.

5.3.9. Fuel Cell Hybrid EV (FCHEV)

Fuel cell vehicles can also be considered as HEVs. In this case, both a primary and a secondary electrochemical power source are used. A rechargeable battery is required to provide the energy and power for starting and for the fuel cell auxiliary equipment, and to provide the peak power necessary during acceleration. Energy is provided by the fuel cell (the primary, refuelable power source) at a more constant power level, closer to the average. That permits downsizing of the fuel cell stack, which is a very important contributor in fuel cell cost. The battery operates under similar conditions to present full hybrid cars. Li-ion battery is the most serious candidate for this application.

5.3.10. Large Hybrid Vehicles: Buses, Light Trucks and Tramways

Public urban transportation and urban delivery light trucks are a very interesting field for application of hybrid propulsion, aiming at both improvement of fuel economy and better environmental conditions. The typical driving pattern includes very frequent stops, leading to needs for electric launch of the vehicle and regenerative storage of braking energy. Hybrid electric bus technologies have been demonstrated in several cities in Europe in recent years.

In hybrid tramways or trolley buses, power is all electric and the "cordless" battery operation allows more flexibility and simplification of infrastructures.

Many battery systems have been experimented, using lead-acid, Ni–Cd or Ni-MH. Practical experience shows that Ni-MH, in spite of its higher cost, is the most suitable system in terms of endurance and reliability. The performance requirements are basically the same as for all hybrids, with higher energy and power. Typical needs are summarized in Table 5.9.

Table 5.9. Typical needs of large hybrid public transportation vehicles.

	Hybrid Bus (ICE or FC)	Hybrid Dual Mode Bus	Hybrid Trolley Bus	Hybrid Tramway
km/cycle in autonomous electric mode	Power assist/no autonomous operation	5–20	1–5	0.4–1
Cycles/hour	20–100	1 or 2	1–3	1–3
Cycles/year	100 000 to 500 000	5000 to 10 000	5000 to 15 000	5000 to 15 000
kWh/cycle	<0.5	1–5	1–7	2–10
kW	50–100	50–100	50–100	50–150
Allowable DOD from battery	1–5%	5–20%	5–20%	5–20%

Source: From Ref. [8].

5.4. More Information on Hybrid Vehicles

When looking for an alternative to a conventional ICE vehicle, HEVs offer several benefits in comparison to other solutions, as evidenced in Table 5.10.

Hybrids have improved fuel efficiency and performance (for example, acceleration), lower emissions and greater convenience (higher range, less refuelling stops). Unlike diesel, hybrids do not sacrifice performance to gain fuel efficiency; unlike bio-fuel vehicles, hybrids do not require special pumps at refueling stations; unlike EVs, hybrids do not limit their driving range. Additionally, consumers should soon have a wide array of model choices [1].

Diesel cars are often mentioned as viable competitors to HEVs. However, as shown in Figure 5.7, HEVs can improve both performance and fuel economy in comparison with diesel vehicles.

Since gasoline hybrids do not use the ICE for starting, and instead draw on the battery, they can utilize the power of their motors to produce instantaneous torque at low rpm (see Table 5.11) [1]. For instance, the Camry hybrid produces 29 kg of torque from 0 to 1500 rpm, more torque than conventional gasoline or

Table 5.10. Comparison of the characteristics of a typical HEV with those of different types of vehicles.

Factor	Hybrid (gasoline)	Diesel	CNG[a]	Flex-Fuel[b]	All Electric
Fuel efficiency	OOO	OO	O	O	OOO
Performance (acceleration)	OOO	OO	O	OO	OO
Emission/air quality	OOO	O	OO	OO	OOO
Model choice/ flexibility	OOO	OO	O	OO	O
Convenience (range, refuelling)	OOO	OO	OO	OO	O
Load capacity	OO	OOO	O	OO	O
Initial cost	O	OO	O	OOO	O
Cost per mile	OOO	OO	OOO	O	OOO
Representative model	Toyota Prius	VW Jetta	Honda Civic	GM Monte Carlo	Toyota RAV4 EV

OOO, excellent; OO, good; O, unfavourable.

[a] CNG, compressed natural gas.

[b] These cars can use different fuels, for example alcohol, gasoline or a combination of the two.

Source: From Ref. [1].

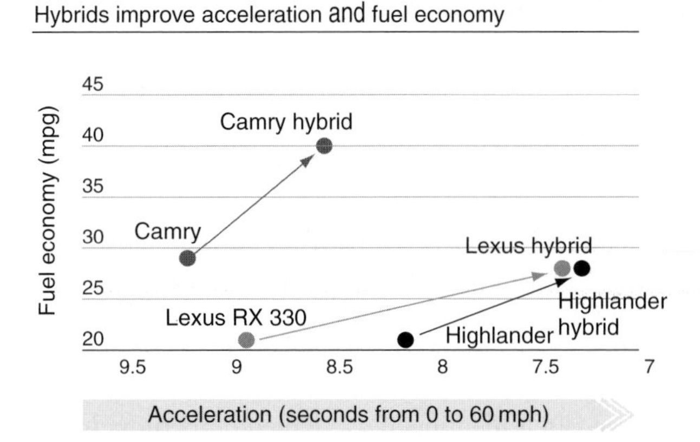

Figure 5.7. Improvement of fuel economy and acceleration in HEVs *vs* corresponding diesel versions. 1 mile, 1.6 km; 1 gallon, 3.8 L.
Source: From Ref. [1].

Table 5.11. Comparison of mid-size cars with different propulsions.

	Diesel (Turbo) (VW Passat)	Gasoline (Toyota Camry)	Hybrid (Gasoline) (Toyota Camry)
Maximum power[a]	134 hp at 4000 rpm	160 hp at 5700 rpm	188 hp at 4000–5000 rpm
Maximum torque[a]	35 kgm at 1800 rpm	23 kgm at 4000 rpm	29 kgm at 0–1500 rpm
Acceleration (s from 0 to 60 mph)	10.5	8.8	8.5
Combined MPG[b] (city/highway)	32.5 (27/38)	29 (24/34)	40 (43/37)

[a] Torque is a measure of rotational force; horsepower (hp) is directly proportional to torque and revolutions per minute (rpm): hp = (torque × rpm)/5250.
[b] MPG: miles per gallon.
Source: From Ref. [1].

diesel vehicles can achieve at those engine speeds. In contrast, gasoline hybrids have lower torque at high rpm than their traditional diesel and gasoline engine counterparts. However, many hybrids counterbalance their lower torque at high rpm with the ability to revolve their engines to high rpm, which enable them to generate higher power (see the table). Hence, hybrids have a better acceleration both in the 0–60 mph range and at high speed (e.g. during passing on the highway) [1].

5.4.1. Present HEV Production and Perspectives

Because of these advantages and the limited drawbacks, the number of hybrid vehicles is expected to grow rapidly over the next decade and eventually reach over 80% of new cars and light trucks sold worldwide (both gasoline-electric and diesel-electric hybrids are included in this estimate). Sales for the period 2001–2007 are reported in Figure 5.8. In Figure 5.24 (Section 5.5.3.3), forecasts for the period 2008–2011 will be reported.

This will certainly be favoured by cost decrease. The main components of hybrid systems are batteries and related electrical systems, electric motors/generators, power-split device and electronics. For a compact or mid-compact full-hybrid, the cost of these components in 2005 was in the range of US$4500–6000. An estimate for 2010 foresees a reduction to US$2000.

In Table 5.12, the possible roadmap to 2015 for the development of hybrids is indicated. In particular, hybrids with Li-ion batteries should be deployed by the end of 2008; PHEVs should become commercial by 2009; automakers should offer hybrid versions of most car models by 2013.

If this roadmap is correct, fuel economy brought about by the larger HEV's commercialization will have a significant impact on oil demand. There is some optimism about that, because after about 10 years of research and development,

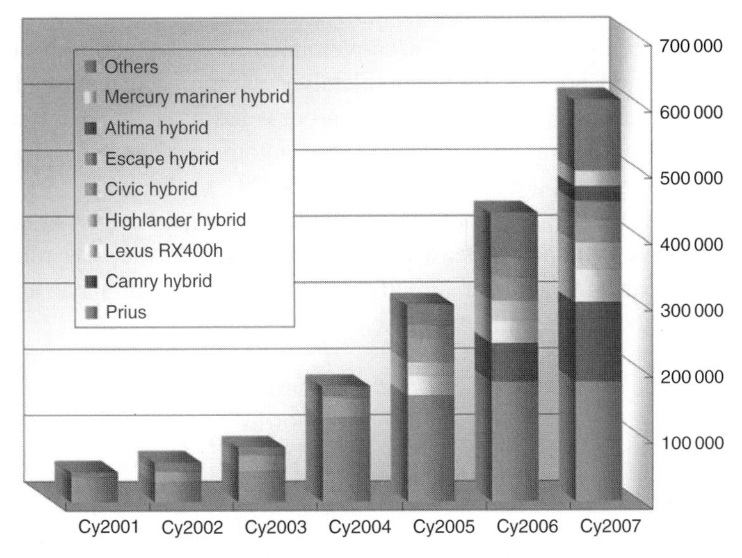

Figure 5.8. Sales of HEVs in the period 2001–2007; the contribution of the most important models is shown. Prius, Camry, Lexus and Highlander by Toyota; Civic by Honda; Altima by Nissan; Escape and Mariner by Ford (See color plate 8).
Source: From Ref. [9].

Table 5.12. A road map to mass adoption of HEVs.

Year	Action
2007	More automakers launch own hybrid systems
2007	Diesel hybrids introduced
2008	Toyota launches third generation Prius, creating new benchmark
2008	Lithium-based hybrids launched
2008	Full hybrids become standard in USA
2009	Plug-ins commercialized using high-power batteries
2010	OEMs will have introduced over 50 hybrid models in USA
2010	Toyota offers hybrids as option on all models
2010	Hybrids take 10.6% share of new sales globally
2011	Plug-ins commercialized with electric driving range of 8–16 km
2012	Toyota introduces fourth generation Prius
2013	Plug-ins driving range increases to 30–50 km
2013	Advanced materials introduced
2013	Automakers offer hybridization as option on most models
2014	Nickel battery displaced
2015	Hybrids take 50% share of new sales globally

Source: From Ref. [1].

the major automakers are ready to mass produce hybrids. Indeed, they will be able to adapt much of their existing capacity to produce hybrid vehicles with acceptable additional expenses.

Assembly of hybrid versions of conventional models could also be simple. Because hybrids do not call for a completely new approach to propulsion and fuelling, adjustments to the manufacturing process are limited. Therefore, hybrid technology will not represent a revolution for automakers, and massive capital-investment requirements are not a limiting factor [1].

Toyota is the leader, but every major automaker has launched or announced plans to launch at least one hybrid vehicle in the near term. A list of major companies that have already produced HEVs, or have announced production within 2009, is given below.

- Toyota
- Honda
- Chrysler
- BMW
- Kia Motors
- Porsche
- Subaru
- Mercedes
- Opel
- GM
- Ford
- Audi
- Hyundai
- Peugot
- Saab
- Volkswagen
- BYD

Ford, GM, Honda and Toyota have been the first to produce several HEVs models since 1997.

It is obviously impossible to give here details on all models. However, some particular concepts will be presented, before describing in some detail the construction and characteristics of the first and more successful HEV, Toyota Prius.

GM is developing a new concept, the E-Flex Electric Vehicle System, and launching the Chevrolet Volt as a first application [10]. "E" stands for electric drive and 'Flex' for different sources of electricity the motor can use. The Volt is a PHEV (even though GM considers it an all-electric vehicle with range extending capacity) using a Li-ion battery and different forms of energy sources for the battery: generator, fuel cell or regenerative braking. The 16-kWh Li-ion battery pack delivers power to a 161 hp (120 kW) electric motor. Plug-in charging is designed for the home (110 V, 15 A) and will take ~6 h.

Acceleration from 0 to 100 km/h takes 8–8.5 s. The basic strategy is to run the vehicle in all-electric mode until the state of charge (SOC) of the battery reaches 30%, this corresponding to a claimed 65-km range. A 53-kW motor generator set allows recharging of the battery while driving. The generator in the present Volt version is driven by a 1-L, three-cylinder, turbocharged bio-fuel engine.

The Volt is endowed with a 12-gallon (45 L) fuel tank, giving the vehicle a total driving range of around 1000 km (presumably, this range would increase with a diesel variant).

In April 2007, GM has presented the E-Flex Hydrogen version of the Chevrolet Volt. The Li-ion battery would reportedly grant a driving range of 483 km. The battery pack is recharged by fuels cells working on 4 kg of H_2 (half the amount needed in the previous GM's fuel cells). A vehicle of this type is fully electric (zero emissions) and has plug-in capability.

By January 2008, the GM Chevrolet Volt was still at the prototype stage, shown in Figure 5.9, but GM's representatives have expressed total confidence in its near commercialization.

Diesel hybrids have been commercialized in the last few years. DaimlerChrysler introduced a hybrid light truck, Dodge Ram, at the end of 2004 as a fleet vehicle. Soon after (January 2005), Opel introduced its Astra Diesel Hybrid, developed thanks to the GM-DaimlerChrysler cooperation. This vehicle has been created with a turbodiesel engine and two electric motors (30 and 40 kW). In March 2008, BMW/Mercedes have presented a mild hybrid SUV (Vision Bluetech Hybrid) with a diesel engine and an electric motor powered by Li-ion batteries. This car should be commercialized by 2009. In March 2008, Volkswagen too has presented a diesel hybrid, the Golf Tdi Hybrid concept, which pairs a 74 hp diesel engine with a 26.6 hp electric motor powered by a 220 V Ni-MH battery. At low speeds, this hybrid will run in EV mode only. This should reduce fuel consumption and bring CO_2 emissions below 90 g/km.

Figure 5.9. Chevrolet Volt plug-in hybrid. See text for details.
Source: Courtesy of GM.

Mercedes has introduced in 2007 a micro-hybrid version of its Smart, the Mhd. This small car, 2.7-m long and with a power of 71 hp, allows saving ~10% fuel vs the conventional version, and features a low CO_2 emission (maximum 103 g/km). Mhd is the smallest micro hybrid and, as any other hybrid, automatically stops its engine when the vehicle comes to a halt.

5.4.2. Toyota Prius

This hybrid car has been produced since 1997 in different versions. Table 5.13 includes data for the models circulating to the end of 2007. NHW10 and NHW11, also referred to as THS1 and THS2, form the first-generation Toyota Prius, while NHW20 is also known as second-generation Prius. This last version (Figure 5.10) uses the so-called Hybrid Synergy Drive (HSD) System. The third generation is expected by the end of 2008 (as a 2009 model). Furthermore, in mid-2007, a plug-in version was also introduced.

The components of the HSD Prius are reviewed here [4, 11]. Technical specifications may be found in Table 5.13 (NHW20).

Electric Motor. This compact, lightweight AC permanent-magnet synchronous motor does not use brushes like a DC motor, this increasing its durability. And, since the permanent magnet needs no electricity to create a magnetic field, it is much more energy efficient. The motor functions as a generator during regenerative braking, converting the vehicle's kinetic energy into electricity to charge the battery (see below).

Battery. The Ni-MH battery, designed especially for electric vehicles, offers high-power density, light weight and long life. Its performance has been further boosted for the Toyota Hybrid System (see table). This allows the battery to provide a great amount of electricity to the motor for start and acceleration.

Table 5.13. Features of the first three models of Toyota Prius.

Feature		Model Code		
		NHW10	NHW11	NHW20
Body style		Four-Door Sedan	Four-Door Sedan	Five-Door Hatchback
First sales		1997	2000	2003
Battery	Modules	40	38	28
	Cells per module	6	6	6
	Total cells	240	228	168
	Volts per cell	1.2	1.2	1.2
	Total volts (nominal)	288	273.6	201.6
	Capacity (Ah)	6.0	6.5	6.5
	Energy (Wh)	1728	1778.4	1310.4
	Weight (kg)	57	50	45
Petrol engine	Power (kW)	43	52	57
	Max rpm	4000	4500	5000
Electric motor	Operating voltage (V)	288	273	500
	Power (kW)	30	33	50
Combined	Power (kW)	?	73	82

Source: Courtesy of Toyota.

Generator. The high-efficiency AC permanent-magnet synchronous generator produces electricity to charge the battery pack and power the motor. By controlling the generator speed, the power split device functions like a continuously variable transmission. The generator also functions as a starter for the engine.

Power Split Device. The power split device (Figure 5.11(left)) splits power from the engine into two routes: mechanical and electrical. Its planetary gear

Figure 5.10. Toyota Prius (Hybrid Sinergy Drive).
Source: Courtesy of Toyota Motor Company.

can transfer power between engine, motor, generator and wheels in almost any combination. It is also called "hybrid transaxle."

Power Control Unit (Inverter). The inverter (Figure 5.11(right)) changes the battery's direct current into alternating current to drive the motor and turns the generator's alternating current output into direct current to charge the battery. It also varies the frequency of the current, depending on motor rpm to maximize efficiency. The inverter unit is water cooled for improved reliability.

Regenerative Braking System. When decelerating or braking, the motor acts as a generator, recovering kinetic energy from the wheels, converting it into electricity and storing it in the battery (Figure 5.12). It is an ideal energy-saving system for city driving. When the driver applies the brakes, both the hydraulic brakes and the regenerative brakes are used, with the priority going to the regenerative braking system to maximize energy recovery.

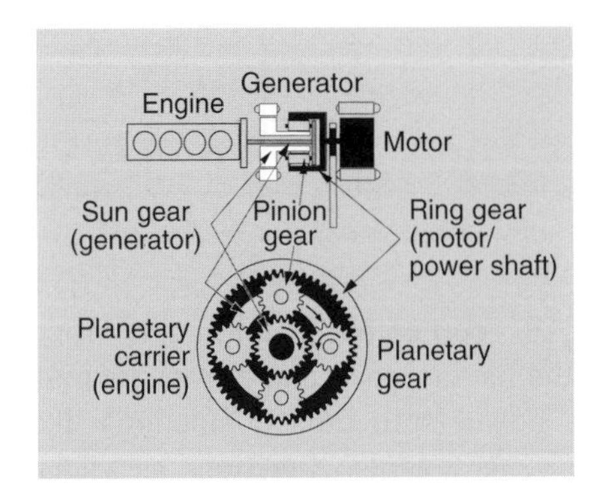

Figure 5.11. Left: scheme of the power split device (engine, generator and motor are also shown); right: power control unit.
Source: From Ref. [4].

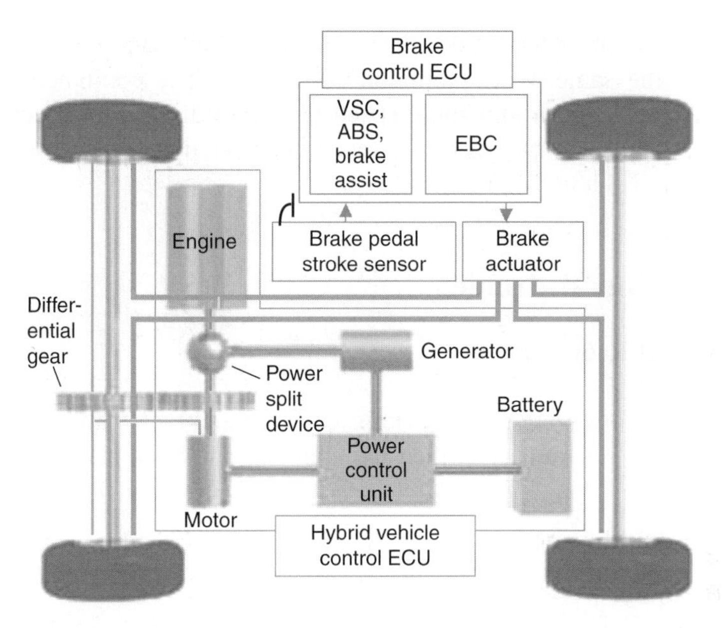

Figure 5.12. Components of the Prius HSD system. Regenerative braking follows the route (in the lower half of the figure): front wheels→electric motor→power control unit→battery. The hydraulic braking components are in the upper half of the figure. ECU, electronic control unit; VSC, vehicle stability control; ABS, anti-lock braking system. *Source*: From Ref. [4].

As already mentioned, the third-generation Prius has been announced to appear by the end of 2008. According to earlier reports, this new hybrid should have been equipped with Li-ion batteries; however, more recent communications have led to suppose that Toyota will delay the use of Li-ion for safety reasons. Probably, this Prius will not be ready until the second half of 2009. At this point, it is really difficult predicting what the third-generation Prius will be. Anyway, Toyota aims at building a Prius which is 20–30% more efficient that the previous model. In the medium term, they plan to have a car allowing a driving range of 40 km/L.

In July 2007, Toyota has introduced the plug-in version of the Prius. Increased battery capacity (13 Ah) gives it a longer electric-motor-only cruising range and a battery-charging device allows users to replenish the batteries using household electricity. These features enable the vehicle to run more often in gasoline-free, electric-only mode, such as on short trips in city driving. Also, charging the battery with less-expensive night time electricity lowers total running costs, providing an economic benefit to owners. It has to be noted, however, that the driving range is only extended from ~2 to ~13 km. Furthermore, Toyota has stated that this version will be tested in Japan, USA and Europe, but will not be commercialized until Li-ion batteries will be available.

The maximum power of the electric motor (50 kW) and that of the engine (57 kW) are the same as the common HSD Prius. The combined maximum power is 100 kW, and a maximum speed of 100 km/h with the electric motor only is anticipated. The battery can be recharged from the electric grid in 1–1.5 h at 200 V or in 3–4 h at 100 V.

5.5. Traction Batteries

5.5.1. General Requirements

In the previous sections, characteristics of EV and HEV batteries have often been mentioned. A summary of their requirements can be in place here [12].

5.5.1.1. EV Requirements

The battery:

- Must have a large capacity for an acceptable range. A typical electric car uses 95–125 Wh/km, depending on the route and the driving style
- Must be capable of regular deep discharge (80% DOD) operation
- Must be designed to maximize energy content and deliver sufficiently high power even with deep discharge to ensure long range
- Must accept very high repetitive pulsed charging currents (greater than 5 C) if regenerative braking is required
- Without regenerative braking, must accept at least continuous 2 C charge
- Must routinely receive a full charge
- Needs a battery management system (BMS)
- Needs thermal management
- Has voltage in the range of 200–350 V
- Has energy in the range of 25–40 kWh
- Accepts continuous discharge currents of up to 1 C and peak currents of 3 C for short durations.

5.5.1.2. HEV Requirements

The battery:

- Is designed to maximize power delivered
- Must deliver high power (up to 40C) in repetitive shallow discharges and accept very high recharge currents
- Has a very long cycle life: some hundred thousands of shallow cycles

- Has operating range between 15 and 40% DOD (in the SOC range of 40–80%) to allow for regenerative braking
- Never reaches full discharge
- Rarely reaches full charge
- Needs thermal management
- Needs a BMS
- Needs interfacing with overall vehicle energy management
- Has a voltage > 100 V
- Has a power > 40 kW (50 hp) for full hybrids
- Has energy in the range of 1–10 kWh depending on the application.

5.5.2. Battery Management System

In EVs or HEVs, battery management is much more demanding than for portable batteries (Chapter 3). It has to interface with a number of other on-board systems, it has to work in real time in rapidly changing charge/discharge conditions as the vehicle accelerates and brakes, and it has to work in harsh and uncontrolled environments [13].

The BMS must manage the system over the entire operating cycle of the EV/HEV vehicle, and has to ensure the following functions [2]:

- Collecting information from sensors in the battery: current, voltage, temperatures, etc. The information is processed by the BMS to ensure correct battery operation.
- Controlling the charger to ensure proper charge of the battery. The charger can be an onboard or outside. The control is designed to react to data from the sensors associated with each battery module (telemetry boards), and responds according to parameters or special algorithms in the computer. The control of the charger is normally accomplished via the vehicle communication bus, which allows dialogue with other on-board equipment.
- Managing cell balance to ensure optimum performance of the battery. Balancing is necessary in multi-cell batteries: the weakest cell limits the total performance of the battery. The BMS controls the balancing electronic devices integrated on each telemetry board according to a predetermined strategy or algorithm.
- Safety control: avoids overcharge or overdischarge or other major anomalies that can occur in case of failure of the battery, auxiliary equipment or surrounding environment. The action of the electronic system can be physical (emergency shut down of the battery) or informational (reporting trouble to the user).
- Reporting the battery state: communication of information (alarms, gauge, etc.) to the user and to other on-board equipment via the communication bus.

- Thermal management of the battery: the BMS monitors the temperature of the cells in all modes of operation (drive, charge, etc.) and controls the pump, fans and heater to manage the temperature of the battery.
- Communication with the vehicle: vehicle computer and BMS exchange data via the communication bus (CAN 2.0B). This bus is a standard of the automotive field.
- Maintenance via the BMS: users will have the ability to connect maintenance and diagnostic tools to undertake necessary operations for the maintenance of the battery.
- Data transfer to a laptop PC, which can monitor and store the battery measurement data collected by the BMS; in this way, the results can be examined using normal software tools such as spreadsheets.

The block diagram of a BMS is shown in Figure 5.13 for a Li-ion battery, but a similar scheme is valid for any battery system.

As shown in the figure, the BMS not only performs battery monitoring and control, but interfaces with the vehicle controller via the CAN bus. The BMS can thus be coupled to other vehicle systems, for example anti-theft devices which disable the battery.

Determining the battery state of charge is particularly critical in an HEV. These batteries require both high-power charge capability (regenerative braking) and high-power discharge capability for start or acceleration. Therefore,

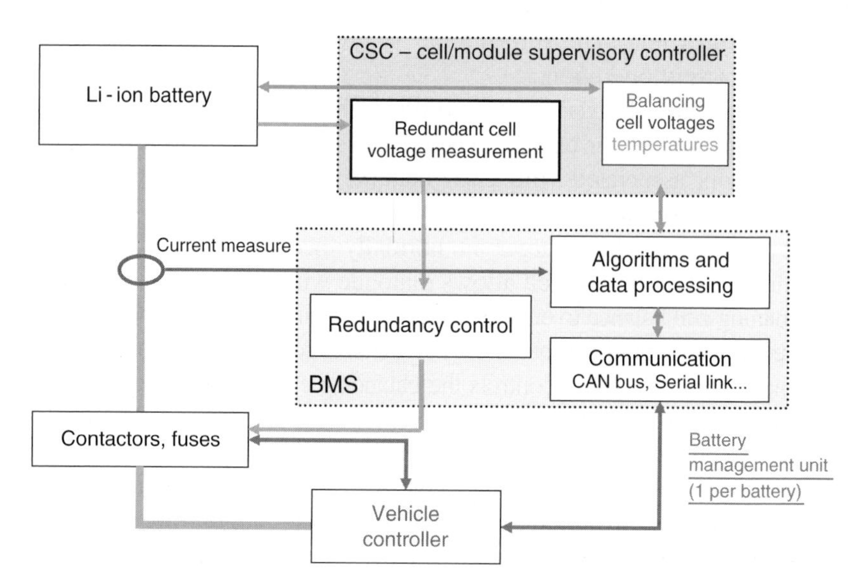

Figure 5.13. BMS developed by Johnson Controls – Saft for an electric vehicle using a Li-ion battery pack.
Source: From Ref. [14].

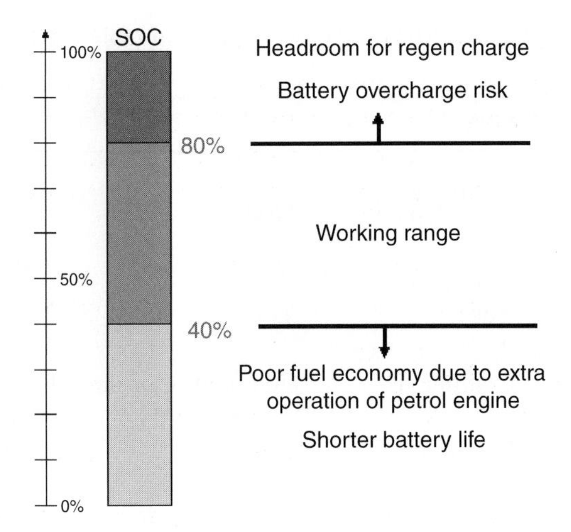

Figure 5.14. Battery SOC and performance.
Source: From Ref. [13].

they must be maintained at a SOC that allows delivering the required power, while still having enough headroom to accept regenerative charge without overcharging the cells. A safe range, as indicated in Figure 5.14, is between 40 and 80% SOC. Any battery may provide higher discharge powers at high DODs; however, for the above reasons, an upper limit of 80% is set in this case. The lower limit is set to optimize fuel economy and also to prevent over-discharge which could shorten the battery's life.

A BMS specifically intended for HEV has recently been described in a series of papers [15–17]. It is based on the so-called extended Kalman filtering (EKF) and aims at determining battery state of charge, power fade, capacity fade and instantaneous available power of the battery pack. The method was applied to a Li-ion polymer battery but can be extended to other chemistries.

The requirements of an efficient BMS for EVs and HEVs are quite different vs those of portable electronics (PEs), as clearly evidenced in Table 5.14. The particularly harsh environment of HEVs motivates the use of advanced management techniques and algorithms. A correct algorithm sequence is the following [15]:

- Initialization (as soon as the vehicle is turned on, the algorithms must be initialized)
- SOC update (voltage, temperature and current are measured)
- SOH update (battery capacity and other parameters are estimated)
- Maximum available power (based on the SOC and a dynamic cell model, the BMS estimates the maximum discharge power)
- Equalization (the BMS determines which cells need a special charge to keep the pack balanced).

Table 5.14. Typical characteristics of HEV, EV and portable electronics (PE) from the viewpoint of BMS.

Characteristic	HEV	EV	PE
Maximum rate	20C	5C	3C
Rate profile	Very dynamic	Moderate	Piecewise constant
SOC estimation	Very precise	Precise	Crude
Predict available power	Yes	Yes	No
Cell balancing	Continuous	Continuous, or on charge only	On charge only
SOH estimation	Required	Required	Not essential
Lifetime	10–15 years	10–15 years	<5 years

Source: From Ref. [14].

This procedure needs a good cell mathematical model, which may allow estimating all the relevant quantities. SOH estimation, in particular, includes battery capacity fade, power fade and self-discharge. The cell model parameters must be adjusted to account for cell ageing [16, 17].

Battery thermal management is also critical for the battery life and the vehicle life and performance. Battery temperature influences the availability of discharge power (for start up and acceleration), energy and charge acceptance during energy recovery from regenerative braking. These factors affect vehicle driveability and fuel economy. Therefore, batteries should ideally operate within a temperature range that is optimum for performance and life. This range varies with battery chemistry, and is usually much narrower than the specified operating range for the vehicle (identified by the vehicle manufacturer). For example, the desired operating temperature for a lead-acid battery is 25 to 45°C; however the specified vehicle operating range could be –30 to 60°C [18].

For thermal batteries, such as ZEBRA and lithium-metal-polymer batteries (see Chapter 2), thermal management is considered an integral part of the battery pack and has been included in the design by the battery manufacturers. For ambient temperature batteries, for example VRLA, Ni-MH and Li-ion, this aspect was not obvious initially, but now EVs and HEVs with these batteries also have efficient thermal management systems (BTMS).

The goal of a BTMS is to make a battery pack working at an optimum average temperature (with life and performance trade-off) with even temperature distribution in the modules and within the pack. However, the BTMS has to meet the requirements of the vehicle as specified by the manufacturer: it must be cheap, compact, lightweight, easily packaged and compatible with location in the vehicle. It must also consume low power, allow the pack to operate under a wide range of climate conditions (from very cold to very hot), and provide ventilation if the battery generates potentially hazardous gases.

A BTMS may use air (Figure 5.15), or liquid (Figure 5.16) . The thermal management system may be passive (i.e. only the ambient environment is used) or active (i.e. a built-in source provides heating and/or cooling at low or high temperatures). The thermal management control strategy is done through the battery electronic control unit.

Heat transfer with air is achieved by directing/blowing the air across the battery modules. Instead, heat transfer with liquid could be achieved: (a) through discrete tubing around each module; (b) with a jacket around the module; (c) submerging modules in a dielectric fluid for direct contact; and (d) placing the modules on a liquid heated/cooled plate (heat sink). Using tubes or jackets, the heat transfer medium could be water/glycol or even refrigerants, which are common automotive fluids. If modules are submerged in the heat transfer liquid, the liquid must be dielectric, such as silicon-based or mineral oils, to avoid electrical shorts.

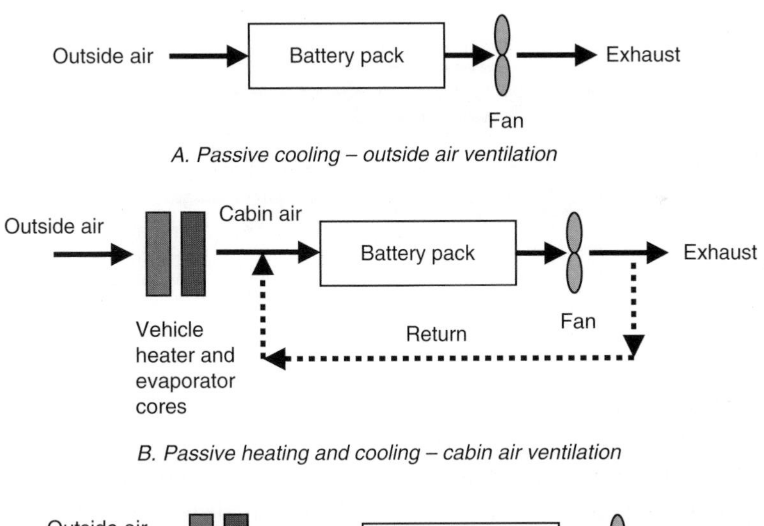

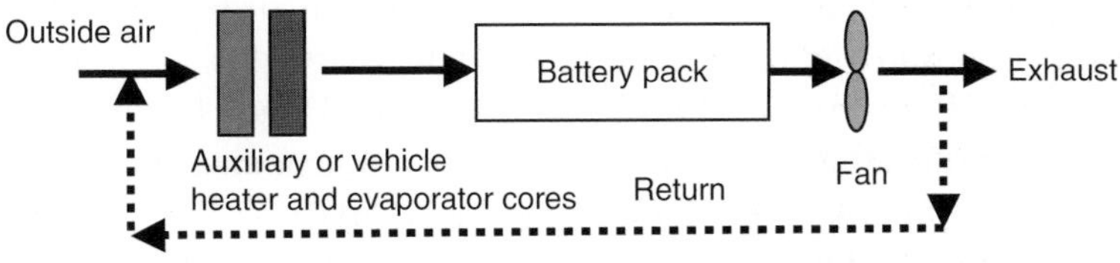

C. Active heating and cooling – outside or cabin air

Figure 5.15. Thermal management using air.
Source: From Ref. [18].

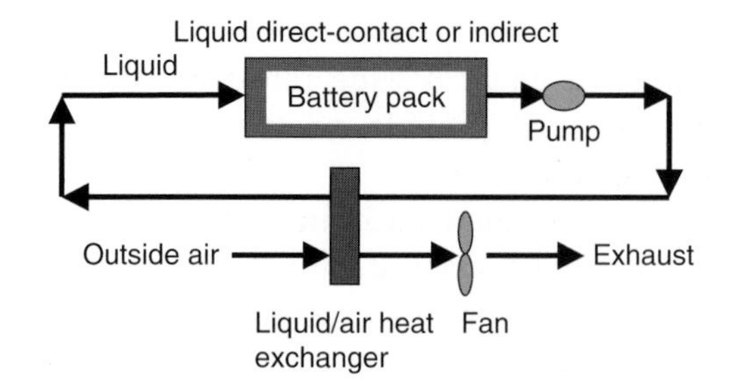

D. Passive cooling – liquid circulation

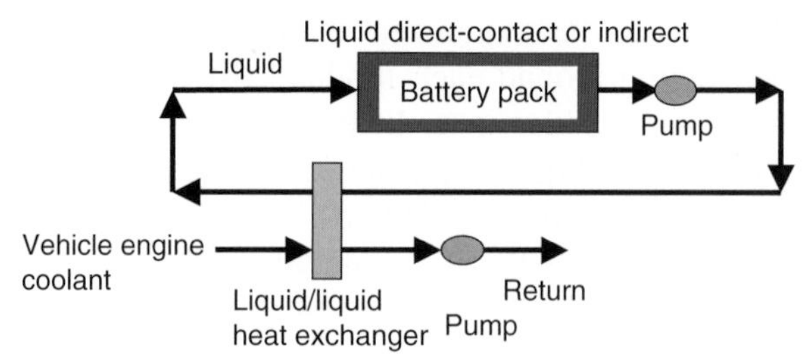

E. Active moderate cooling/heating – liquid circulation

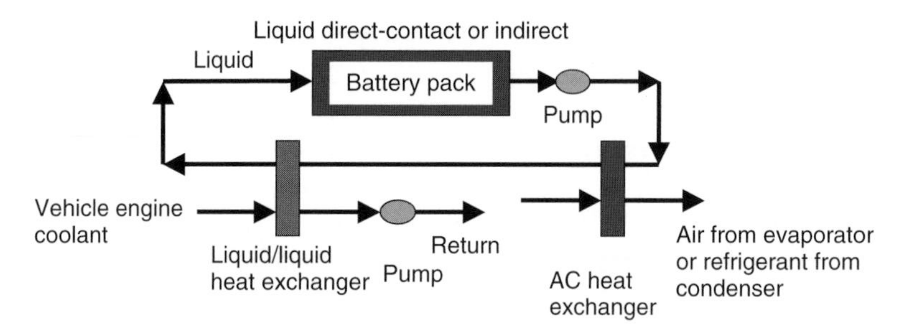

F. Active cooling and heating – liquid circulation

Figure 5.16. Thermal management using liquid.
Source: From Ref. [18].

For the same flow rate, the heat-transfer rate for most practical direct-contact liquids, such as oil, is much higher than with air because of the thinner boundary layer and higher fluid thermal conductivity.

Current HEVs, for example Toyota Prius and Honda Insight, use cabin air for cooling and heating the pack: in this case, ambient air is heated and cooled by the air-conditioning (AC) system (Figure 5.15B).

As shown in Figure 5.16D, ambient air can be used for heat rejection in a passive liquid system; however, this is only possible if the ambient temperature is in the range of 10–35°C. Outside of these conditions, active components such as evaporators, heating cores, engine coolant, etc., are needed.

The performance of the Ni-MH battery is rather sensitive to temperature; therefore, these batteries need a more involved battery management control. This is also evident from various efforts to use the more effective liquid cooling for these batteries. Li-ion batteries also need a good thermal management system because of safety and low-temperature performance concerns.

5.5.3. Battery Technologies

5.5.3.1. Lead-Acid Batteries

Micro-hybrid vehicles (see Section 5.3.4) are the best opportunity today for lead-acid to be introduced in passenger electric vehicles, besides such applications as forklift trucks, golf carts and other specific vehicles where they have long been used [2]. Improved products adapted to this application that should strongly expand in the near future are being developed by all the main lead-acid battery manufacturers.

An AGM VRLA battery that can be used in the micro-hybrid application is presented in Figure 5.17. A VRLA battery performs much better than one with liquid electrolyte (flooded battery), as indicated by start and stop cycling tests (see Figure 5.18). In these tests, the battery undergoes thousands of shallow cycles (1.3% DOD) starting from 100% SOC. After 60 000 such cycles,

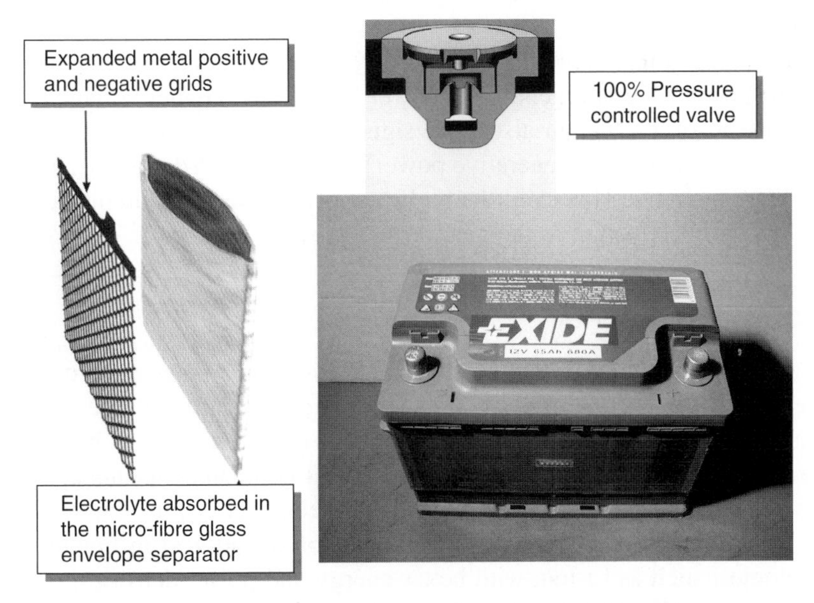

Figure 5.17. Example of AGM VRLA battery for micro hybrid vehicles.
Source: Courtesy of Exide Technologies.

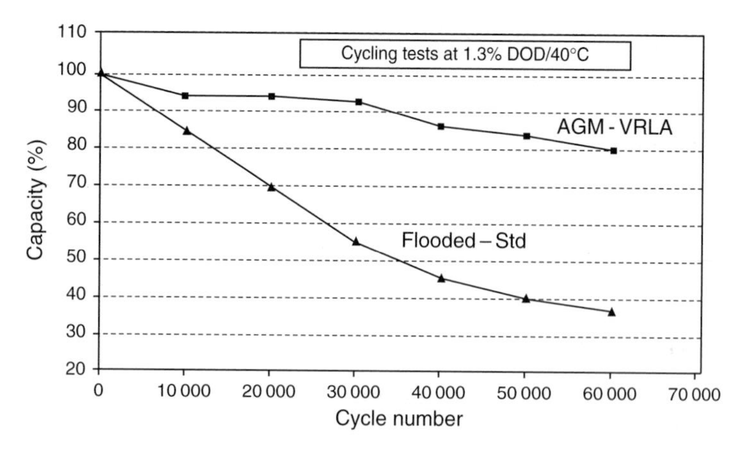

Figure 5.18. Capacity evolution of AGM VRLA cell during start and stop cycling tests, compared to conventional flooded lead-acid.
Source: From Ref. [19].

the VRLA battery still retains 80% capacity, while the flooded battery has collapsed to 37%.

When more power and better cycle life are required, a spiral wound cell configuration with thin electrodes is more appropriate. Such a design, which has been developed by most of the lead-acid battery manufacturers, provides better cycling at deep DOD, higher cranking power (the power needed to start the ICE) and charge acceptance.

This type is more suited for the soft hybrid application (see Section 5.3.5) requiring acceptance of regenerative power on braking. Higher voltage batteries (36 V instead of 12 V) can also be designed. Typical specific power of these batteries (cranking and regenerative power) is up to 500 W/kg on discharge and 200 W/kg on charge. However, the high cost of this electrical architecture does not allow its introduction in the near term.

5.5.3.2. Ni-MH Batteries

As soon as Ni-MH batteries became available, they were considered for replacing Ni–Cd batteries in EV applications because of the significant improvement in energy density. Considerable effort, supported by programs funded by the USABC, was devoted by several companies to this application in the 1990s, and many EV prototypes have been successfully built, although never commercialized. These projects were soon abandoned in favour of new technologies, such as Li-ion, with better energy and power densities.

When the HEV concept was developed into a commercial product, in the second half of the 1990s, Ni-MH was the advanced battery technology selected.

Mass produced for several years, this battery system is presently the only one used in commercially available full hybrid cars.

The main manufacturers of Ni-MH batteries for HEV are PEVE (a joint venture of Panasonic/Toyota, Japan), SANYO in Japan, JCS Advanced Power Solutions (a joint venture of Johnson Controls Inc/Saft) in USA/Europe and Cobasys in the USA [2].

Both cylindrical and prismatic shapes exist in this application and are adopted in commercialized vehicles. Each cell shape has advantages and disadvantages. The electrical characteristics (energy density, power density) are essentially the same. The choice of a specific design is mostly oriented by geometrical constraints on the complete battery system, heat control systems, etc.

The characteristics of Ni-MH batteries used in the most popular full hybrid, the Hybrid Synergy Drive by Toyota, are described here. The battery is presented in Figure 5.19, and the module characteristics are given in Table 5.15 [20]. For other characteristics, see Table 5.13. The expected life of this battery, based on laboratory tests, is 250 000 km.

As power is the main property for a HEV battery, all parameters that can lead to a reduction in cell resistance must be carefully studied. In the latest design, the Panasonic electric vehicle energy (PEVE) module provides a 50% power increase vs the first configuration, in addition to a 20% volume reduction obtained with a specific cell-to-cell connection.

The present module design consists of six prismatic cells attached side by side. Because of the power increase, the number of modules in the battery system could be reduced from 38 to 28 and the energy reduced from 1.8 to 1.3 kWh (see Table 5.13).

Channels are provided to allow air cooling by forced air produced by fans. A metallic casing, although heavier than the polymeric one, provides much easier heat transfer. Prismatic design for the Ni-MH cell is also preferred by other

Figure 5.19. Ni-MH battery pack used in Toyota HSD Prius.

Table 5.15. Main characteristics of Ni-MH high power modules (Panasonic) for Toyota Prius.

	Cylindrical	First-Generation Prismatic	Second-Generation Prismatic
Nominal voltage (V)	7.2	7.2	7.2
Nominal capacity (Ah)	6	6.5	6.5
Output power [a] (6-cell module) (W)	872	1050	1350
Mass (kg)	1.09	1.05	1.04
Dimension (cm)	1.96 (W) 10.6 (H) 28.5 (L)	1.96 10.6 28.5	1.96 10.6 27.6
Battery pack height (cm)		17.4	17.6

[a] 10 s, 50% SOC, 23°C.
Source: From Ref. [20].

companies, for example Johnson Controls-Saft Advanced Power Solutions and Cobasys [2]. On the other hand, in two HEVs by Honda, Civic and Accord, cylindrical cells, based on the same design of consumer products, were preferred.

As previously mentioned, in full-hybrid applications, the Ni-MH battery pack undergoes a large number of shallow cycles. During driving, the battery is continuously charged and discharged, but the average state of charge must be maintained within specified limits, as already pointed out. Figure 5.20 illustrates the power output/input variations during actual use, and the corresponding state of charge variations [21] (see also Figure 5.14.)

Ni-MH batteries can also be used in soft and mild hybrids, and in hybrid buses and tramways [2]. The battery module for large vehicles has a typical power of 4.2 kW vs 1.3 kW for full hybrids.

5.5.3.3. Li-Ion Batteries

The term Li-ion battery includes a family of batteries using various active materials (see Table 2.6) to build systems that can also be used in vehicular application (see Table 2.8). Indeed, due to their high energy and power, these batteries are well suitable for powering both EVs and HEVs.

Sony built the first prototype of an EV battery for Nissan in Japan, using $LiCoO_2$ as a positive material [22]. Saft chose from the beginning a nickel-based positive material, because of its good performance and the high price of

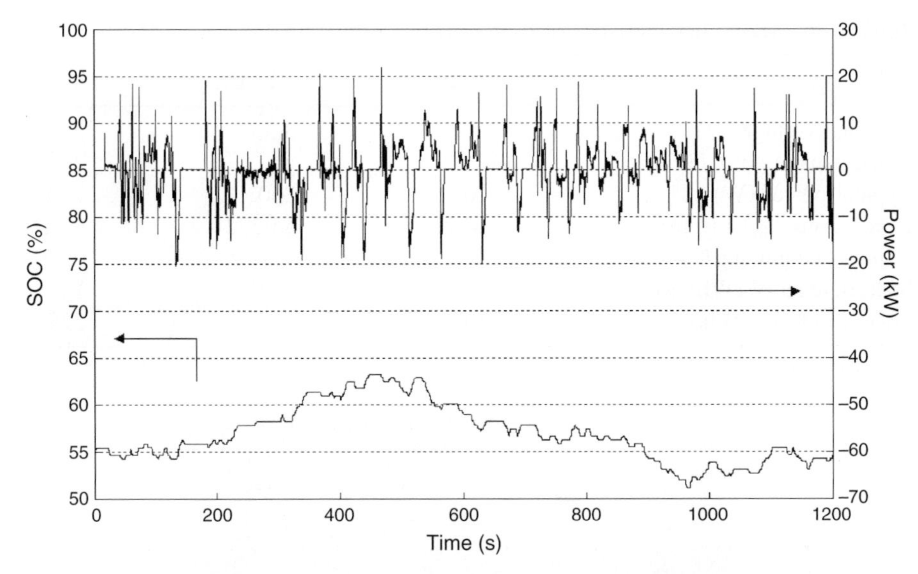

Figure 5.20. Examples of power and SOC variations during Prius HEV urban driving. *Source*: From Ref. [21].

cobalt [23]. The manganese spinel $LiMn_2O_4$, which offers a potentially lower cost, has also been largely studied for this application, especially in Japan [24]. The drawback of poor battery life due to the slight solubility of Mn is being tackled by partially substituting Mn with other ions (see again Table 2.8). $LiFePO_4$ has been more recently considered: it features an improved stability on overcharge (increased safety), a very high power, and the potential for lower cost due to the use of iron [25].

Batteries for EVs

Cylindrical cells or round-edge prismatic cells with wound electrodes are preferred in EV applications. Their capacities are in the range of 25–100 Ah. As an example, the main characteristics of the high-energy cells manufactured by Johnson Controls-Saft are described in Table 5.16.

As already mentioned, EV batteries must possess high energy for a long driving autonomy. Typical energy is in the range of 25–40 kWh, with a voltage of 300 V or more. A number of modules with energy in the range of 1–2 kWh are necessary. An example of EV module comprising six cells in series is shown in Figure 5.21. This module includes an electronic board that transmits data to the battery controller, and can also manage the cell temperature.

Table 5.16. Main characteristics of high-energy Li-ion cells for EVs.

Cells	VL45E	VL41M	VL27M
Nominal voltage (V)	3.55	3.55	3.55
Rated capacity at C/3 (Ah)	45	41	27
Typical power (W)	710	850	760
Dimensions (⊘/h) (mm)	54/222	54/222	54/163
Typical weight (kg)	1.07	1.07	0.77
Specific energy (Wh/kg)	150	135	130
Energy density (Wh/L)	310	285	275

Source: From Ref [2].

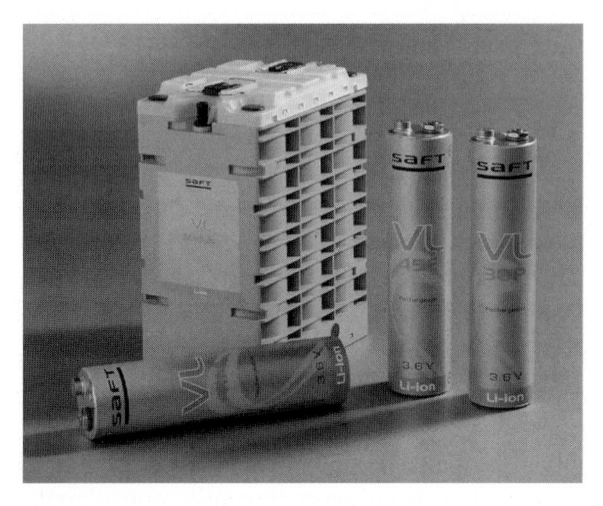

Figure 5.21. EV Li-ion module formed by 6 VL41M cylindrical cells in series. *Source*: From Refs. [2, 14].

Large batteries of this type are endowed with subsystems, managed by the battery controller, such as [2]:

- A cooling system that may be limited to fans in the case of air cooling, or a pump, hydraulic manifolds, a thermal regulation valve and a liquid-to-air heat exchanger in the case of liquid cooling.
- A contactor to insulate the battery from the vehicle when parked. This contactor may also be connected to a vehicle impact detector to insulate the battery in case of an accident.
- A ground fault detector, which is mandatory for high-voltage batteries.

Batteries for HEVs

By reducing the electrode thickness to increase the capability of sustaining high currents, Li-ion batteries can provide very high power and are therefore a good candidate for HEV batteries. Their high voltage, low weight, small volume and potentially long life make them very attractive. With decreasing cost as production volume increases [1], they are considered to be economically competitive to replace the presently used Ni-MH [26]. However, commercialization of HEVs using Li-ion has been lately announced several times and then postponed: problems connected with safety, cost and, perhaps, technical issues still exist.

In the past few years, several innovations have been presented in material chemistry and morphology. Some of the most promising proposals are here summarized.

Johnson Control-Saft has selected the Ni-based positive material $LiNi_xCo_yAl_zO_2$ already studied for high-energy cells [27]. In Table 5.17, characteristics of high-power cells are reported. A battery for a hybrid demonstration that can be assembled with modules formed by 10 VL7P cells is shown in Figure 5.22 (up-right).

Johnson Controls-Saft is also evaluating batteries for plug-in HEVs. As already pointed out, batteries for these vehicles must have enough energy to allow a large driving range before recharging. Therefore, they must be built on high-energy cells, like the ones of Table 5.16. Figure 5.22 (down) shows one such battery – see caption for details.

EnerDel has proposed a new chemistry: instead of carbon, they use $Li_4Ti_5O_{12}$ as a negative electrode, and modified manganese spinel, $LiMn_2O_4$, as a positive [28]. The advantage of using the titanate vs carbon is a lower structural stress when Li^+ is inserted, this increasing battery longevity. This battery is claimed to allow a volume reduction of 50% and a weight reduction of 35% vs an equivalent Ni-MH battery. Other claimed features include: very limited capacity loss upon cycling (5% after 1000 cycles), capability to operate between 90 and 10% DOD, and excellent low-temperature performance (at –30°C, 90% of the room temperature capacity would be maintained).

Table 5.17. Characteristics of high-power Li-ion cells.

	VL7P	VL20P	VL30P
Capacity (Ah)	7	20	30
Max continuous current (A)	100	250	300
Peak current (A)	250	500	500
Energy (Wh) at C/3	25	71	107
Power (W)	670	1130	1250
Diameter (mm)	41	54	54
Length (mm)	145	163	222
Weight (kg)	0.37	0.80	1.1

Source: From Refs. [2, 14].

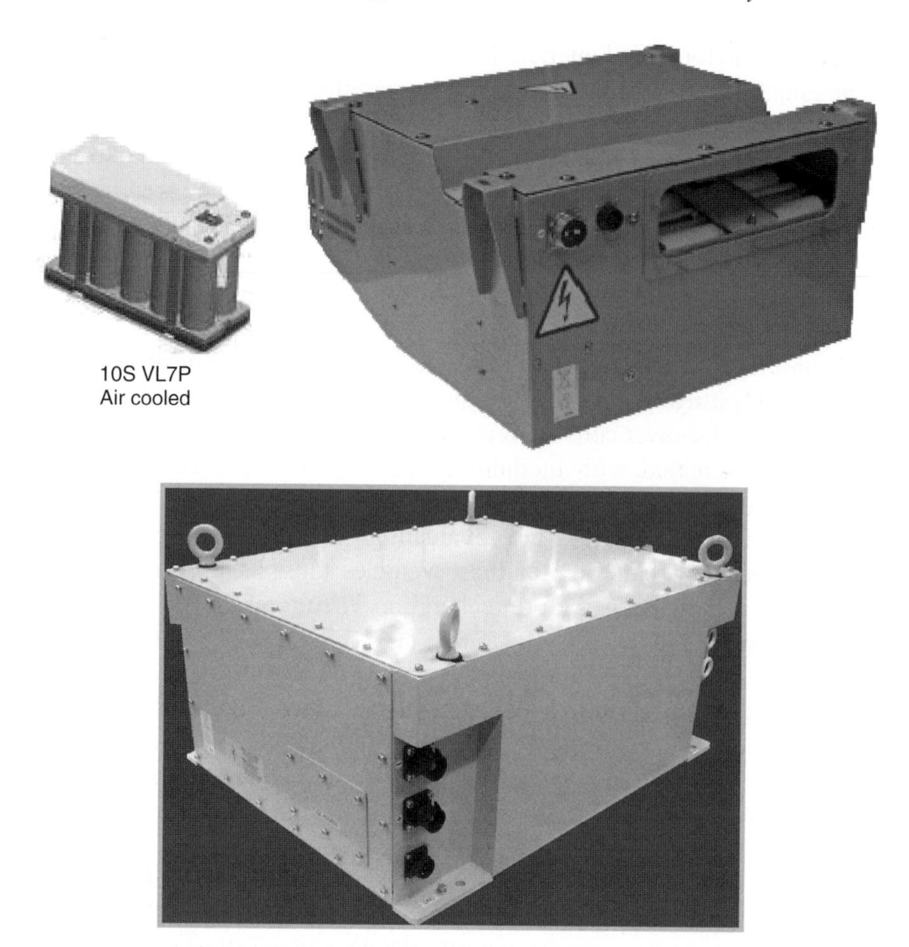

Figure 5.22. Up left: module formed by 10 VL7P cells in series; Up right: hybrid demonstration vehicle battery; down: battery for plug-in hybrid (12 modules formed by 6 VL41M cells (see Figure 5.21), 259 V, 10.6 kWh, 61 kW at 50% SOC for 30s, liquid cooled). *Source*: From Ref. [14].

This chemistry has been tested in a Toyota Prius and, according to a contract with Think Global, EnerDel's batteries should be supplied for the Think City EV in 2008 (see Section 5.2.1).

Remarkable steps forward, especially as far as the power is concerned, are reported by several companies.

Altair Nanotechnologies reports the use of nano-sized titanate-negative electrode (its composition is supposed to be the same as of EnerDel) [29]. Traditional Li-ion batteries allow a 3–5 year operation and 1000 cycles. This firm claims batteries that can last more than 20 years with 25 000 cycles, that is 400 000 km.

A123Systems is developing a Li-ion with a positive material based on nano-sized $LiFePO_4$ (see also Section 3.4.1) [30]. This material has already demonstrated, for example in power tools, extraordinary power capabilities. In March 2007, this company has exhibited a plug-in Toyota Prius with its battery. Combined with the Prius's existing battery, an electric continuous drive of 50–55 km is reported. Large Li-ion batteries for hybrid trucks and buses have also been shown.

Valence has also developed nano-sized phosphate for traction Li-ion batteries [31].

Hitachi Vehicle Energy has reported recent improvements as regards manganese spinel as a positive electrode. Some kind of (undisclosed) doping allows limiting the main drawback of this material, that is capacity fade upon cycling [32].

ElectroEnergy is proposing a new cell geometry: the wafer Li-ion cell [33]. This company claims significant reduction in manufacturing cost, 20% increase in energy and power density vs prismatic batteries, 15–35% volume reduction, flexible footprint and increased reliability.

In conclusion of this survey of Li-ion batteries for traction, a direct comparison with Pb-acid and Ni-MH may be in place [34]. As shown in Figure 5.23, Li-ion batteries have superior energy and power output, and can accept higher regenerative currents (faster charge). They have weak points in low-temperature performance, cost and manufacturing maturity. The introduction of small-particle materials can improve the low-temperature behaviour; cost is decreasing and will decrease even more once mass production starts.

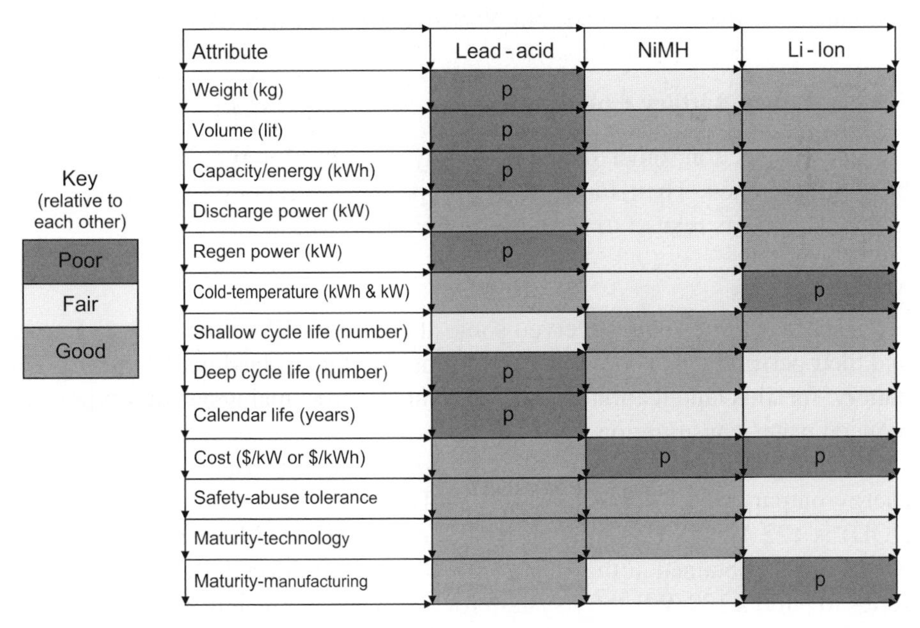

Figure 5.23. Qualitative comparison of large-format battery technologies for plug-in HEVs. "P" denotes "Poor".
Source: From Ref. [34].

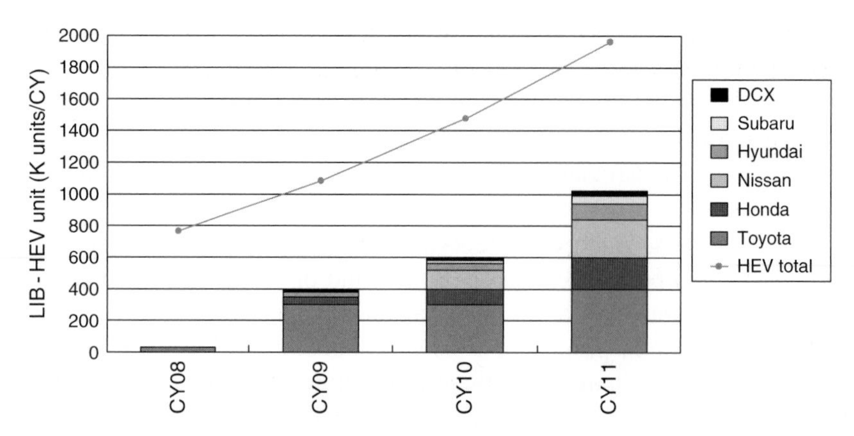

Figure 5.24. Near-term market expectations for HEVs with Li-ion batteries *vs* total number of HEVs sold. The legend corresponds to the rectangles of each bar, from bottom to top. DCX: DaimlerChrysler.
Source: From Ref. [14].

The favourable properties of Li-ion batteries, stressed throughout this chapter, allow positive forecasts in spite of the recent deployment delays by some car makers. As shown in Figure 5.24, by 2011 half the circulating HEVs are expected to have an electric motor powered by a Li-ion battery. Toyota will continue to be the front runner, but Nissan and Honda will grow at a faster rate.

5.5.3.4. Other Battery Chemistries

In this section, other batteries mainly used in prototype vehicles will be briefly mentioned. Their fundamental properties are reported in Chapter 2, so here only aspects related to their use in EV/HEV will be highlighted.

Ni–Cd batteries

These batteries have received some attention for EVs especially in Europe and their performance is obviously confronted to that of lead-acid batteries. The battery for this application is of the vented "low maintenance" type with reduced water consumption [2].

A 6-V battery module (5 cells in series) has a capacity of 100 Ah and energy outputs of 55 Wh/kg and 88 Wh/L, while the specific power (at 80% DOD) is 122 W/kg. This module has an excellent cycle life: more than 3000 cycles can be obtained at the 1C rate (to 80% DOD). Modules are connected in series to build a 132-V battery system for EV. Typical range is ~80 km in urban drive.

This battery has been tested on several cars, but none of them is produced at present, with the exception of small specific models. Reasons for this failure are relatively low driving range, high selling price and the availability of batteries with higher energy densities (Ni-MH and Li-ion).

Zn-air batteries

As anticipated in Section 2.4.1.4, these batteries have been proposed for large vehicles, that is buses and vans. A typical module has 47 cells, an operating voltage range of 57–40 V, a capacity of 325 Ah, a specific energy of 200 Wh/kg, and an energy density of 220 Wh/L (total energy, 17.4 kWh) [36]. A transit bus is powered by three trays of six modules each, for a total of 312 kWh.

These batteries are not electrochemically rechargeable: indeed, once the modules are exhausted, new ones are used. Exhausted modules are sent to a regeneration plant for re-use, but this complicate procedure raises questions about the real application of this technology.

Li-metal-polymer batteries

A typical EV battery module, working in the temperature range of 60–80°C, has a specific energy of 120 Wh/kg, a specific power of 260 W/kg and can sustain ~800 cycles (see also Table 2.13). For HEV applications, the modules have energy of 850 Wh, power of 20.5 kW and specific power of 1.4 kW/kg (10 s pulses at 80% DOD) [37]. Tests of EV modules have given performances very near to the USABC's criteria for mid-term (commercialization) batteries.

LMP batteries have been tested with good results in experimental cars. For instance, Avestor's batteries have powered such cars as the Ford Think City, Honda Insight and Cleanova [38]. However, in 2005 Avestor stopped its production of prototypes for EVs. This Canadian company has been acquired by Bolloré Group, which is going to use LMP batteries in its forthcoming Blue Car (see also Section 5.2.1).

Zebra batteries

As reported in Section 5.2.1, these batteries will be used in the new Think City produced in Norway by Think Global, while waiting for Li-ion batteries. Zebra batteries (see section 2.4.2 for details on their chemistry) are noted for safety and resistance to overcharge and overdischarge. Lifetime of a battery pack should correspond to about 160 000 km.

5.6. The Vehicle Control Systems

Road vehicles have long been almost entirely mechanical devices. However, over the last ~30 years several systems and functions have been put under electronic control. Electronic components provide more precise control, greater reliability and higher efficiencies. They are also smaller, lighter and, often, less expensive. In 1977, the average value of electronic components per vehicle was \$110, well under 5% of the total cost of materials and components [39]. By comparison, today's conventional vehicles contain \$1400–\$1800 in electronic components, about 10–20% of the total cost [40]. In full-hybrid models, the cost of electronics amounts to 40–50% of the total.

Vehicle electrification will certainly continue. Industry experts claim that 80–90% of automotive innovation is based on electronics; they also foresee that electronic innovation may lead to "smart" vehicles capable of acting autonomously: this would result in safer, more comfortable and efficient driving [1]. The demand for electrical power within vehicles is increasing at the rate of about 4% per year, putting increasing pressure on the present 14-V architecture [38].

The electronic control units (ECUs) of a vehicle take care of several devices, for example engine, brakes, battery, locks, navigation system, airbags, lights, stereo, etc. New vehicles may contain up to 80 ECUs. In hybrid vehicles, an additional electronic control unit serves as the "brain" of the hybrid, managing the power flow among the battery, generator, motor and transmission or power-split device.

In a vehicle, the embedded electronic control systems work in these main areas: driving, body and information system controls [41].

Driving Control System. It includes these subsystems:

- *Engine Control Unit.* This was the first embedded system used in a standard vehicle and remains the most important one. It allows great precision in ignition's time and control of the amount of fuel injected into each cylinder. It can also control other aspects of the internal combustion process, such as variable valve timing. A standard unit is based on these components: processor and operating system, sensors, actuators, converters and buses.
- *Electronic Stability Control (ESC).* It ensures a safe driving in dangerous conditions, for example on a wet and slippery road. To do so, the ESC uses systems like ABS and traction control. Traction control prevents the tyres from slipping by controlling engine power and/or braking.
- *Electric Power Steering.* The microcontroller of the electric power steering system determines how much power the assist motor must give.

Body Control System. It includes:

- *Dashboard Meters.* In newer cars, all dashboard meters are driven by electric motors controlled by an instrumentation controller.
- *Control Switches.* More and more switches (telephone, navigation system, audio system, etc.) tend to appear. In order to avoid having one wire per switch, a "local interconnect network" has been adopted.

- *Power Windows.* A control system provides safety locking.
- *Intelligent Keys* (e.g. for door opening). These keys ask for wireless command signals together with radio transmitters, receivers and antenna's built into the door.
- *Adaptive Front Lights.* In bad light conditions, a microcontroller can turn the headlights for a better view when turning.
- *Tire Pressure Monitoring.* This affects not only tyre and fuel consumption, but also driver's safety. When tyre pressure is too low, the driver can be alerted.
- *Airbag Control.* Accelerometers (normally based on MEMS) are the primary impact detection sensors.
- *Automatic Windscreen Wipers.* A sensor detecting the reflection of LED light being irradiated on the windscreen detects whether or not rain is on the windscreen and how much.
- *Central Monitor.* It informs the driver about the vehicle condition, for example low battery, broken lamp, etc.

Driving Information System. In a common car, it coincides with the navigation system. It is becoming a multimedia console, for example with audio features and computational capabilities.

Details of the electronic control in car with a combustion engine are shown in Figure 5.25. The operation of a number of components, for example fuel injector, throttle control valve, oil and water pump, radiator fan, etc., is controlled by the electronic management unit (ECM).

In Figure 5.26, the set of safety, comfort and entertainment devices present in a newer car is shown. Obviously, the power they require is remarkable (see later) and would be very onerous for a traditional 12 V battery/14 V generator system. Air bag, tyre pressure monitoring and telematics emergency system (see also Section 4.6.3) are especially critical for driver's safety [42].

The tendency towards a growing electronic control of the power train and other car functions has been stressed above. "X-by-wire" is a generic term describing the displacement of mechanical systems by electronic components such as sensors, controllers, power circuits, motors and actuators [1]. In Table 5.18, the displacement of mechanical functions by electronics is described. As will be later discussed, using so many power-hungry systems needs envisioning a battery pack of higher voltage and power.

In Table 5.18, the advantages brought about by car electrification are outlined. In general, the driver will find easier driving a car, while, as is typical of electric systems with respect to mechanical systems, the need for repairing or substituting will be decreasing.

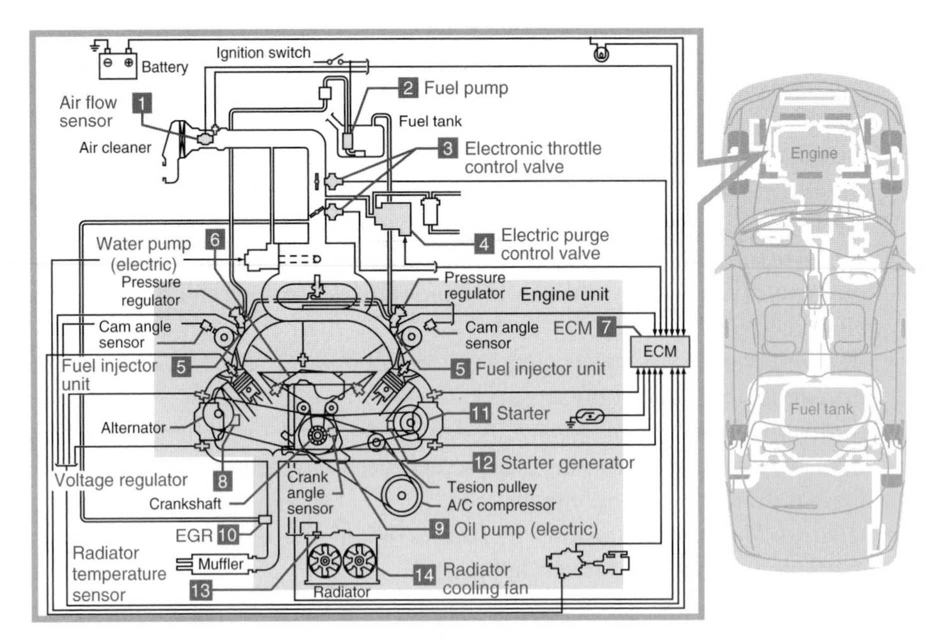

Figure 5.25. Block diagram of ECM of an ICE vehicle. The engine unit is enclosed in the light-grey area.
Source: Courtesy of TDK.

The benefit of electronic systems in two functions of great concern from the safety standpoint, that is braking and steering, are particularly relevant.

Braking-by wire allows an independent control on the four wheels: the speed of each wheel is measured, and the force applied to a wheel is a function of its speed, so to equalize the braking action. The ESC may be integrated in this system.

The steer-by-wire technology allows independent turns of the wheels. In a conventional car, both wheels are mechanically linked, this generating a slip angle upon turning and increasing the car resistance. With the new technology, both wheels move on a circular path with the same centre point: this may become possible only if the wheels turn under a different angle.

As anticipated above, a high electrification level is possible if the car is powered by a high-voltage battery, as in hybrid vehicles. In Figure 5.27, a block diagram highlighting the presence of electronic controls in an HEV is shown. In particular, the electronically controlled transmission (ECT) is evidenced. This transmission, except for the valve body and speed sensor, is virtually the same as a fully hydraulic one, but also includes sensors and actuators [43]. An electronic transmission improves efficiency and performance, and is already implemented in Toyota's hybrids.

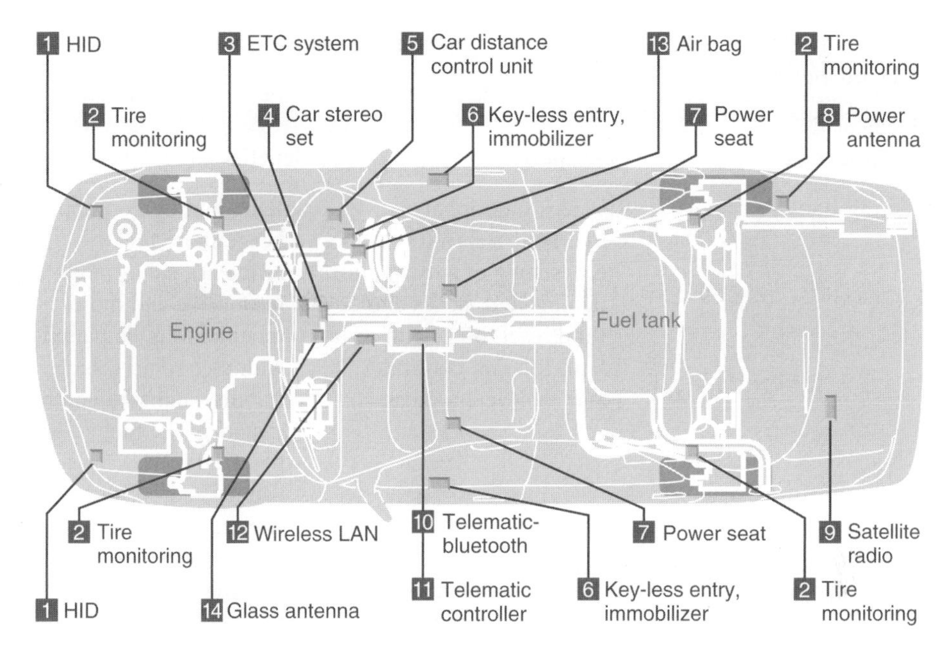

Figure 5.26. Examples of devices for safety, comfort and entertainment that are under electronic control. IDE: high discharge density lamp; ETC: electronic toll collection. *Source*: Courtesy of TDK.

Hybrids with high-voltage batteries can have electrical architectures supporting 300–500 and even 600 V [44]. For instance, Ford's Escape hybrid works off 330 V, Toyota Prius 500 V (see Table 5.13), and Lexus' hybrid SUV 650 V. 500-V architecture affords ~4 kW of accessory power vs 1.5 kW for a conventional vehicle. Some automotive engineers forecast for the near future 8–10 kW or even more.

With a power of 1.5 kW in today's conventional vehicles, adding such features as heated seats, electrically-assisted steering systems, electric air conditioning and pre-heated catalysts can hardly be accomplished.

For this reason, since 1999 many automotive groups supported the idea of shifting from 12/14 to 36/42 V architectures. However, thus far, efforts in this direction have failed, especially for cost considerations.

In high-voltage hybrids these problems do not exist, but as is obvious, cost issues cannot be neglected. Higher voltages allow to have steer-by-wire and brake-by-wire, but many industries are still discussing the cost-benefit trade-offs of that technology. This is why engineers are trying to add new features to conventional 12/14 V architectures, such as hybrid steering and regenerative braking.

Table 5.18. Displacement of mechanical functions in vehicles with a high level of on-board electronics.

Mechanical Displacement	Description	Benefits
Steering-by-wire	Replaces traditional hydraulic steering system (consisting of a hydraulic pump, fluid, hoses and fluid reservoirs) with a fault-tolerant controller and electrical motors.	Smaller and lighter (and hence more fuel efficient), safer, more responsive and more reliable. May be paired with electronic vehicle-stability and skid-control systems. The motors are powered by the car's battery and not the engine, which also improves fuel efficiency.
Throttle-by-wire	Replaces cable that connects throttle valve and gas pedal with an electrical connection.	More efficient than traditional throttle systems and safer; provides electronic stability and traction control.
Braking-by-wire	Replaces hydraulic-braking system with an electrical-component system that connects the four brake corners with the brake pedal and each other.	Enhances driver control in braking, provides greater uniformity in distribution of brake force, and enhances stability and traction control.
Shift-by-wire	Replaces conventional transmissions with electronically controlled variable transmissions.	Improves efficiency and performance.
Suspension-by-wire	Replaces mechanical damping with active electronic control of damping stiffness.	Improves vehicle stability by active support of wheel traction and by counter adjustments to pitch and roll when braking or cornering.
Integrated starter/alternator	Combines the alternator and starter functions into an electrical generator.	Reduces part count, weight and cost, and offers fast start times, stop/start and regenerative braking capabilities. Improves fuel efficiency and lowers emissions.

Table 5.18. (*Continued*)

Mechanical Displacement	Description	Benefits
Variable valve control	Replaces cams on an engine-powered camshaftwith electrical control from the motor; each cam and perhaps each valve can be powered separately.	Improves fuel efficiency and performance; reduces emissions.
Electric catalytic converters	Replaces the mechanical heating of catalysts needed to convert emissions into safe by-products.	Reduces emissions by more quickly achieving high temperatures.
Electric devices	Replaces belts connecting air conditioners, oil pumps, cooling fans and water pumps to the engine for power; provides power from electric motor/generator.	Virtually no impact on performance, but better fuel efficiency.
Electrical accessories (heated windshields, advanced seats and personal settings)	Replaces mechanical processes with electrical power and adds new functionality.	Greater comfort, efficiency and integration of systems.

Source: From Ref. [1].

In Table 5.19, an electrical load list including the main power-consuming devices/functions is presented. Even if only some devices were simultaneously used, the present electrical architecture with a power of 1.5 kW would hardly be adequate.

From the safety viewpoint, the high-voltage of hybrids is potentially dangerous for rescue workers in case of serious accidents. However, with cutoff switches in place, when an accident causes airbag deployment, the vehicle excludes the main voltage lines. Instead, lower voltage lines can be maintained so that the telematics system may send an alarm (automatic crash notification). A low-voltage system may be based on primary Li batteries, for example Li/MnO_2 or $LiSOCl_2$ (see Section 4.6.3).

These batteries are also used in another safety device, the direct tyre pressure monitor system (TPMS). In the United States, all new cars must have

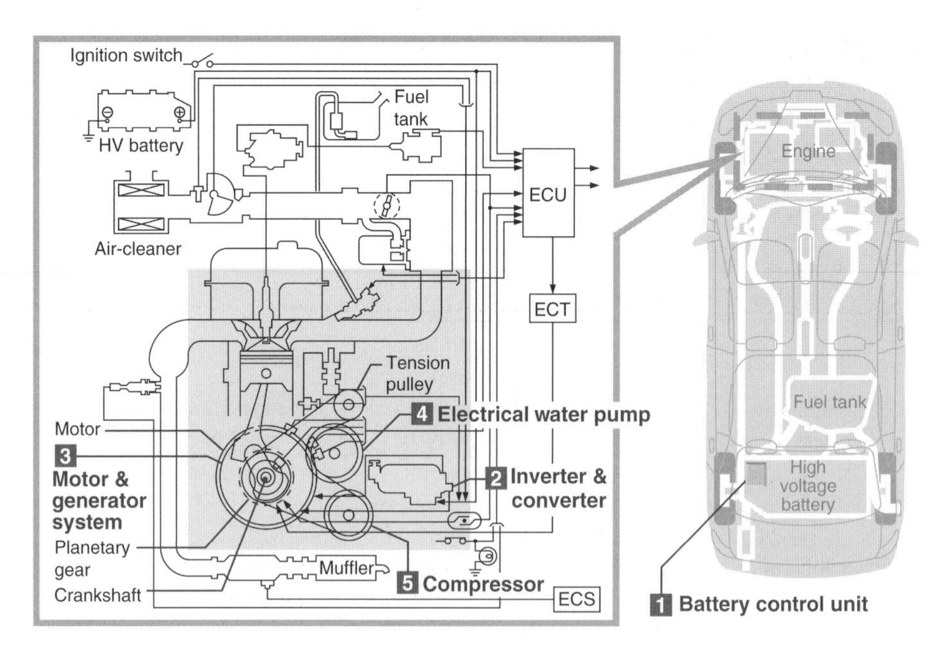

Figure 5.27. Block diagram showing electronic control in a hybrid vehicle. ECT, electronic control transmission; ECS, exhaust control system.
Source: Courtesy of TDK.

Table 5.19. List of components that can be powered by the vehicle's electric architecture, and maximum power they would require.

Electrical Load List	Max Power (kW)
Electrical AC compressor	3.8–4.0
Electrically actuated valve train	3.0–3.2
Pre-heated catalyst	3.0
Windshield heating	2.5
Brake-by-wire	2.0
Steer-by-wire	1.8
Mobile AC power outlet	1.5
Electric power steering	1.0
Electro-hydraulic brakes	0.9
Lights	0.6
ABS pump	0.6
Power windows (front)	0.5–0.7
Power windows (rear)	0.5–0.7
Heated front seats (left and right)	0.5
Heated rear seats (left and right)	0.5
High-power stereo	0.3
Front wiper motor	0.2–0.3

Source: Courtesy of Vishay Intertechnology Inc.

a TMPS (direct, indirect, or battery-less), which alerts the driver if the pressure decreases below 25% of the required level [42]. In the direct method, a battery-powered pressure sensor is located inside each tire. Therefore, flat and small batteries (coin cells), able to work in a wide temperature range and to deliver current pulses in the mA range, must be used.

5.6.1. Recent Developments in Automotive Lead-Acid Batteries

This section regards the benefits that improved automotive (not traction) batteries may bring about to cars with a combustion engine.

There is a growing demand for batteries with greater cycle stability and storage capacity. Indeed, there is a considerable effort to reduce fuel consumption, which, by 2015, should be brought to half the actual values. This is thought to be possible only with the introduction of technologies that will further increase the electrical content. Few examples, some of them already discussed, are: automatic transmission, automatic switch gear (ASG), automatic idling stop, regenerative braking, electrical support for the combustion engine at the start and at low rpm, and optimum fuel mixing in the engine by electromagnetic valve actuation (EMVA) [45].

A dual battery system is considered to cope with these increasing electrical demands. Such solution is already exploited in sports cars and some SUVs. The system guarantees the capability to start the combustion engine and to maintain the mobility of the vehicle even in extreme operating conditions and in case of failure of the power-supply system. The 42 V architecture is losing consent, so splitting the battery seems to be a more convenient solution. Two batteries may give some redundancy and may provide the average power demands for which the system is designed. However, even this solution would not prove satisfactory: only if the battery has been monitored and its state has been determined can the energy flow be managed appropriately. This asks for an intelligent integration of battery management into the more general concept of vehicle energy management [45, 46].

In the dual battery system, one battery would keep the task of engine cranking (start), whereas the second would power most of the electrical load. The former would be optimized for power output (for example a small-capacity high-CCA (cold cranking amps) flooded lead-acid battery), whereas the latter would be optimized for cycling capability (for example an AGM battery with low or moderate power ratings). To reduce costs, using two identical starter-type batteries may be considered [47]. Whether the two batteries are equal or different, both batteries have a nominal voltage of 12 V and are connected in parallel during normal driving (charging).

As mentioned above, the 42 V architecture has been set aside for its high cost, and this applies to a hybrid 14/42 V system as well: the cost associated with the generation and distribution of a second voltage level would be prohibitive. Furthermore, some components of the conventional 14 V system have been considerably improved. For example, enhanced 14 V claw pole alternators were developed that can continuously generate an electric power output of 3 kW and more.

AGM batteries offer at least three-fold longer shallow cycle life, compared to conventional SLI batteries [47]. Their use in start/stop applications and regenerative braking has already been discussed. Currently, an AGM battery costs twice as much a SLI battery; in the long term, and with high-volume production, the cost could decrease to 1.2–1.3 times.

The well-established SLI Lead-acid battery with flooded electrolyte is expected to be improved in its shallow-cycle life: in this way, more than the actual 150 capacity turnovers should be obtained at lower cost than those of AGM VRLA batteries. On the other hand, since the cost per charge delivered during battery life outperforms the SLI battery technology, AGM batteries may become standard if medium-high cycle life is required. Furthermore, the development of high-temperature resistant AGM batteries could lead to their broad introduction, because they could replace SLI batteries in the engine bay, where heat irradiation is high [47].

A final example of recent Lead-acid applications is in wire-based systems, such as brake-by-wire and steer-by-wire, where the power demand is high. In this case, generator (alternator) and battery provide double redundancy, but triple redundancy with a small reserve battery, of the VRLA type, is requested [48]. In this dual-battery system, the auxiliary VRLA has a small capacity (e.g. 14 Ah), and its SOC and SOH are constantly monitored to determine the battery's ability to carry out the duty cycle if required.

5.7. Electric Bikes

Electric bikes (e-bikes) have been thus far the most successful electric vehicles, with a cumulative production of 30 million pieces [49]. They have been traditionally powered by VRLA batteries, whose limits in energy and power (and hence driving range) are known. Alternative chemistries have also been considered: Ni–Cd, Ni-MH and, especially, Li-ion.

Both bicycles and scooters can be powered by these batteries (see Figure 5.28), and specifications are reported in Table 5.20.

VRLA battery packs consist of 3–4 modules of 12 V, with 12–20 Ah capacity, for a total voltage of 36–48 V and energy of 0.4–1.0 kWh. They are of the AGM type and can be deeply discharged (down to ~20% SOC).

Li-ion battery packs have voltages in the range of 24–37 V and capacities varying in a broad range (5–60 Ah).

Figure 5.28. Bicycle-style and scooter-style electric bikes.
Source: From Ref. [50].

Table 5.20. Power system characteristics of e-bikes.

Specifications	Bicycle	Scooter
Total battery pack (kWh)	0.4–0.6	0.8–1.0
Maximum current (A)	15	20–30
Voltage (V)	36	48
Modules/pack (typical)	3	4
Cells in series	18	24
Peak motor power (kW)	0.24	0.50–1
Maximum DOD (%)	80	80

Source: From Ref. [50].

Advances in VRLA technologies have made e-bikes efficient and affordable. On the other hand, the advantages of Li-ion in terms of specific power and energy are well known.

The characteristics of VRLA and Li-ion batteries are compared in Table 5.21. The batteries are sized for an average 48 V scooter with 60 km range (0.90 kWh) and 350 W motor. This type of e-bike is comparable in performance to a 50-cm^3 gasoline scooter. An energy consumption of 14 Wh/km, and average travel distance of 15 km/day were assumed in making the battery comparison.

The choice depends on the user's preferences: either a less expensive, but heavy and relatively short-lived bike with a VRLA battery, or an expensive, light and durable one with a Li-ion battery.

Other factors to be considered are VRLA batteries tend to exhibit more scattering performance from module to module; Li-ion batteries have higher safety issues and need adequate management.

Table 5.21. Comparison of VRLA and Li-ion batteries for a scooter-style e-bike.

	VRLA	Li-ion
Cost ($)	75	424
Mass (kg)	26	8
Lifetime (years)	3	9
Volume (L)	10	5
Max. theor. power (kW)	6.2	2.9
Charge safety	High	Low
Temperature effect	Moderate	High
Specific energy (Wh/kg)[a]	35	110
Energy density (Wh/L)[a]	86	170
Specific power (W/kg)[a]	240	350
Cycle number[a]	300	800

[a] Authors' assumptions based on their knowledge.
Source: From Ref. [50].

Li-ion batteries may use different materials as positive electrodes: $LiCoO_2$ and its derivatives, $LiMn_2O_4$ and $LiFePO_4$. The penalty in specific energy of the last two materials is balanced by their higher safety and lower cost (see Sections 2.3.3 and 5.5). A pack for a bicycle-style e-bike, made with cells utilizing the Mn spinel, has the following specifications: 37 V, 10 Ah, maximum discharge rate: 2C, maximum charge rate: C/2, ~500 cycles, weight: 3.8 kg.

References

1. A. Bernstein, 'Ending Oil's Stranglehold on Transportation and Economy – The Emergence of Hybrid Vehicles', Report 2006.
2. M. Broussely, in *Industrial Applications of Batteries. From Cars to Aerospace and Energy Storage'*, M. Broussely and G. Pistoia, Eds., Elsevier, Amsterdam, 2007.
3. M. Broussely, in *Lithium Batteries – Science and Technology*, G.A. Nazri and G. Pistoia, Eds., Kluwer Academic Publishers, Boston, USA, 2004.
4. Toyota Motor Corporation, 'A Guide to HSD', March 2007.
5. Union of Concerned Scientists, 'Hybrids Under the Hood (Part 2)', 2007.
6. T. Kameda, *Proceedings of the AABC Conference*, Hawaii, USA, 2005.
7. F. Kalhammer, *Extended Abstracts, 46th Battery Symposium in Japan*, Nagoya, Japan, 2005.
8. P. Ulrich and J.L. Liska, *Proceedings of the 20th Electric Vehicle Symposium (EVS20)*, Long Beach, USA, 2003.
9. HIEDGE Institute, 'HEV/EV Market Report 2007'.
10. Green Car Congress, 'GM Introduces E-Flex EV System; Chevrolet Volt the First Application', January 2007.
11. Toyota New Zealand, 'HSD Overview', 2006.
12. MPower, 'Traction Batteries for EV and HEV Applications', 2005.

13. MPower, 'Battery Management Systems', 2005.
14. M. Andrew, 'Li-ion: Enabling a Spectrum of Alternate Fuel Vehicles', *CARB ZEV Symposium,* Sacramento, CA, September, 2006.
15. G. Plett, *J. Power Sources* 134 (2004) 252.
16. G. Plett, *J. Power Sources* 134 (2004) 262.
17. G. Plett, *J. Power Sources* 134 (2004) 277.
18. A. Pesaran, 'Battery Thermal management in EVs HEVs: Issues Solutions', *Advanced Automotive Battery Conference,* Las Vegas, Nevada, February, 2001.
19. F. Trinitad, C. Hernandez, M. Lacadena, B. Jacq, G. Fossati and R. Johnson, *Proceedings of the AABC Conference*, Hawaii, USA, 2005.
20. H. Miyamoto, T. Asahina, T. Matsuura, S. Hamada and T. Eto, *Proceedings of the 21th Electric Vehicle Symposium (EVS21),* Monaco, 2005.
21. S. Nagata, H. Umeyama, Y. Kikuchi and H. Yamashita, *Proceedings of the 20th Electric Vehicle Symposium (EVS20)*, Long Beach, USA, 2003.
22. T. Miyamoto, M. Touda and K. Katayama, *Proceedings of the EVS13 Symposium*, Vol. 1, 37–44, Osaka, Japan, 1996.
23. M. Broussely, G. Rigobert, J.P. Planchat and G. Sarre, *Proceedings of the EVS13 Symposium*, Vol. 1, Osaka, Japan, 1996.
24. K. Nakai, T. Aiba, K. Hironaka, T. Matsumura and T. Horiba, *Abstracts of the 41st Battery Symposium in Japan*, Nagoya, Japan, 2000.
25. V. Biancomano, 'Lithium-Ion Batteries Make a Grab for Power', *Power Management Design Line*, May 2006.
26. M. Broussely, P. Blanchard, P. Biensan, G. Rigobert, J.L. Liska, P. Genin and F. Barsacq, *Abstracts of the 46th Battery Symposium , 3E*, Nagoya, Japan, 2005.
27. M. Broussely, K. Nechev, G. Chagnon, J.P. Planchat, G. Rigobert and G. Sarre, *Abstracts of the 43rd Battery Symposium*, Kyushu, Japan, 2002.
28. Green Car Congress, 'EnerDel to Market Automotive Li-Ion by End of 2008', December 2007.
29. V. Evan House and F. Ross, 'How to Build a Battery that Lasts Longer Than a Car', *Power Management DesignLine*, August 2007.
30. The Energy Blog, 'A123 Systems Li-ion Battery in Prius', March 2007.
31. Green Car Congress, 'Valence Technology Introducing New Large Format Li-Ion Battery Technology at EVS-23', November 2007.
32. T. Horiba, T. Maeshima, T. Matsumura, M. Koseki, J. Arai and Y. Muranaka, *J. Power Sources* 146 (2005) 107.
33. ElectroEnergy, 'Advanced Lithium Ion Technology for PHEV and HEV', *Lithium Mobil Power 2007*, San Diego, CA.
34. Pesaran, 'Battery Choices for Different Plug-in HEV Configurations', *Plug-in HEV Forum and Technical Roundtable,* Diamond Bar, CA, July 2006.
35. H. Takeshita, Worldwide Market Update on NiMH, Li Ion and Polymer Batteries for Portable Applications and HEVS' *24th International Seminar on Primary/Secondary Batteries,* Fort Lauderdale, FL, March 2007.
36. Electric Fuel Corp., 'Zinc Air Fuel Cell Technology', 2004.
37. C. St-Pierre, R. Rouillard, A. Bélanger, B. Kapfer, M. Simoneau, Y. Choquette, L. Gastonguay, R. Heiti and C. Behun, "Lithium-Metal-Polymer Battery for Electric Vehicle Hybrid Electric Vehicle Applications", www.avestor.com/rtecontent/document/evs16.
38. G. Pistoia, Chapter 1 in *Industrial Applications of Batteries. From Cars to Aerospace and Energy Storage*, M. Broussely and G. Pistoia, Eds., Elsevier, Amsterdam, 2007.
39. G. Leen and D. Heffernan, 'Expanding Automotive Electronic Systems', *Computer*, January 2002.

40. A. Bindra, ‘Electronics Redefines the Auto of the Future’, *Electronic Design*, December 2000.
41. Wikibooks, Embedded Control Systems Designs, 2007.
42. H. Yamin, in *Industrial Applications of Batteries. From Cars to Aerospace and Energy Storage*, M. Broussely and G. Pistoia, Eds., Elsevier, Amsterdam, 2007.
43. Toyota Motor Corporation, Electronic Control Transmission (ECT).
44. C.J. Murray, ‘Turning on the Juice’, *Design News*, February 2006.
45. E. Meissner and G. Richter, *J. Power Sources* 144 (2005) 438.
46. E. Meissner and G. Richter, *J. Power Sources* 116 (2003) 79.
47. E. Karden, P. Shinn, P. Bostock, J. Cunningham, E. Schoultz and D. Kok, *J. Power Sources* 144 (2005) 505.
48. G.J. May, D. Calasanzio and R. Aliberti, *J. Power Sources* 144 (2005) 411.
49. J. Wang, *China International Battery Forum*, Beijing, China, 2006.
50. J.X. Weinert, A.F. Burke and X. Wei, *J. Power Sources* 172 (2007) 938.

LIST OF ACRONYMS

AAC: Advanced Audio Coding
ABS: Anti-lock Braking System
ACM: Acoustic Current Meter
ACN: Automatic Crash Notification
A/D or ADC: Analog to Digital Converter
ADCP: Acoustic Doppler Current Profiler
AED: Automated External Defibrillator
AGM: Absorptive Glass Mat
AI: Artificial Intelligence
ALABC: Advanced Lead-Acid Battery Consortium
AM: Amplitude Modulated
AMR: Automatic Meter Reading
AP: Access Point
APU: Auxiliary Power Unit
ARGOS: Advanced Research and Global Observation Satellite
ASG: Automatic Switch Gear
ASIC: Application Specific Integrated Circuit
ASIMO: Advanced Step in Innovative Mobility
AUV: Autonomous Underwater Vehicle
AVI: Automated Vehicle Identification

BCB: Battery Control Board
BESS: Battery Energy Storage System
BMS: Battery Management System
BTMS: Battery Thermal Management System

CAES: Compressed Air Energy Storage
CCA: Cold Cranking Amps
CC-CV: Constant Current - Constant Voltage
CCD: Charge-Coupled Device
CCTV: Closed Circuit TV
CDMA: Code Division Multiple Access
CEV: Crew Exploration Vehicle
CMOS: Complementary Metal Oxide Semiconductor
Codec: Compressor-Decompressor
CORK: Circulation Obviation Retrofit Kit
COSPAS: *Russian acronym meaning* Space System for Search of Vessels in Distress
CPLD: Complex Programmable Logic Device
CPR: Cardio-Pulmonary Resuscitation
CPU: Central Processing Unit
CPV: Common Pressure Vessel
CRIEPI: Central Research Institute of Electric Power Industry

D/A or DAC: Digital to Analog Converter
DART: Deep Ocean Assessment and Reporting of Tsunamis
DCP: Digitally Controlled Potentiometer
DDR: Double Data Rate
Demux: De-multiplexer
DMB: Digital Multimedia Broadcasting
DOD: Depth of Discharge

DOE: Department of Energy
DPGS: Differential Global Positioning System
DSC: Digital Still Camera
DSP: Digital Signal Processor
DVB: Digital Video Broadcasting
DVD: Digital Versatile Disc

ECG (or EKG): Electrocardiogram
ECM: Engine Control Management
ECS: Exhaust Control System
ECT: Electronic Control Transmission
ECU: Electronic Control Unit (or Engine Control Unit)
EDL: Entry, Descent and Landing
EDLC: Electric Double Layer Capacitor
EEG: Electroencephalogram
(E)EPROM: (Electrically) Erasable and Programmable ROM
EGR: Exhaust Gas Recirculation
ELT: Emergency Locator Transmitter
EMVA: Electromagnetic Valve Actuation
EPIRB: Emergency Position Indicating Radio Beacon
EPRI: Electric Power Research Institute
EPS: Emergency Power Supply
ESA: European Space Agency
ESD: Electrostatic Discharge
ETC: Electronic Toll Collection
EV: Electric Vehicle
EVA: Extra Vehicular Activity

FCHEV: Fuel Cell Hybrid Electric Vehicle
FCVT: FreedomCAR and Vehicle Technology
FDD: Floppy Disk Drive
FET: Field Effect Transistor
FLO: Forward Link Only
FPGA: Field Programmable Gate Array
FreedomCAR: CAR *stands for* Cooperative Automotive Research

GEO: Geostationary Earth Orbit
GIF: Graphics Interchange Format
GLONASS: GLObal NAvigation Satellite System
GOES: Geostationary Operational Environmental Satellite
GPRS: General Packet Radio Service
GPS: Global Positioning System
GSM: Global System for Mobile Communication

HEV: Hybrid Electric Vehicle
HF: High Frequency
HID: High-Intensity Discharge (Lamp)
HLC: Hybrid-Layer Capacitor
HST: Hubble Space Telescope

IC: Integrated Circuit
ICD: Implantable Cardioverter Defibrillator
ICE: Internal Combustion Engine
I/F: Interface
I/O: Input/Output
IP: Internet Protocol
IPV: Individual Pressure Vessel
IrDA: Infrared Data Association
ISS: International Space Station

JPEG: Joint Photographic Experts Group

LAN: Local Area Network
LCD: Liquid Crystal Display
LDO: Low Dropout (Regulator)
LED: Light Emitting Diode
LEO: Low Earth Orbit
LIBES: Lithium Battery Energy Storage (Technology Research Association)

LMP: Lithium-Metal-Polymer (Battery)
LVAD: Left Ventricular Assist Device
LWD: Logging While Drilling
MAC: Medium Access Controller
MAV: Micro Air Vehicle
MCU: Micro Controller Unit
MEMS: Micro Electromechanical System
MEO: Mid-altitude Earth Orbit
METI: Ministry of Economy Trade and Industry
MFL: Magnetic Flux Leakage
MMR: Mobile Micro Robot
MOSFET: Metal Oxide Semiconductor FET
MPEG: Motion Picture Experts Group
MP3: MPEG-1 Audio Layer 3
MWD: Measurement While Drilling

NASA: National Aeronautics and Space Administration
NEDO: New Energy and Industrial Technology Development Organization
NOAA: National Oceanic and Atmospheric Organization
NTSC/PAL: National TV Standards Committee/Phase Alternating Line

OBS: Ocean Bottom Seismometer
OEM: Original Equipment Manufacturer
OLED: Organic Light Emitting Diode

PCI: Peripheral Component Interconnect
PCS: Power Conversion System
PCU: Power Conditioning Unit
PDA: Personal Digital Assistant
PEVE: Panasonic Electric Vehicle Energy
PHES: Pumped Hydroelectric Energy Storage
PIC: Programmable Interrupt Controller
PIG: Pipeline Inspection Gauge
PLB: Personal Locator Beacon
PLC: Programmable Logic Controller
PLD: Programmable Logic Device
PMDC: Permanent Magnet Direct Current
PMEL: Pacific Marine Environmental Laboratory
PMP: Portable Media Player
PNGV: Partnership for a New Generation of Vehicles
PQ: Peak Quality
PS: Peak Shaving
PSTN: Public Switched Telephone Network
PTC: Positive Temperature Coefficient
PTT: Platform Transmitter Terminal
PV: Photovoltaic
PWM: Pulse Width Modulation

RAM: Random Access Memory
RF: Radio Frequency
RFID: Radio Frequency Identification
RGB: Red, Green, Blue
ROHS: Reduction of Hazardous Substances
ROM: Read Only Memory
RTC: Real Time Clock

SARSAT: Search and Rescue Satellite Aided Tracking
SBS: Smart Battery System
SD: Secure Digital
SDRAM: Synchronous Dynamic RAM
SEI: Solid Electrolyte Interface
SIP: Session Initiation Protocol
SLA: Sealed Lead-Acid (Battery)

SLI: Starting, Lighting, Ignition
SMbus: System Management Bus
SMES: Superconducting Magnetic Energy Storage
SMPS: Switch-Mode Power Supply
SMS: Short Message System
SoC: System on Chip
SOC: State of Charge
SOH: State of Health
SOI: Silicon on Insulator
SPLB: Sub-marine Personal Locator Beacon
SRAM: Static Random Access Memory
SSPS: Space Solar Power Station
STC: Standard Test Conditions

TAH: Total Artificial Heart
TCM: Trajectory Control Manoeuvre
TCU: Telematics Communication Unit
TFT: Thin Film Transistor
TPMS: Tire Pressure Monitoring System

UART: Universal Asynchronous Receiver/Transmitter
UAV: Unmanned Air Vehicle
UCAV: Unmanned Combat Air Vehicle
UMPC: Ultra Mobile Personal Computer
UMTS: Universal Mobile Telephone System
UPS: Uninterruptible Power Supply
USABC: United States Advanced Battery Consortium
USB: Universal Serial Bus
USCAR: United States Council for Automotive Research
UUV: Unmanned Undersea Vehicle
UWB: Ultra Wide Band

VoIP: Voice over Internet Protocol
VRB: Vanadium Redox Battery
VRLA: Valve Regulated Lead-Acid (Battery)
VSIP: Video Security over Internet Protocol
VSLI: Very Large Scale Integration
VSC: Vehicle Stability Control

WAN: Wide Area Network
WAV: WAVeform Audio Format
WEEE: Waste from Electrical and Electronic Equipment
Wi-Fi: Wireless Fidelity
Wi-MAX: Worldwide Interoperability for Microwave Access
WLAN: Wireless Local Area Network
WPAN: Wireless Personal Area Network
WSN: Wireless Sensor Network

ZEBRA: Zero Emission Battery Research Activity
ZEV: Zero Emission Vehicle

3G: 3rd Generation

μC: Microcontroller
μP: Microprocessor

Index

Plate 1. An E-book reader – Kindle by Amazon. Size: 19.1 × 13.5 × 1.8 cm; weight: 292 g. See text for other characteristics. (See Figure 3.8, p. 82).

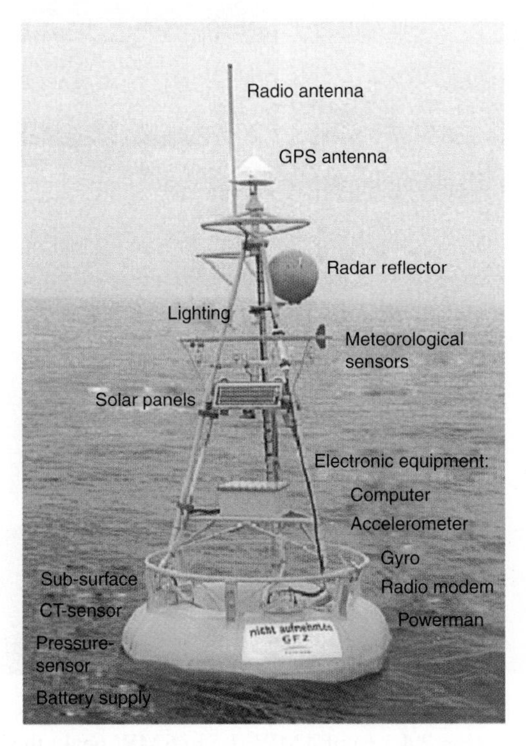

Plate 2. Off-shore moored buoy. (See Figure 4.26, p. 203).

Plate 3. Left: Portable weather station for measuring different meteorological parameters; right: fire station. (See Figure 4.52, p. 237).

Plate 4. Left: Mars Exploration Rover (Spirit and Opportunity are twin rovers); Right: Phoenix Lander. (See Figure 4.55, p. 248).

Plate 5. 32 MW wind farm in Sapporo, Japan. The building houses a Vanadium Redox Battery (VRB, see Section 2.4.1.6) of 4 MW, 1.5 h, 6 MW peak; the battery can smooth power output fluctuations due to wind variations. (See Figure 4.84, p. 291).

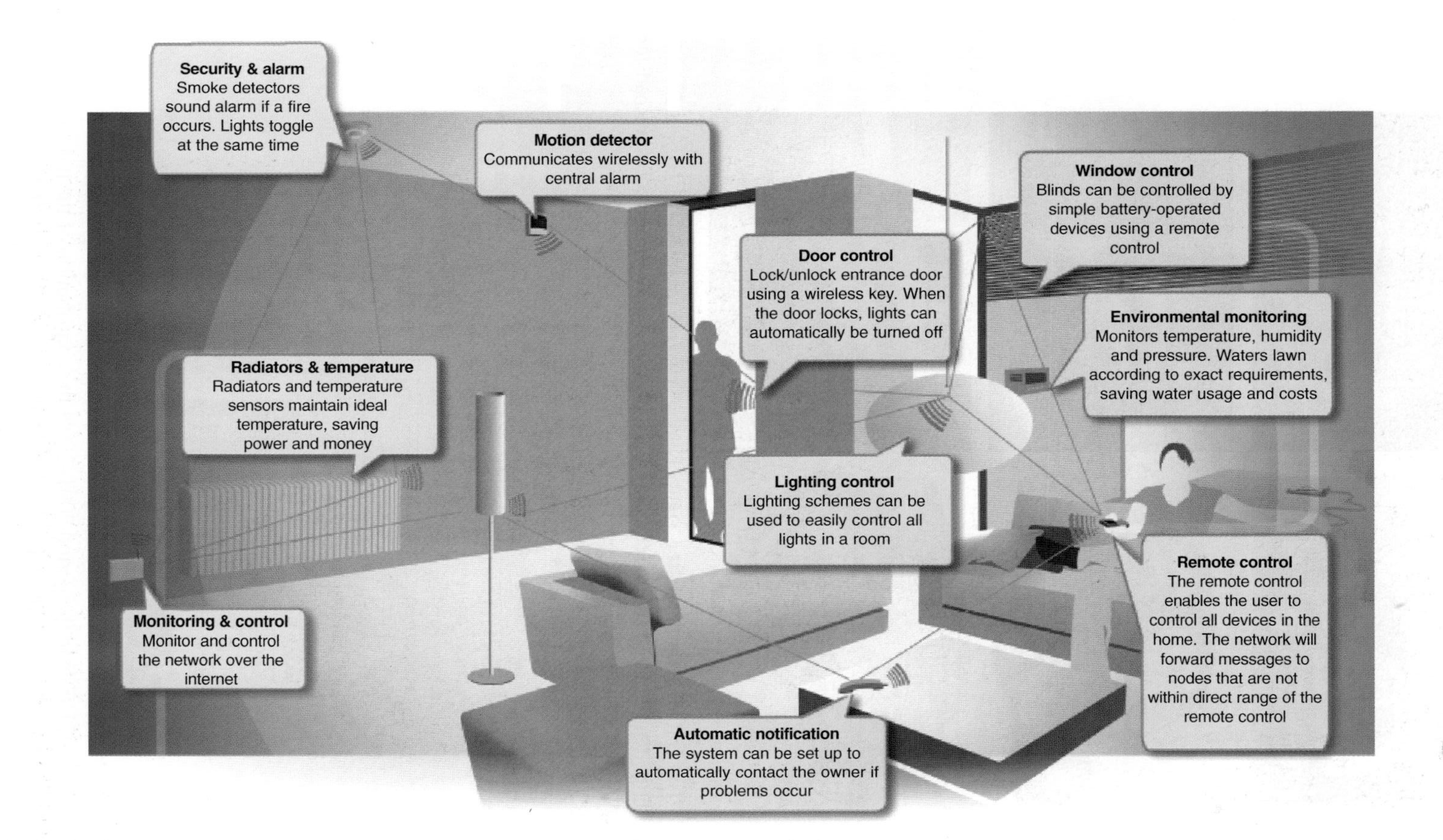

Plate 6. An example of home wireless networking using ZigBee technology. (See Figure 4.98, p. 313).

Plate 7. Top left: Cleanova II; top right: Nissan Mixim; middle left: Nissan Pivo 2; middle right: Think City; bottom left: Blue Car (Bolloré); bottom right: Tesla Roadster. (See Figure 5.4, p. 328).

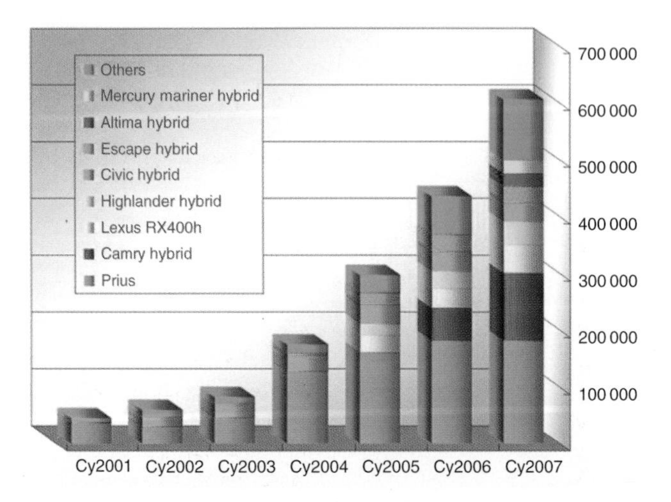

Plate 8. Sales of HEVs in the period 2001–2007; the contribution of the most important models is shown. Prius, Camry, Lexus and Highlander by Toyota; Civic by Honda; Altima by Nissan; Escape and Mariner by Ford. (See Figure 5.8, p. 341).

RECEIVED
APR - 3 2009